TOWARD A H[ISTORY] OF EPISTEMIC THINGS

Synthesizing Proteins in the Test Tube

WRITING SCIENCE

EDITORS Timothy Lenoir and Hans Ulrich Gumbrecht

TOWARD A HISTORY OF EPISTEMIC THINGS

Synthesizing Proteins in the Test Tube

Hans-Jörg Rheinberger

STANFORD UNIVERSITY PRESS

STANFORD, CALIFORNIA 1997

Stanford University Press
Stanford, California
© 1997 by the Board of Trustees of the
Leland Stanford Junior University
Printed in the United States of America

CIP data are at the end of the book

For Ineke

Contents

✤✤✤

TOWARD A HISTORY OF EPISTEMIC THINGS

Synthesizing Proteins in the Test Tube

Prologue

With this book, I pursue three lines of inquiry. First, I expound my view on the arrangements that twentieth-century laboratory scientists refer to as their "experimental systems."[1] In a post-Kuhnian move away from the hegemony of theory, historians and philosophers of science have given experimentation more attention in recent years. This book is an attempt at an epistemology of contemporary experimentation based on the notion of "experimental system." Second, my aim is to emphasize the dynamics of research as a process of the emergence of "epistemic things." Such an endeavor involves unfathoming the basic question of how novel objects come into existence and are shaped in the empirical sciences. The consequences of such a shift of perspective from the actors' minds and interests to their objects of manipulation and desire lead us toward a history of epistemic things. Third, I intertwine a case study with these epistemic and historiographic issues. I take a microscopic look at a particular laboratory—that of Harvard-trained oncologist and biochemist Paul Charles Zamecnik and his colleagues—located in a specific setting, the Collis P. Huntington Memorial Hospital of Harvard University at the Massachusetts General Hospital in Boston. Over a period of fifteen years—between 1947 and 1962—this group developed an experimental system for synthesizing proteins in the test tube that put Zamecnik and his coworkers at the forefront of those leading biochemistry into the era of molecular biology.

Having worked on this project for several years, I came to see that pursuing these three lines of argument, reflection, and reconstruction would require a special form of presentation. I realized that casting the epistemic and historiographic issues into a coherent framework would interfere with presenting the case study in appropriate detail and from a perspective that was sufficiently nonanticipatory such that the microscopic contingencies involved in generating scientific objects would become visible. I have tried to solve this problem of representation by dividing the chapters of the book into two categories. Alternating between a more reflective and a more narrative mood, they oscillate between the recursive view of an epistemologist and the—alas—intractable and foreclosed perspective of a "historical participation" in the endeavor. The chapters refer to each other and draw upon each other's arguments, but they can also be regarded as two separate embodiments of a play alternating between reenactment and recounting. This book is written for both scientists and historians as well as philosophers of science. For none of them will it be easy reading. Those interested mainly in philosophical issues can concentrate on the epistemic and historiographic chapters while consulting the historical material to the extent that they want to go into the details. Those interested mainly in the case study can concentrate on the chronologically ordered historical chapters while consulting the reflective parts to the degree they wish to. The readers from both camps will have to judge whether this concept is viable.

The Epistemology

I have sketched the lines of this study's argument in two previous papers, which contain the epistemic message of this book *in nuce*.[2] Discussions with many colleagues have helped me clarify what sounded confusing to them. The price of such disentanglement is the inevitable attenuation of an enthusiasm that arose precisely from my hybrid existence between the laboratory bench and the community of epistemologists when I started this enterprise. Nevertheless, I hope the text still bears the marks of the early excitement.

The following is a brief summary of matters of epistemology. First, experimental systems (Chapter 2) are the genuine working units of contemporary research in which the scientific objects and the technical conditions of their production are inextricably interconnected. They are, inseparably and at one and the same time, local, individual, social, institutional, technical, instrumental, and, above all, epistemic units. Experimental systems are thus impure, hybrid settings. It is in these "dynamic bodies" that experimenters shape and reshape their epistemic things. My approach in this book is certainly biased toward the epistemic aspect. There is only one excuse for that: nobody can claim that all scientific activity is locally constrained without accepting that this holds true for the activity of a historian and philosopher as well. My decade-long entanglement in an experimental system, at the laboratory bench, has left its marks on my bottom-up epistemological idiosyncrasies and predilections. Others may add where they see gaps and lacunae.

Second, experimental systems must be capable of differential reproduction (Chapter 5) in order to behave as devices for producing scientific novelties that are beyond our present knowledge, that is, to behave as "generator[s] of surprises."[3] Difference and reproduction are two sides of one coin; their interplay accounts for the shifts and displacements within the investigative process. To be productive, experimental systems have to be organized in such a way that the generation of differences becomes the reproductive driving force of the whole experimental machinery. Third, experimental systems are the units within which the signifiers of science are generated. They display their meanings within spaces of representation (Chapter 7) in which graphemes, that is, material traces such as fraction patterns or arrays of counts, are produced, articulated, and disconnected and are placed, displaced, and replaced. Scientists think within such spaces of representation, within the hybrid context of the experimental systems at hand. More precisely, scientists create spaces of representation through graphematic concatenations that represent their epistemic traces as engravings, that is, generalized forms of "writing." Finally, it is through conjunctures and bifurcations that particular experimental systems get linked into experimental ensembles, or experimental cultures (Chapter 9). As a rule, conjunctures and bifurcations are the result of unprecedented and unanticipated events. Such local condensation points in the economy of scientific practice are the attractors around which research communities and, in the long run, scientific disciplines become organized.[4]

The Historiography

In matters of historiography, first and foremost, I claim with Bruno Latour that "there is a history of science, not only of scientists and there is a history of things, not only of science."[5] In this cascade from scientists to science to (scientific) things, it is on the level of things that my historical perspective is anchored. Call it a biography of things, a filiation of objects, not as pictures of an exhibition, but as records of the process of their coming into existence. In this respect, I have been inspired by art historian George Kubler's thoughts about the temporal forms of artistic production, especially his "remarks on the history of things." Exploring what he calls "formal sequences" of works of art, of experiments, of tools, and of technical constructs, Kubler notes, "The value of any rapprochement between the history of art and the history of science is to display the common traits of invention, change, and obsolescence that the material works of artists and scientists both share in time."[6] The rapprochement I have in mind concerns the history of the material culture of the sciences and the experimental arrangements in which they dwell. Instead of reading a history of objectivity from concepts, I embark on reading a history of *objecticity* from material traces.

Two coupled notions coined by Jacques Derrida, "*différance*" (Chapter 5) and "historiality" (Chapter 11),[7] will help me to frame my thoughts on histories of science as histories of traces and of things. *Différance*, here, stands for the peculiar displacing power of experimental systems, whereas historiality designates their intrinsic spatiotemporal load. If experimental systems have a "life of their own,"[8] they have also a time of their own. The notion of *différance* will allow me, if not to escape, at least to perforate any nontemporal dichotomy between the "true" and the "false," between fact and artifact, that pitfall of naive realism. The notion of historiality will be instrumental in avoiding any nonspatial dichotomy between past and future, that trap of naive historicism. Thus, I will follow what might be called a Derridean principle of supplementarity. Stated simply, we deal here with an economy of epistemic displacement, such that everything intended as a mere substitution or addition within the confines of a system will reconfigure that very system—sometimes beyond recognition. The movement of supplementation carries with it a translinguistic idea of semantics: that which displaces, the very materiality of the trace, overturns the meaning of that which is displaced.

This book is no exception to that rule. When I started to write down shreds of what later turned into this text, I did not know that the notion of experimental system would take over. "Experimental system" first was a supplement to "empirical reasoning," and it acquired genuine displacing powers—intrinsic ones as well as extrinsic ones. It allowed me to anchor my errant thoughts both with respect to my subject matter and the audience I came to address as a scientist turning historian.

The Case Study: From Microsomes to Ribosomes, from Soluble RNA to Transfer RNA

Whoever immerses him- or herself in the details of the cancer-related and biochemical studies of protein synthesis during the decade after World War II gains the impression of a maze from which there seems no easy way out. Gunther Stent has pointed to the derogatory attitude of the first generation of molecular biologists toward the messy and at times literally bloody procedures of traditional biochemistry, which were of no appeal to former physicists turned biologists.[9] In his autobiography, Mahlon Hoagland has spoken of a "wide gulf" that in the 1950s and even in the early 1960s, separated the experimental culture of biochemists from those who regarded themselves as molecular biologists.[10] Yet, as my case study of the group at the Massachusetts General Hospital will highlight, much of the laborious work that established the molecular details of what later became codified as the process of replication, transcription, and translation of genetic information between 1953 and 1963 was the result of biochemical endeavors that had not at all been set up from the perspective of molecular genetics. And the success of what Stent called the "dogmatic phase" of molecular biology largely rested on the fact that, in the late 1950s, many molecular biologists, among them James Watson's laboratory at Harvard, resorted to the previously denounced procedures and experimental systems of biochemistry in elucidating the details of the so-called molecular flow of genetic information.

The issue of the relation between biological chemistry and the emerging discipline of molecular genetics is a complex one, and it has taken amazingly different shades and shapes according to local research traditions, institutional structures and disciplinary affiliations, national research policies, and philosophical commitments. In this respect, my book aims at adding one more facet to a long-standing and ongoing discussion among historians of molecular biology.[11]

I will look at and focus on a particular experimental system that came to occupy a central place in clarifying the mechanism of protein synthesis. I will reconstruct the intricacies of a research trajectory, with its quirks, its distractions, and its inherently unpredictable output. In doing so, I try to avoid making cheap use of the benefits of retrospection. Yet, I am aware that no historian can, or should, completely abstain from the opiate of hindsight. As with other drugs, its effect, beneficial or deleterious, lies with the dosage.

The case study part of the book focuses in considerable detail on the construction of an in vitro system of protein biosynthesis and the laboratory emergence of what molecular biologists today call *transfer RNA*. This molecule, on the one hand, "transfers" amino acids from the cytoplasm to the sites of their assembly, the ribosomes. On the other hand, it "transfers," according to its coding properties, the genetic message from its nucleic acid form into its protein form.

The group whose work at the Collis P. Huntington Memorial Hospital of Harvard University I will follow included, besides Paul Zamecnik, a medical doctor with an M.D. from Harvard, especially the following members: Mary Stephenson, who worked with Zamecnik right from the beginning as a technical assistant and later earned a Ph.D. in biochemistry; Harvard M.D. Ivan Frantz and Robert Loftfield, who held a Ph.D. in organic chemistry from Harvard and had been a research associate at MIT, both early postwar companions of Zamecnik; Philip Siekevitz, who joined the group in 1949 as a biochemistry Ph.D. from the University of California, Berkeley; Elizabeth Keller, who came as a biochemistry Ph.D. from Cornell in 1949; Mahlon Hoagland, a Harvard M.D. who had worked with Fritz Lipmann before joining Zamecnik in 1953; John Littlefield, an M.D. from Harvard, who began to work as a clinical and research fellow at Massachusetts General Hospital (MGH) in 1954; Jesse Scott, with an M.D. from Vanderbilt, who was an associate biophysicist at Huntington during the 1950s; Liselotte Hecht, with a Ph.D. in oncology from Wisconsin, who came in 1957; and Marvin Lamborg, with a Ph.D. in biology from the Johns Hopkins University, who joined the group as a research fellow in biochemistry in 1958.

Zamecnik can be regarded as the grand old man of protein synthesis. Yet he and the members of his group never belonged to the handful of molecular biologists whose names were bandied about as the heroes of a new discipline. To tell a tale of heroes, or geniuses, even by way of depreciation, was not the thing I wanted to accomplish. This is the history of a collaborative enterprise whose proficiency was exceptional. But, on the other hand, it can also be viewed as an exemplar of mid-twentieth-century biomedicine-biochemistry-molecular biology. I wanted to retrace a particular pathway of investigation characteristic of post–World

War II life sciences. In doing so, Francois Jacob's account of his laboratory experience at the Pasteur Institute, and his reflections thereupon, have been an invaluable companion to me.[12] I read his autobiography at the time I began this project. It gave me the necessary confidence that it would be worthwhile to structure my narrative from the viewpoint of an experimental system.

The history of the early test-tube representation of transfer RNA as soluble RNA is illuminating in many respects. Its details will be dispersed throughout the following chapters. I will show that the emergence of this molecule was embedded in a background of cancer research and bio-chemistry that had nothing to do with molecular biology at the begin-ning. Toward the end of the 1950s, however, transfer RNA produced a major conjuncture between classical biochemistry and molecular biology and by the early 1960s came to occupy a central place within the scaffold-ing of molecular biology. It became instrumental for dealing with the genetic code, and it bridged the gap between DNA, which is said to store the genetic information, and the proteins, which are said to represent the code's biological meaning. With the assessment of the different steps of the life of this molecule as an epistemic thing, I intend to exemplify a more general concept of how the history of science can be set in perspec-tive as a history of experimental events.

This is not to say that a chronicle is intended here. I do not claim to give a complete account of the historical development of protein synthesis research, much less so of the genetic code. Despite the reflections of a few participants in the endeavor, such an account is still lacking.[13] Rather, my focus on a single laboratory notwithstanding, I am aiming at what, in a Foucauldian sense, would be called the "archaeology" of transfer RNA, that is, the analysis of those dispositions and depositions through which the culture of protein synthesis of the 1950s can be reread and reenacted. Thus, this account is also not biographical. My purpose has been to show how, from a standard research situation, an investigative enterprise took shape that surreptitiously pervaded molecular biology by way of capillary diffusion.

[Archaeology] does not try to restore what has been thought, wished, aimed at, experienced, desired by men in the very moment at which they expressed it in discourse. [It] does not try to repeat what has been said by reaching it in its very identity. It does not claim to efface itself in the ambiguous modesty of a reading that would bring back, in all its purity, the distant, precarious, almost effaced light of the origin. It is nothing more than a rewriting: that is, in the preserved form of exteriority, a regulated transformation of what has already been written. It is not a return to the innermost secret of the origin; it is the systematic description of a discourse-object.[14]

This "discourse-object" is what I call an epistemic thing. It is my contention, therefore, that epistemic things—things embodying concepts—deserve as much attention as generations of historians have bestowed on disembodied ideas.

In Chapter 1, I sketch the coordinates of description to which my narrative is committed. Chapter 2 presents some basic thoughts on experimental systems and epistemic things and situates them in the current debate over the role of experimentation in understanding the dynamics of the modern sciences. In Chapters 3 and 4, I focus on the establishment of a first test-tube system for protein synthesis based on rat liver tissue, between 1947 and 1952, by Paul Zamecnik's group at MGH. Chapter 5 picks up the thread of Chapter 2 and reflects on the differential reproduction of experimental systems. These thoughts are exemplified in Chapter 6, where I follow the fractionation of the protein synthesis system between 1952 and 1955. Chapter 7 concentrates on the issue of representation in scientific practice. Representing protein synthesis as a cascade of intermediate steps in a metabolic reaction chain is the topic of Chapter 8, which covers the period from 1954 to 1956, and in which the concept of amino acid activation took shape. How developing experimental systems can be set into a broader context will be discussed in Chapter 9, which deals with conjunctures, bifurcations, and experimental cultures. The emergence of a particular epistemic thing from the in vitro protein synthesis system will be related in Chapter 10. Soluble RNA came as an unprecedented finding, and it connected protein synthesis research to molecular biology. Chapter 11 centers on historiographical deliberations on the emergence of unprecedented events. The transformation of soluble RNA into transfer RNA, the transformation of microsomes into ribosomes, and the transformation of what for a long time had been "microsomal template," into messenger RNA will be traced in Chapters 12 and 13. Chapter 13 ends with another unprecedented event, the deciphering of the genetic code on the basis of a bacterial in vitro system of protein synthesis. As an epilogue, a short remark on science and writing concludes the book.

Here and there throughout the text, I have resorted to materials and reflections that I have published during the past few years. I wish to cite especially the following: *Experiment, Differenz, Schrift* (Basiliskenpresse, 1992), "Experiment, Difference, and Writing," I and II (*Studies in the History and Philosophy of Science* 23, 1992, 305–31, 389–422), "Experiment and Orientation" (*Journal of the History of Biology* 26, 1993, 443–71), "Experimental Systems: Historiality, Narration, and Deconstruction" (*Science in Context* 7, 1994, 65–81), and "From Microsomes to Ribosomes: 'Strategies' of 'Representation' 1935–55" (*Journal of the History of Biology* 28, 1995, 49–89). The permission of the publishers—Basiliskenpresse, Pergamon, Kluwer, and Cambridge University Press—is gratefully acknowledged. I thank Nancy Bucher, Ivan Frantz, Liza Hecht-Fessler, Mahlon Hoagland, John Littlefield, Robert Loftfield, Heinrich Matthaei, Philip Siekevitz, Mary Stephenson, and above all Paul Zamecnik for patiently answering my boring questions and for offering valuable comments on earlier drafts of different chapters of this book. I am especially grateful to Heinrich Matthaei and to Paul Zamecnik, who gave me access to the notebooks of their laboratories; to Richard Wolfe, the librarian of the Francis A. Countway Library of Medicine in Boston; and to Penny Ford-Carleton from the Research Affairs Office of MGH, who made archival sources from that institution available to me. Intellectually, I am indebted to Jacques Derrida and François Jacob, whose works have served as a great source of inspiration for me in preparing this text. I thank Richard Burian, Soraya de Chadarevian, Lindley Darden, Peter Galison, Jean-Paul Gaudillière, David Gugerli, Michael Hagner, Klaus Hentschel, Timothy Lenoir, Ilana Löwy, Peter McLaughlin, Everett Mendelsohn, Robert Olby, Paul Rabinow, Johannes Rohbeck, Joseph Rouse, Bernhard Siegert, Hans-Bernd Strack, Bettina Wahrig-Schmidt, Norton Wise, and especially Marjorie Grene, Yehuda Elkana, and Lily Kay for countless discussions, thoughtful critiques, and constant encouragement. Last but not least, I greatly enjoyed the expert editorial advice of the staff of Stanford University Press, in particular Nathan MacBrien, and of my copy editor, Paul Bodine.

I started to work on this book during a sabbatical spent at the Program in the History of Science at Stanford University with Timothy Lenoir, generously granted to me by Heinz-Günter Wittmann, the late director of the Max Planck Institute for Molecular Genetics in Berlin-Dahlem. I am grateful to Dietrich v. Engelhardt, the director of the Institute for the History of Medicine and Science at the University of Lübeck, for a congenial research atmosphere that left me ample space to continue this project. Finally, a fellowship at the Institute for Advanced Study, Berlin, under its rector, Wolf Lepenies, during the academic year 1993–94 gave me the opportunity to complete this book much more quickly than would have been possible otherwise. It took a considerable amount of time nonetheless. The University of Salzburg granted me a leave, which I spent at the Max Planck Institute for the History of Science, Berlin, in the summer of 1995 to make the manuscript ready for press. But there is somebody I have to thank for *giving* me that time I took away from her.

CHAPTER 1

After All: An Epistemology of the Beginning

❖❖❖   [It is] the vague, the unknown that moves the world.
                   —Claude Bernard, *Philosophie: Manuscrit inédit*

Forewords are afterthoughts. They are exemplars of the paradox of writing, hence of creating history. By definition, they precede the text, but, as a rule, they do so by deferment and insinuation. They try to impose a closure on what until then could be considered work in progress. Yet by this very gesture they keep the game going. The sciences are characterized by a permanent process of reorientation and reshuffling of the boundary between what is thought to be known and what is beyond imagination. Such a reorientation has at times been characterized as an approximation, an adjustment to "truth" or "reality" as a presupposed *terminus ad quem*. In contrast, there are reasons to pretend that the sciences are much better understood as continuously *engendering* what Gaston Bachelard has called the "scientific real," thereby reshaping their agenda through their own action.[1] This recurrent action applies to the historian's practice as well: he or she continuously reorients the historical understanding of the process of reorientation we see at work in the sciences. There is no end to this endeavor, just as there is no end to the empirical endeavor on which it bears.

To Begin With

To read Freud as an epistemologist might seem surprising. I will not enter into a discussion of the discourse of psychoanalysis, although I will touch upon it here and there in the following chapters, nor am I guided by the conceptual apparatus of that science. Rather, I am interested in finding out what Freud, both from a general perspective and from within his own analytical practice, considers to be "the true beginning of scientific activity"—scientific activity tout court, as he seems to suggest in passing. I will have to question this assertion of universality as well as the hope for a "true beginning." But it serves me as a preliminary starting point. The passage to which I refer reads as follows:

> We have often heard it maintained that sciences should be built up on clear and sharply defined basic concepts. In actual fact no science, not even the most exact, begins with such definitions. The true beginning of scientific activity consists rather in describing phenomena and then in proceeding to group, classify and correlate them. Even at the stage of description it is not possible to avoid applying certain abstract ideas to the material in hand, ideas derived from somewhere or other but certainly not from the new observations alone. Such ideas—which will later become the basic concepts of the science—are still more indispensable as the material is further worked over. They must at first necessarily possess some degree of indefiniteness; there can be no question of any clear delimitation of their content. So long as they remain in this condition, we come to an understanding about their meaning by making repeated references to the material of observation from which they appear to have been derived, but upon which, in fact, they have been imposed. Thus, strictly speaking, they are in the nature of conventions—although everything depends on their not being arbitrarily chosen but determined by their having significant relations to the empirical material, relations that we seem to sense before we can clearly recognize and demonstrate them. It is only after more thorough investigation of the field of observation that we are able to formulate its basic scientific concepts with increased precision, and progressively so to modify them that they become serviceable and consistent over a wide area. Then, indeed, the time may have come to confine them in definitions. The advance of knowledge, however, does not tolerate any rigidity even in definitions. Physics furnishes an excellent illustration of the way in which even "basic concepts" that have been established in the form of definitions are constantly being altered in their content.[2]

This text is both clear and bewildering, if not contradictory. In "scientific activity," the "grouping," "classification," and "correlation" of phenomena, cannot proceed from "clear and sharply defined basic concepts." Nonetheless, scientific activity presupposes "certain abstract ideas [from] somewhere or other," which, however, cannot be derived from what, as a new experience, is still wanting. If they are to allow "new observations" at all, they must "at first necessarily possess some degree of indefiniteness." But they dare not be so indeterminate that they are unable to settle into the "material of observation." As soon as one starts to apply these ideas to the material of experience, understanding them further will depend on how the material is handled. This in turn makes it "appear" that the ideas had been "derived" from the material, although, "in fact," just the opposite has taken place: they have been "imposed" upon it. Thus, on the one hand, such notions are "conventions," and, on the other hand, they must not be "arbitrarily chosen" if they are to allow "new" observations on the material of experience. But how can we guarantee their nonarbitrariness? In the end, we can only do so because they bear "significánt relations to the empirical material," relations we "seem to sense before we can clearly recognize and demonstrate them." This draws us inevitably into a circle where moving on is only possible by further convolution.

What Freud invites us to ponder here is the ineffable trace of scientific action or, as I would like to call it, the *experimental situation*. It appears as if the relationship of "deriving" ideas from the material of observation and of "imposing" ideas upon that material represented the focal point of the argument. In this game of deriving from / imposing upon, the contours of what will, perhaps one day, constitute the basic concepts of a science emerge in a gradual fashion. The solidus is meant to indicate that both processes are inextricably interconnected, in fact, that they become effective only in one and the same movement. The deriving from / imposing upon construction is a precarious one whose ongoing reorientation can be stopped only by destroying it. The tension in this movement demonstrates the ambivalence of what we like to call a "concept": it is the product of a scientific activity, and it is a compelling instrument of performing that same activity. It causes some irritation to observe Freud's attempts to cope—using the traditional opposition of concept and reality, theory and empirical foundation, idea and matter—with a movement that resists this scheme to the extent that it continually forces him to invent new formulations. What is more disquieting is that there is another level of reasoning beneath the deriving from / imposing upon movement of concept and reality. This schematism does not aim—despite its outward appearance—at *the* "basic concept" as the point at which it would come to rest; on the contrary, it aims at "new observations." In working through the material, it gives hope of rendering accessible what until now was not accessible, of making explicit what up to now could only be "sensed." Freud writes,

[I] am of opinion that that is just the difference between a speculative theory and a science erected on empirical interpretation. The latter will not envy speculation its privilege of having a smooth, logically unassailable foundation, but will gladly content itself with nebulous, scarcely imaginable basic concepts, which it hopes to apprehend more clearly in the course of its development, or which it is even prepared to replace by others. For these ideas are not the foundation of science, upon which everything rests: that foundation is observation alone. They are not the bottom but the top of the whole structure, and they can be replaced and discarded without damaging it.[3]

Somewhat later, in Ludwik Fleck's writings, we encounter the expression of "pre-ideas" or "proto-ideas" as guidelines for the development of knowledge in the sense of a heuristics of "somewhat hazy" concepts.[4] Yehuda Elkana has found the notion of "concepts in flux" to be an appropriate description for the early development of thermodynamics.[5] In a different, immunological context, Ilana Löwy has used the notion of "boundary concept" and emphasized its function in organizing research fields and in creating "federative experimental strategies."[6] I deliberately slip over the nuances of these attempts at what might be called, with Hans Blumenberg, an epistemology of the "non-conceptual" in science.[7] Suffice to say here that what Freud calls the "nebulous, scarcely imaginable basic concepts," are not, in his or in all these related instances, the gemmules of an anticipated future theory. They are more like auxiliary organs of touch in an otherwise impalpable space of experience that, on occasion, one will be ready to "replace by others." As Freud sees it, in the edifice of science the concepts, whatever their degree of elaboration, form "the top of the whole structure, and they can be replaced and discarded without damaging it."

This does not imply considering them as mere ephemeral intermediates in a thoroughgoing empirical process of establishing facts. Freud is not, as the vocabulary might suggest, speaking in favor of a shallow empiricism; what he is attempting is to emphasize the primacy of scientific experience-in-the-making, for which conceptual indeterminacy is essential, over its conceptually defined and consolidated results. He is trying to characterize that state of affairs that Michel Serres has claimed nearly defies any possibility of a description—that "free-wheeling, fluctuating, non-determinate time in which those who do research are not yet quite sure what they are looking for, and yet blindly do know what they are after." "No," Serres continues, "he who does research, does not know. He gropes about, he tinkers, lingers, keeps his options open. No, he does not construct, thirty years before its realization, the calculator of the day after tomorrow. He does not foresee it, as we, who know and use it might conclude." And Serres compares the history of science, which does not unfold in time but on which time acts as an operator, with a "meandering river, its many streamlets and tributaries, whose fluxes adventurously clash against obstacles, barriers, impediments, and icings, and force themselves into narrow passes, defiles and crevasses. Not to mention the turbulences and the more stable currents, the countercurrents that reverse the river's flow, its blindly ending ramifications, its dead backlashes."[8]

Serres, of course, is not the first to compare the scientific endeavor in its unruliness with a meandering river. Fleck used the metaphor but also pointed to its limits: yes, rivers meander, they may change trajectories, but, "provided enough water flows in the rivers and a field of gravity exists, all rivers must finally end up at the sea." Fleck has gone one step further and contended that for science in genesis and development "there is no such thing as the *sea as such*," where its movement may finally come to rest.[9] The process of investigation does not and cannot come to an end, for the very reason that there is no possibility of anticipating the future objectal constellations that accrue from it. Instead of following the entropic track toward a stable equilibrium, the material activity of pattern formation we call research obeys the rules of a never ending ramification. It is in this latter sense only that the comparison of science with a meandering river seems appropriate.

Coordinates of Description

The passages cited in the previous section point to the hinge that holds the following chapters together. Alternately more reflective and more narrative, they revolve and evolve around the experimental groundwork of the empirical sciences—molecular biology in the given case. In the past decade and a half, philosophers and historians of science have begun to focus on experimentation as a field of investigation in its own right. The role of experiments in the sciences has begun to be recognized and reconceptualized as extending far beyond their seemingly unproblematic acceptance as instances of verification, of corroboration, of refutation, or of the modification of theories—in short, beyond their function as mere empirical instances in the evaluation of theoretical propositions. Viewed in a restricted sense and still largely within the framework of classic epistemological thinking, this means basically—to use Hans Reichenbach's terms—that experimentation has been removed from the "context of justification" and inserted into the "context of discovery."[10] But such a displacement does not remain without consequences for the logical functioning of the opposition between justification and discovery itself from which it proceeded.

This dislocation also affects the accompanying opposition between the realm of logic and that of a psychology of mind. By going a step further and by transforming the psychological space of discovery into a space of experimental manipulation, we also transcend the ideal of a creative genius, of a free play of individual mental faculties, bent and domesticated only by the stringency of their own performance. By plunging into the spaces of experimental manipulation, we find ourselves confronted with a rhizomic network of recurrent epistemic practices, a filigree of "investigative operations."[11] Correspondingly, we need a "pragmatogony" of scientific action.[12]

After the Kuhnian move from continuity and verity to discontinuity and relativity of scientific knowledge, a move decisively radicalized by Paul Feyerabend, we are witnessing another turn—from Kuhn's predilection for viewing science as theory dominated to a post-Kuhnian engagement with the experimental aspects of the sciences.[13] A philosophical landmark in this move has been Ian Hacking's *Representing and Intervening.* Hacking reminded philosophers of science that "experimentation has a life of its own."[14] Historian of science Peter Galison has taken up the challenge and argued in favor of "granting a measure of autonomy to the practices of experimentation and instrumentation" that renders them respectable subjects of inquiry. Galison calls for "a history of the laboratory without idolatry and without iconoclasm," a history of experimentation that "accords that activity the same depth of structure, quirks, breaks, continuities, and traditions that we have come to expect from theory."[15] Accordingly, Galison has proposed a "brick model" of scientific development and change. In it, on the one hand, the three levels of theory construction, experimental traditions, and instrument building are allowed to unfold their own developmental potentials and to follow their own time requirements in relative autonomy. On the other hand, they are expected to generate new coherences by "intercalation" within the ever changing framework of local, situated research programs.

Timothy Lenoir has suggested locating the relation between experiment and theory in the proper realm of experimental practice itself, as a process of "packaging" practices and concepts: "The very construction of the concepts is intertwined with the practices which operationalize them, give them empirical reference, and make them function as tools for the production of knowledge."[16] The size and shape of these packages, however, depends on the qualification of what is to be taken as practices, and here the scope extends from the more restricted realm of epistemic practices, passing through all kinds of intermediates, to the all-encompassing conception of practice as a "form of life."[17]

Thus, once theory as the distinguishing and distinctive feature of the scientific enterprise is put back into the context of practice, the quest for the social context of practice commences. Today, a growing history and philosophy-of-science industry is carrying on that move and amalgamating it with what has come to be labeled "science as practice and culture."[18] "Social construction of science" has become a shibboleth for those wishing to be members of the club. Actors, interests, politics, power, and authority have acquired the status of key terms in a "strong program" to treat science on a par with any other cultural activity.[19] That Thomas Kuhn is "among those who have found the claims of the strong program absurd: an example of deconstruction gone mad"[20] might not surprise and might perhaps be put aside as a matter of taste. I will not go into detail here on the variants and certainly not into an extensive evaluation of social constructivist endeavors. They are of concern to my purpose insofar as they—like Steve Shapin and Simon Schaffer in their *Leviathan and the Air Pump*[21]—give voice to the relation of science and society as mutually coproductive, rather than burying the "generative entrenchment" of both under the cover sheet of sociologists' shoptalk about dominance and subordination, expression and influence.[22] From a fundamental epistemological point of view it is perhaps Latour who most explicitly and most radically has called attention to an impasse in "science and society" studies from which there seems to be no easy escape. To put it sharply: what do we gain by substituting social conditions for what, long enough, have been taken to be the natural referents of scientific activity? If, in the perspective of social construction, we have lost the illusion of an ultimate referent called "nature," what do we gain by trying to compensate for this loss with the mirror image of "society" as a new and insurmountable foundation?[23] From where do we hope to derive the epistemic legitimation of this move? Andrew Pickering, in his most recent book, raises similar arguments in the performative idiom of the "mangle of practice."[24]

With the tetragonic coordination of theory and practice, nature and society, we remain, despite all rotation of competences, within the confines of a conceptual framework that Jacques Derrida has qualified as the logocentric legacy of occidental metaphysics. "It could perhaps be said that the whole of philosophical conceptualization, which is systematic with the nature/culture opposition, is designed to leave in the domain of the unthinkable the very thing that makes this conceptualization possible."[25] In discussing Lévi-Strauss's anthropological writings, Derrida sees him conserving "all these old concepts within the domain of empirical discovery while here and there denouncing their limits, treating them as tools which can still be used. No longer is any truth value attributed to them; there is a readiness to abandon them, if necessary, should other instruments appear more useful. In the meantime, their relative efficacy is exploited, and they are employed to destroy the old machinery to which they belong and of which they themselves are pieces."[26] Derrida himself prefers to question the whole conceptual framework by working with/on its limits. His endeavor has come to be known as "deconstruction."[27]

Latour's model of science in action presents itself as a "hybridogony" of networks that are "simultaneously real, like nature, narrated, like discourse, and collective, like society."[28] To mention only one aspect, but in any event a very central one of his model, it tries to capture how, in the process of engendering scientific "things," conditions are created for what I would like to call unprecedented events. Latour ponders,

No matter how artificial the setting, something new, independent of the setting, has to get out, or else the whole enterprise is wasted. It is because of this "dialectic" between fact and artefact, as Bachelard puts it, that although no philosopher defends a correspondence theory of truth it is absolutely impossible to be convinced by a constructivist argument for more than three minutes. Well, say an hour, to be fair.[29]

Thus, no debunking of the sciences in the name of another authority, but

instead of taking in their objectivity, their truth, their coldness, their exterritoriality—qualities they have never had, except after the arbitrary withdrawal of epistemology—we retain what has always been most interesting about them: their daring, their experimentation, their uncertainty, their warmth, their incongruous blend of hybrids, their crazy ability to reconstitute the social bond.[30]

The following chapters try to convey a feeling for this quandary. The concepts that will accompany my narrative do not claim to posit themselves beyond the tetragon of the conceptual coordinates just mentioned—theory and practice, nature and society—at least not immediately and not from the beginning. Their purpose is humbler and more modest. I hope to help, through a series of displacements, crack open this age-worn framework that has become so deeply entrenched in our minds. I will stick to the Derridean program of reworking these oppositions from within, of trying to perfuse and defer their limits. I will try to widen the gaps in which the very effort of tracing these distinctions has buried them, to the extent that we are getting a sense for what they have been hiding before us. Once again with Freud, philosophy and history of science, too, must derive their own "abstract ideas" from somewhere other than the "new," still pending, "observations alone." Instead of remaining categorial beyond question, these notions have to become involved in a process in which "the material is further worked over" so that perhaps, one day, they can be "replaced" and "discarded without damage." Only the work of displacement itself will be able to tell us whence, when, and whether this will happen. I therefore start from the more narrowly conceived type of move from theory to experiment within the realm of scientific activity characterized by Hacking and Galison, and try to develop a framework in which experimentation takes meaning as a set of epistemic practices that constitute a specific kind of material culture. Nevertheless, this project is ambitious. It tries to characterize as incredibly prolific hybrids those structures that are recalcitrant to being classified as belonging to either realm, the natural or the social, the theoretical or the practical alone.

That "Scarcely Imaginable Basic Concept"

The concept around which this book has taken shape initially was itself a kind of "nebulous, scarcely imaginable basic concept." It is the *experimental system* as a point of orientation for the historian in the overly complex happenings of the modern empirical sciences. Its "somewhere" can be located in the everyday practice of the sciences at issue in this book: mid-twentieth-century biochemistry and molecular biology. To explore its historical emergence in biology and to trace its disciplinary range remains a task that is beyond the limits of this book. Be this as it may, the notion of experimental system is frequently used by scientists in biomedicine, biochemistry, biology, and molecular biology to characterize the space and scope of their research activity. Whoever asks a contemporary laboratory bioscientist what he or she is doing will be told about his or her "system" and the things that happen there. It is, in the first place, a practitioner's notion, not an observer's. Just to mention one example among scores that can be found in the research literature, Mahlon Hoagland speaks of the "selection of a good system" as a key to success on the "itinerary into the unknown."[31]

Only very recently have historians of science started to become aware of the descriptive and historiographical potentials of this hazy and fuzzy concept.[32] David Turnbull and Terry Stokes have used the notion of "manipulable systems" in their case study on malaria research at the Walter and Eliza Hall Institute of Medical Research in Melbourne.[33] Robert Kohler, in dealing with Drosophila, Neurospora, and the rise of biochemical genetics, has spoken of "systems of production."[34] By now, the term *experimental system* is becoming fashionable.[35] The concept, as I use it, does not derive its justification from an a priori definition of how the empirical sciences ought to be shaped and of what they ought to be about. Neither must the concept of experimental system, despite the connotations commonly attributed to the notion of a system, be placed within the framework of systems theories. Nowhere do I argue on the basis of a systems theory. It will be easy to see, however, that notions such as "differential reproduction" (see Chapter 5) play an important role in sociological accounts of science on a systemic level.[36] For me, the notion of experimental system marks a point of departure for "further working over" that "material" of experience with which the historian of the empirical sciences is confronted. To try to do so is inevitably a deconstructive endeavor because one has to work with concepts whose appropriateness is the very point at issue.

How far this notion will take us, and above all what it is able to carry along with it, will be explored from different perspectives in the following chapters. In a very general sense, this book is concerned with experimental reasoning. But this expression has to be used cautiously because it can easily be misunderstood. For its grammatical structure presupposes reasoning as the *genus proximum*, whose specific difference it is to be guided by experimentation. What is at stake, however, is just the opposite. I try to circumscribe and identify a kind of movement oriented and reoriented by generating its own boundary conditions, within which reasoning displays itself as a dynamic interaction between material entities swept off by tracing. "In science, an idea can become substance only if it fits into a dynamic accumulating body of knowledge."[37] The dynamic body of knowledge, the network of practices structured by laboratories, instruments, and experimental arrangements, is a reasoning machinery in its own right. Gaston Bachelard has spoken of the instruments of modern research as "theories materialized" and has concluded, "contemporary science thinks with/in its apparatuses."[38] These instruments embody the heavy load of knowledge taken for granted at a particular time. Correspondingly, by using a kind of mirror image, we might characterize theories, when come of age, as "machines idealized." I agree with Bachelard that we need to know more about this peculiar kind of "scientific real," whose "proper noumenal contexture it is to be able to orient the axes of the experimental movement."[39] In analogy to Wittgenstein's well-known expression, we could call this a "tracing-game." Wittgenstein says: "I shall also call the whole, consisting of language and the actions into which it is woven, the 'language-game.' "[40] And he continues: "Our mistake is to look for an explanation where we ought to look at what happens as a 'proto-phenomenon.' That is, where we ought to have said: *this language-game is played*."[41] Just as we are never able to locate ourselves behind this weaving as language users, we are not able to locate ourselves behind the conceptual tracing-game as scientific practitioners. Thus, I am not looking for a "logic" in the relationship between theory and experiment. I am not looking for a logic behind experiment. Rather, I am grappling with what must be seen, irreducibly, as the "experimental situation." In this situation, which is irrevocably local and situated in space and time, there are scientific objects and the technical conditions of their coming into existence, there is differential reproduction of experimental systems, there are conjunctures of such systems, and graphematic representations. All these are notions related to the process of producing what I shall call epistemic things, and all will be explored further in the following chapters.

Within these complex, tinkered, and hybrid settings of emergence, change, and obsolescence, scientific objects continually make their appearance and eventually recede into technical, preparative subroutines of an ongoing experimental manipulation. As a result, there is again a continuous generation of new phenomena, which need not have anything to do either with the preceding assumptions or with the presupposed goals of the experimenter. They usually begin their lives as recalcitrant "noise," as boundary phenomena, before they move on stage as "significant units."

Concatenations

Throughout this exposition as well as in the chapters that follow, I frequently refer to a series of French-speaking philosophers, scientists, and historians of science, from Gaston Bachelard, Georges Canguilhem, Michel Foucault, Louis Althusser, Jacques Lacan, and Jacques Derrida to Michel Serres and Bruno Latour; from Claude Bernard to François Jacob, Isabelle Stengers, and Ilya Prigogine. To speak of a series here is not to imply a genealogy. Such a genealogy does not exist. What exists is a finespun network of demarcation lines. The notion of network emphasizes the links by which the pieces of a system are held together.[42] One might accuse me of producing a *contradictio in adjecto* by introducing the term "*network of demarcations.*" I do it to stress that what makes each of these figures count in my account is just what makes them *different*. In contrast to looking first and foremost on what holds the threads of this web together, I am interested in what differentiates its texture. Therefore, I also avoid speaking of a tradition of structuralism, of poststructuralism, of deconstructionism, or postmodernism. What I see is an endless series of displacements, a concatenation of attempts to answer, from widely different starting points and domains of experience, the basic epistemological questions raised by the *sciences* of our century. If there is a glue that pastes them together, it is the transpositivistic challenge to objectivity.

Not without recourse to Nietzsche, Freud, and Martin Heidegger, a move reverberates throughout this web that touches the roots of the occidental *episteme*. What is at stake is the fissure between knowledge and truth, the fragmentation of the unity of knowledge through the sciences themselves, in space and time. What is at stake is the grand project of modernity, the instantiation of Kant's rationalist credo that we understand only what we can make in terms of our conceptualizations. In his *Critique of Judgment*, at the end of the section on the analytics of teleological judgment, Kant states: "[When] we study nature in terms of its mechanism, we keep to what we can observe or experiment on in such a way that we could produce it as nature does, at least in terms of similar laws; for we have complete insight only into what we can ourselves make and accomplish according to concepts."[43] Meanwhile, that credo has taken on a very non-Kantian appearance. What we can ourselves make and accomplish, we always only know in the form in which we locally do it, and not even this completely. "The grand narrative legitimation of the history of science as a history of rationality, progress or the search for truth must go, but so too must the debunking of science which too often motivates the repudiation of such modernist narratives. Take legitimations of scientific practices and beliefs always to be partial, and to take place in specific contexts, for particular purposes, to which large-scale (de-)legitimation has little relevance."[44] No remedy has ever been raised against the weeds that always spring up eventually.[45] We would not have the incommensurable plurality of the sciences as we experience—and fear—them today if their movement were not excessive, if they were not continuously producing a surplus that is beyond what we may have wanted, beyond what we might have been able to imagine. In this way, time and again they prevent the closure that a whole epoch of philosophical systems, from Descartes to Hegel, from Gottlob Frege to Rudolf Carnap, has tried to impose on them. Do we finally begin to understand Jacques Lacan's "strange remark," in his seminar on "Science and Truth," that "our science's prodigious fecundity is to be examined in relation to the fact, sustaining science, that science doesn't want-to-know-anything about the truth as cause"?[46]

Consequently, on an epistemological level, we need a "philosophy of epistemological detail."[47] Traditional philosophy has seen in this situation a necessary and intrinsic limitation of empirical knowledge—whence that longing for an "integral philosophy" that would restore the Gordian knot cut through by the positive sciences. One can see it differently. One can conceive of this presumed limitation as being the prerequisite for the occurrence of events that cannot be anticipated. As a rule, the new is the result of spatiotemporal singularities. There is reason to assume that this is especially the case for matters of knowledge. Indeed, experimental systems are arrangements that allow us to create cognitive, spatiotemporal singularities. They allow us to produce, in a regular manner, unprecedented events. In this sense such systems are "more real" than reality. The "scientific real," therefore, is not that ultimate referent to which all knowledge must finally accommodate itself. The reality of epistemic things lies in their resistance, their capacity to turn around the (im)precisions of our foresight and understanding. As Michael Polanyi says,

this capacity of a thing to reveal itself in unexpected ways in the future, I attribute to the fact that the thing observed is an aspect of reality, possessing a significance that is not exhausted by our conception of any single aspect of it. To trust that a thing we know is real is, in this sense, to feel that it has the independence and power for manifesting itself in yet unthought of ways in the future.[48]

CHAPTER 2

Experimental Systems and Epistemic Things

❖❖   [The] meticulous care required to connect things in
     unbroken succession.
         —Goethe, "The Experiment As Mediator between
                 Object and Subject"

At a symposium on the structure of enzymes and proteins in 1955, Paul
Zamecnik read a paper on the "Mechanism of Incorporation of Labeled
Amino Acids into Protein." When, in the ensuing discussion, Sol Spiegelman reported his own experiments on the induction of enzymes in yeast
cultures, Zamecnik responded, "we would like to study induced enzyme
formation, too; but that reminds me of a story Dr. Hotchkiss told me of a
man who wanted to use a new boomerang but found himself unable to
throw his old one away successfully."[1]

What Does It Mean to Do Experiments?

Better than any lengthy description, the opening anecdote illustrates an
essential feature of experimental practice. It expresses an experience familiar to every working scientist: the more he or she learns to handle his
or her own experimental system, the more it plays out its own intrinsic
capacities. In a certain sense, it becomes independent of the researcher's
wishes just because he or she has shaped it with all possible skill. What
Lacan states for the structuralist human sciences holds here, too: "The
subject is, as it were, internally excluded from its object."[2] It is this "intimate exteriority," or "extimacy,"[3] captured in the image of a boomerang,
that we may call virtuosity.

     Virtuosity creates pleasure. When Alan Garen once asked Alfred
Hershey for his idea of scientific happiness, he answered: "To have one
experiment that works, and keep doing it all the time." As Seymour

Benzer wrote later, this became known as "Hershey Heaven" among the first generation of molecular biologists.[4]

In his autobiography, François Jacob has formulated the same experience from the perspective of being engaged in an ongoing research process:

> In analyzing a problem, the biologist is constrained to focus on a fragment of reality, on a piece of the universe which he arbitrarily isolates to define certain of its parameters. In biology, any study thus begins with the choice of a "system." On this choice depend the experimenter's freedom to maneuver, the nature of the questions he is free to ask, and even, often, the type of answer he can obtain.[5]

Thus, we have, at the basis of biological research, the choice of a system, and a range of maneuvers that it allows us to perform. Which is, if I see it correctly, a specific reformulation of Heidegger's claim that to "open up a sphere," and to "establish a procedure" is what modern research, considered as representing the "essence" of occidental science, is all about.

In his "The Age of the World Picture," Heidegger states with respect to the modern sciences:

> The essence of what we today call science is research. [But] in what does the essence of research consist? In the fact that knowing establishes itself as a procedure within some realm of what is, in nature or in history. Procedure does not mean here merely method or methodology. For every procedure already requires an open sphere [*offener Bezirk*] in which it moves. And it is precisely the opening of such a sphere that is the fundamental event in research.[6]

To open a sphere and to establish a procedure: such ought to be the grounding feature of the modern sciences, viewed from a Heideggerian point of view.

With respect to intent and context, these quotations are utterly different. Zamecnik, Garen, Jacob, and Heidegger speak about experimentation in the light of acquaintance, satisfaction, constraint, and conquest. But in another respect they coincide: they all identify a research setting, or experimental system, as the core structure of scientific activity. Such a view, if taken seriously, entails epistemological as well as historiographical consequences. If we accept the thesis that *research* is the basic procedure of the modern sciences, we are invited to explore how research gets enacted at the frontiers between the known and the unknown. If we accept that biological research in particular begins with the choice of a system rather than with the choice of a theoretical framework, it will be in order to focus attention on the characterization of experimental systems, their structure, and their dynamics. To speak of the "choice" of a system here does not mean that such arrangements are there from the beginning. To arrive at an experimental system is itself a laborious process, as my case study of the group at MGH will show. My emphasis is on the materialities of research. Therefore, as my point of departure I will not directly address the theory and practice issue and the relation between theory and practice, the theory-ladenness of observation, or the underdetermination of theory by experiment. My approach tries to escape this "theory first" type of philosophy of science perspective. For want of a better term, the approach I am pursuing might be called "pragmatogonic." I would like to convey a sense of what it means for the participants in the endeavor to be engaged in epistemic practices, that is, in irrevocably experimental situations. Here I claim, with Frederick Holmes, "it is the investigations themselves which are at the heart of the life of an active experimental scientist. For him ideas go in and come out of investigations, but by themselves are mere literary exercises. [I]f we are to understand scientific activity at its core, we must immerse ourselves as fully as possible into those investigative operations."[7]

In this chapter, I first turn to some structural characteristics of such investigative operations on the level of relatively *longue durée*. Let me recall an episode from the end of the eighteenth century. When Goethe was performing the optical experiments that led to his theory of colors, he wrote, in 1793, a remarkable essay entitled "The Experiment as Mediator between Object and Subject."[8] In this essay, Goethe addresses his problem in a similar, but still different, vein, neither with respect to virtuosity nor to pleasure, but—conforming to what Friedrich Kittler has called the "Aufschreibesystem 1800"[9]—with respect to the duty of the scientist. The central sentence reads as follows: "To follow every single experiment through its variations is the real task of the scientific researcher." Goethe compares what he calls "Versuch" with a point from which light is emitted in all possible directions. Through the step-by-step exploration of all of them, a research network is built up that eventually will come into contact with neighboring networks. Establishing such fields, according to Goethe, is the primary task of the experimentalist; disciplinary junctures may be the final outcome of his endeavor. "Thus when we have done an experiment of this type, found this or that piece of empirical evidence, we can never be careful enough in studying what lies next to it or derives directly from it. This investigation should concern us more than the discovery of what is related to it."[10] Five years later, Goethe asked Schiller to comment.[11] In his reply, Schiller immediately pointed to the core of the argument: "It is quite obvious to me how dangerous it is to try to demonstrate a theoretical proposition directly by experiments."[12]

Experimental Systems

According to a long-standing tradition in philosophy of science, experiments have been seen as singular, well-defined empirical instances embedded in the elaboration of a theory and performed in order to corroborate or to refute certain hypotheses. In the classical formulation of Karl Popper, "the theoretician puts certain definite questions to the experimenter, and the latter, by his experiments, tries to elicit a decisive answer to these questions, and to no others. All other questions he tries hard to exclude."[13] Despite the radical shift in perspective in which social studies of science have attempted to deny the naked experiment its ability to decide scientific controversies, the familiar notion of the experiment as a test of a hypothesis is still virulent in them. Even Harry Collins's argument from the "experimenter's regress" embraces, in its very rejection, a view of the experiment as an ultimate arbiter.[14]

What does it mean to speak of experimental systems, in contrast to this clear-cut rationalist picture of experimentation as a theory-driven activity? Ludwik Fleck, Popper's long neglected contemporary, has drawn our attention to the manufacture of scientific practices in twentieth-century biomedical sciences and has argued that—contrary to Popper's claim—scientists usually do not deal with single experiments in the context of a properly delineated theory. "Every experimental scientist knows just how little a single experiment can prove or convince. To establish proof, an entire *system of experiments* and controls is needed, set up according to an assumption or style and performed by an expert."[15] A researcher thus does not, as a rule, deal with isolated experiments in relation to a theory, but rather with a whole experimental arrangement designed to produce knowledge that is not yet at his disposal. What is even more important, the experimental scientist deals with systems of experiments that usually are not well defined and do not provide clear answers. Fleck even goes so far as to claim that "if a research experiment were well defined, it would be altogether unnecessary to perform it. For the experimental arrangements to be well defined, the outcome must be known in advance; otherwise the procedure cannot be limited and purposeful."[16] These remarks are not to be taken as a trivial characterization of a de facto imperfection of a particular research activity. They are to be taken as a profound re-orientation of our view of the inner workings of this process, a process "driven from behind,"[17] a genuinely polysemic procedure defined by ambiguity, not one just limited by finite precision.

Experimental systems are to be seen as the smallest integral working units of research. As such, they are systems of manipulation designed to give unknown answers to questions that the experimenters themselves are not yet able clearly to ask. Such setups are, as Jacob once put it, "machines for making the future."[18] They are not simply experimental devices that generate answers; experimental systems are vehicles for materializing questions. They inextricably cogenerate the phenomena or material entities and the concepts they come to embody. Practices and concepts thus "come packaged together."[19] A single experiment as a crucial test of a properly delineated conception is not the simple, elementary, or basic situation of the experimental sciences. The inverse holds. Any simple case is the "degeneration" of an elementarily complex experimental situation. As Bachelard reminds us, "simple always means simplified. We cannot use simple concepts correctly until we understand the process of simplification from which they are derived."[20] It is only in the process of making one's way through a complex experimental landscape that scientifically meaningful simple things get delineated; in a non-Cartesian epistemology, they are not given from the beginning. They are the inescapably historical product of a purification procedure.[21] This is, again and again, the experience we find when we look at autobiographical science narratives.[22] But this is also what we find when we try to follow particular cases in the history of the modern life sciences. One of them will be expounded in Chapter 3 and traced throughout the rest of the book.

Epistemic Things, Technical Objects

In inspecting experimental systems more closely, two different yet inseparable elements can be discerned.[23] The first I call the research object, the scientific object, or the "epistemic thing." They are material entities or processes—physical structures, chemical reactions, biological functions—that constitute the objects of inquiry. As epistemic objects, they present themselves in a characteristic, irreducible vagueness. This vagueness is inevitable because, paradoxically, epistemic things embody what one does not yet know. Scientific objects have the precarious status of being absent in their experimental presence; they are not simply hidden things to be brought to light through sophisticated manipulations. A mixture of hard and soft, like Serres's veils, they are "object, still, sign, already; sign, still, object, already."[24] With Bruno Latour, we can claim it to be characteristic for the sciences in action that "the new object, at the time of its inception, is still undefined. [At] the time of its emergence, you cannot do better than explain what the new object is by repeating the list of its constitutive actions. [The] proof is that if you add an item to the list you *redefine the object*, that is, you give it a new shape."[25]

To enter such a process of operational redefinition, one needs an arrangement that I refer to as the experimental conditions, or "technical objects." It is through them that the objects of investigation become entrenched and articulate themselves in a wider field of epistemic practices and material cultures, including instruments, inscription devices, model organisms, and the floating theorems or boundary concepts attached to them. It is through these technical conditions that the institutional context passes down to the bench work in terms of local measuring facilities, supply of materials, laboratory animals, research traditions, and accumulated skills carried on by long-term technical personnel. In contrast to epistemic objects, these experimental conditions tend to be characteristically determined within the given standards of purity and precision. The experimental conditions "contain" the scientific objects in the double sense of this expression: they embed them, and through that very embracement, they restrict and constrain them.[26] Superficially, this constellation looks simple and obvious. But the point to be made is that within a particular experimental system both types of elements are engaged in a nontrivial interplay, intercalation, and interconversion, both in time and in space. The technical conditions determine the realm of possible representations of an epistemic thing; and sufficiently stabilized epistemic things turn into the technical repertoire of the experimental arrangement.

Take the following example, to which I will return in detail in Chapter 13: When Heinrich Matthaei and Marshall Nirenberg, in their bacterial in vitro system of protein synthesis, introduced synthetic polyuridylic acid, among other ribonucleic acids, as a possible template for polypeptide formation, the genetic code assumed the quality of an experimental epistemic thing. When the genetic code was solved, the polyuridylic acid assay, within the same in vitro system, was turned into a subroutine for the functional elucidation of the protein synthesizing organelles, the ribosomes. To add one more example, less than twenty years ago, enzymatic sequencing of DNA was a scientific object par excellence. It was a new possible mode of primary structure determination among older ones.[27] A few years later, it became a procedure that had been adopted by the leading DNA laboratories around the world. In the early 1980s, it was transformed into a technical object with all the characteristics of such a "translation." Today, every biochemical laboratory may order a sequence kit, including buffers, nucleotides, and enzymes from a biochemical company, and perform the sequence reaction routinely in a semi-automatic machine. Latour has spoken of "black boxing" in this context.[28] Unfortunately, this expression mainly reflects one particular aspect of the process: its "routine" nature after the event. Perhaps at least as important, however, is its impact on a new generation of emerging epistemic things. Black boxing does not mean just setting aside.

Through this kind of recurrent determination, certain sets of experiments become clearer in some directions but at the same time less independent because they more and more rely on a hierarchy of established procedures. "Once a field has been sufficiently worked over so that the possible conclusions are more or less limited to existence or nonexistence, and perhaps to quantitative determination, the experiments will become increasingly better defined. But they will no longer be independent, because they are carried along by a system of earlier experiments and decisions."[29]

The difference between experimental conditions and epistemic things, therefore, is functional rather than structural. We cannot once and for all draw such a distinction between different components of a system. Whether an object functions as an epistemic or a technical entity depends on the place or "node" it occupies in the experimental context. Despite all possible degrees of gradation between the two extremes, which leave room for all possible degrees of hybrids between them, their distinctness is clearly perceived in scientific practice. It organizes the laboratory space with its messy benches and specialized local precision services as well as the standard scientific text with its specialized sections on "materials and methods" (technical things), "results" (halfway-hybrids) and "discussion" (epistemic things).[30]

If both types of entities are engaged in a relation of exchange, of blending and mutual transformation, why then not cancel the distinction altogether? Does it not simply perpetuate the traditional, problematic distinction between basic research and applied science, between science and technology? If science in action should not be conceived in terms of an asymmetric relation from theory to practice, why then uphold a gradient between epistemic and technical objects? Why then construct a division whose only effect is that it permanently has to be undone? The answer is:

because it helps to assess the game of innovation, to understand the occurrence of unprecedented events and with that, the essence of research.

A Little Note on Technoscience

My remarks in the preceding section suggest that the notion of "technoscience" often used in science studies to characterize contemporary scientific large-scale enterprises needs to be handled with caution.[31] The tendency to lump together what should be understood in its interaction is already virulent in the notion of "phenomeno-technology," which, according to Bachelard, "takes its instruction from construction."[32] Technoscience suggests an identity of science and technology that, I argue, tends to disguise the essential tension of the research process—no matter whether we are concerned with big or little science, hard or soft. It subscribes to the domination of the "theme" (in my words, epistemic objects) by the "method" (in my words, technical objects) that Heidegger, twisting around a sentence of Nietzsche, has characterized by the following words:

In the sciences, not only is the theme drafted, called up [*gestellt*] by the method, it is also set up [*hereingestellt*] within the method and remains within the framework of the method, subordinated to it [*untergestellt*]. The furious pace at which the sciences are swept along today—they themselves don't know whither—comes from the speed-up drive of method with all its potentialities, a speed-up that is more and more left to the mercy of technology. Method holds all the coercive power of knowledge. The theme is a part of the method.[33]

In this passage, Heidegger sees the sciences as tending to become subordinated to and finally swallowed by technology. Heidegger claims that "from the point of view of the sciences, it is not just difficult but impossible to see this situation."[34] Let me claim, in contrast, that it is exactly the viewpoint of opposing philosophy to technoscience and identifying scientific knowledge with "technowledge" that finally leads to the exile of the "theme" and to its surrender to Heideggerian "thinking." I perceive thinking as remaining a constitutive part of experimental reasoning, conceived as an embodied disclosing activity that transcends its technical conditions and creates an open reading frame for the emergence of unprecedented events.

Mahlon Hoagland, like many of his fellow molecular biologists, sees scientific activity basically as a "generator of surprises" on the "itinerary into the unknown."[35] Research produces futures, and it rests on differences of outcome. In contrast, technical construction aims at assuring presence, and it rests on identity of performance. How could it fulfill its purposes otherwise? If the momentum of science gets absorbed into technology, we end up with "extended present."[36] A technical product, as everybody expects, has to fulfill the purpose implemented in its construction. It is first and foremost an answering machine. In contrast, an epistemic object is first and foremost a question-generating machine.[37] It is not technical in itself, although it grants the "goings-on of technics," as Samuel Weber appropriately has translated *Wesen* in the context of Heidegger's "Questing After Technics": "The goings-on of technics are ongoing, not just in the sense of being long-standing, staying in play, lasting, but in the more dynamic one of moving away from the pure and simple self-identity of technology. What goes on, in and as technics, its *Wesen*, is not itself technical."[38]

Yes, technical tools define any system of investigation—"any study thus begins with the choice of a 'system.'" They circumscribe the boundaries of an experimental system. Proper fluctuation and oscillation of epistemic things within an experimental system require appropriate technical and instrumental conditions. Without a system of sufficiently stable identity conditions, the differential character of scientific objects would remain meaningless; they would not exhibit the characteristics of epistemic things. We are confronted with a seeming paradox: the realm of the technical is a prerequisite of scientific research. On the other hand and at any time, the technical conditions tend to annihilate the scientific objects in the sense attributed to this notion. The solution to the paradox is that the interaction between scientific object and technical conditions is eminently nontechnical in its character. Scientists are, first and foremost, *bricoleurs* (tinkerers), not engineers. In its nontechnicality, the experimental ensemble of technical objects transcends the identity condition of its parts. According to the same pattern, established tools can acquire new functions in the process of their reproduction. Their insertion into a productive or consumptive process beyond their intended use may reveal characters other than the original functions they were designed to perform.[39]

A Word on Historiography

Research systems are tinkered arrangements that are not set up for the purpose of repetitive operation but for the continuous reemergence of unexpected events. Experimentation, as a machine for making the future, has to engender unexpected events. However, it also channels them, for their significance ultimately derives from their potential to become, sooner or later, integral parts of future technical conditions. This movement implies that, in the last resort, it is the future integration into the realm of the technical that grants the scientific objects their "legitimate position" within the history of knowledge. No historiography of science can escape this movement of recurrence implanted into its very subject matter: the epistemic things. For every new scientific object sheds a "recurrent light" on those by which it was preceded.[40] A historiography that blindly streamlines this movement *post festum* has been criticized as "whiggish."[41] This is not the place for tracing the arguments against whiggish history and the subsequent critique of this notion.[42] But it is the place to emphasize that no historiography of science—including my own—can escape what might be called a position of "reflected anachronicity."

The Case Study and Its Context

The present investigation, in its case study chapters, concentrates on the history of molecular life science in the formative years between 1947 and 1962. More precisely, it focuses on a particular experimental system, an in vitro system for the biosynthesis of proteins. Even more narrowly, it looks at a particular research group based at the Collis P. Huntington Memorial Hospital of Harvard University at the Massachusetts General Hospital (MGH) in Boston. The work of Paul Zamecnik, Mahlon Hoagland, and their colleagues originated from a cancer research program and, over a period of fifteen years, was transmuted into one of the core systems of the new biology.

Molecular biology, as I hope to show, must be regarded as the result of an extraordinarily complex development that can by no means be described in an adequate fashion through, for example, the fusion of already existing biological disciplines, such as microbiology, genetics, or biochemistry. Nor is it simply another biological discipline supplementing the historically established canon of biological disciplines.[43] Above all, what could be called, with Foucault, the epistemic and technical "formation" of the discourse of molecular biology, is not the straightforward product of the efforts of a few research schools led by a few prominent figures, such as the phage group of the California Institute of Technology in Pasadena and Cold Spring Harbor, the Cavendish crew in Cambridge, and the Pasteur *équipe* in Paris. This is a myth created by some Fest-schriften dedicated to the "members of the club."[44] Neither is it the result of an all-encompassing, paradigmatic theory based on the notion of information. Richard Burian even goes so far as to deny that there exists a unifying theory of molecular biology at all. To assert this, however, is not equivalent to claiming that it was constituted by a mere "battery of techniques."[45] Generally speaking, what we today call molecular biology emerged from and was supported by and constructed of a multiplicity of widely scattered, differently embedded, and loosely (if at all) connected biochemical, genetical, and structural research systems. But all of them, in one way or the other, sought to characterize living beings down to the level of biologically relevant macromolecules. By implementing different modes and models of technical analysis, these systems helped to create a new epistemotechnical space of representation in which the concepts of molecular biology, increasingly revolving around the metaphor of information, gradually became articulated. In terms of what could be called its historical "eventuation," this process is still poorly understood. It appears that we still have to find an appropriate level of analysis through which the key features of the all pervading dynamics of the new biology might become obvious.

In the following chapters, I propose that we turn away from the perspective of a more or less well-defined disciplinary matrix of twentieth-century biology and move toward what scientists are inclined to call their experimental systems. Such systems, I repeat, are hybrid constructions: they are at once local, social, technical, institutional, instrumental, and epistemic settings. As a rule, and insofar as they are research systems, they do not respect macrolevel disciplinary, academic, or national boundaries of science policy and research programs. Insofar as they orient research activity, they may also prove helpful for the orientation of the historian. If experimental systems have a life of their own, precisely what kind of life they have remains to be determined.

In following the development of epistemic things rather than that of concepts, topics, problems, disciplines, or institutions, boundaries have to be crossed, boundaries of representational techniques, of experimental systems, of established academic disciplines, and of institutionalized programs and projects. In following the path of epistemic things, classifications have to be abandoned. Does this study belong to the history of cancer research? of cytomorphology? of biochemistry? of molecular biology? Is it a prehistory of protein synthesis? All of these—and none. My path takes its starting point from protein synthesis research as part of a cancer research program. By way of differential reproduction, by way of the implementation of skills, tracing techniques, and instruments, such as laboratory rats, radioactive amino acids, biochemical model reactions, centrifuges, and technical expertise, it gained a momentum of its own. In the rapidly changing landscape of the new biology, it became disconnected from cancer research where it had been rooted. Instead, through several unprecedented shifts, it ended up with transfer RNA, which provided one of the experimental handles for solving the central puzzle of molecular biology: the genetic code.

Most of the material analyzed in this book has so far received little attention from historians of biology or medicine.[46] There are reasons for this. The material cannot easily be categorized as belonging to either fundamental science or technology, to biology or to medicine: it is situated at their intersection. And it cannot easily be reconstructed in terms of paradigmatic conceptual shifts, which renders it resistant to a historiography oriented toward theoretical breakthroughs. Instead, the breakthroughs I am describing lie in the disseminating power of epistemic things that eventually became transformed into technical things. They lie in the structure of a particular experimental culture of representation, of rendering biological processes manipulable in vitro, which is so characteristic of the life sciences of our century.

Historians of physical specialties are confronted with new technologies in the first place when analyzing such shifting experimental cultures. And they tend to think of instruments in terms of devices that become more and more sophisticated, eventually ending up as large-scale machinery. In biochemistry and in molecular biology, this is not necessarily the case. No ultracentrifuge was instrumental in establishing the in vitro protein synthesis system to be described in the next chapter—although this instrument became crucial for its subsequent development. In fact, the most efficient biochemical and molecular biological instruments are those that accommodate themselves to the level of the analysis—that is, ultimately to the level of molecules. In the establishment of the in vitro protein synthesis system, radioactive amino acids happened to play this role of molecular tools, or instruments. Of course, it goes without saying that there is no routine purchase of isotopes without big machinery such as cyclotrons.[47] But the organic synthesis of amino acids from these isotopes can be, and was, at the beginning, performed with the moderate equipment of an organic chemistry laboratory. On the one hand, such tracer molecules are technical devices for following particular metabolic pathways. But insofar as they are integral parts of the scientific object under scrutiny, it is not easily possible, in the given case, to draw a clear-cut distinction between the scientific object and the technical conditions of its evaluation. It depends largely on the experimental context whether radioactive tracing is to be considered a technical means of analysis or whether it constitutes the epistemic things that are the objects of research.

In addition, instruments by themselves are not the moving forces of the experimental "goings-on." It is their embedment in experimental systems that counts. Instruments display their power only as integral parts of what I call spaces of representation.[48] Without a space of tracing, things cannot be treated as part of the "scientific real."[49] Representations are epistemic things in the first place, they are traces deriving from things like "tracers" rather than concepts. The fractional partition of a cell homogenate and the corresponding radioactivity pattern constitute a representation of the cytoplasm upon which it is possible to act: a material space of signification.

My case study here of the group at MGH shows that it is not the overall orientation of an institutional setting or the initial formulation of a research program, or the sheer introduction of new technology that ultimately determines that program's subsequent direction and scientific productivity. Thus, there is no possibility of a deterministic account, be it socially, theoretically, or technically motivated. Experimental systems grow slowly into a kind of scientific hardware within which the more fragile software of epistemic things—this amalgam of halfway-concepts, no-longer-techniques, and not-yet-values-and-standards—is articulated, connected, disconnected, placed, and displaced. Certainly they delineate the realm of the possible. But as a rule, they do not create rigid orientations. On the contrary, it is the hallmark of productive experimental systems that their differential reproduction leads to events that may induce major shifts in perspective within or even beyond their confines. In a way, they proceed by continually deconstructing their own perspective. Experimental systems, in fact, do not and cannot tell their story in advance.

Let me conclude this chapter with a quotation from Brian Rotman. It is a note on the xenogenesis of texts, and I find it very appropriate for describing an experimental system: "What [a xenotext] signifies is its capacity to further signify. Its value is determined by its ability to bring readings of itself into being. A xenotext thus has no ultimate 'meaning,' no single, canonical, definitive, or final 'interpretation': it has a signified only to the extent that it can be made to engage in the process of creating an interpretive future for itself."[50]

Experimental systems give laboratories their special character as particular cultural settings: as places where strategies of material signification are generated.[51] It is not, in the end, the scientific or the broader culture that determines "from outside" what it means to be a laboratory, a manufactory of epistemic things becoming transformed, sooner or later, into technical things, and vice versa. It is "inside" the laboratory that those master signifiers are generated and regenerated that ultimately gain the power of determining what it means to be a scientific—or a broader—culture.

CHAPTER 3

Out of Cancer Research, 1947–50

❖❖ If you are interested in a clinical problem, start back about
three steps from where you really want to work. Then you
will find out something, although it may not be what you
were hoping for.

—Edwin J. Cohn

When I first met Paul Charles Zamecnik, in March 1990 at the Worcester
Foundation for Experimental Biology in Shrewsbury, Massachusetts, I
encountered a man in his late seventies who had plunged himself into the
study of viruses and their inhibition by so-called "antisense" oligonucleotides. I wanted to know more about the early days of protein synthesis in
which he had been so profoundly involved some forty years before. He
wanted to explain to me the details of his ongoing research on the human
immunodeficiency virus and a possible cure for AIDS.[1]

Paul Zamecnik graduated from Dartmouth College and received his
M.D. from Harvard Medical School in 1936. During the following years,
he was at the Collis P. Huntington Memorial Hospital in Boston, at
Harvard Medical School, and at the University Hospitals in Cleveland. In
the course of his internship in Cleveland, 1938–39, he became interested
in the regulation of growth. He recalls:

On inquiry from a number of medical faculty, I found one who knew of a
scientist at the Rockefeller Institute who was studying protein synthesis. This was
Max Bergmann, an organic chemist who had recently come from Germany, and
who was synthesizing peptides, using a new procedure. Inside cells he found
enzymes capable of hydrolyzing these peptides in very specific ways. He sus-
pected that the same enzymes might catalyze the synthesis of peptides and proteins. I applied for a fellowship to work with him, and offered to try to grow
tissue cultures, in which to study protein synthesis and the role of his enzymes in
this process. Dr. Bergmann stated that he had only organic chemists in his laboratory, and that if I were serious I should get more disciplined training in chemistry,
then apply to him again in a year or two.[2]

To learn more about protein chemistry, Zamecnik spent the next year at Carlsberg Laboratories in Copenhagen, with Kai Linderstrøm-Lang, a noted protein physical chemist. Then, driven out of Denmark by the German occupation in 1940, he returned to the United States via Capri and worked with Max Bergmann for a year. "This time Bergmann took me (1941–1942). I really knew very little more, but stardust from the famous Carlsberg Laboratories had fallen on my shoulders."[3]

Back at MGH, Zamecnik became involved in studies on the toxic factors in experimental traumatic shock. This was a wartime project that the Office of Scientific Research and Development (OSRD) had asked Huntington's director, Joseph Charles Aub, and his associates to work on.[4] In fact, Zamecnik did not return to his own research agenda until the war was over.

The aim of this and the following chapters is neither institutional nor biographical.[5] No hagiography is intended. Behind a prominent name, and behind a prominent institution, there is the world of the laboratory bench, the world of those who are scarcely mentioned in public. Nor will I try to confer logical, *post festum* coherence to a work that none of the protagonists knew initially would lead into the heart of molecular biology. Rather, I intend to show, from a detailed screening of the published record, from personal recollections of some of the actors, and from note-books as well as other laboratory and archival materials, how a particular experimental system arose from the pursuit of cancer research in a medical setting—the Huntington Laboratories at MGH—how it gained a bio-chemical momentum, and how it was, in the course of the years, trans-formed into a system for pursuing questions of molecular biology. Al-though Zamecnik and other members of the group have given historical accounts on the in vitro protein synthesis system on several occasions, they are less explicit about the details of the early phase of their work.[6]

The Local Context: A Cancer Research Setting

By the end of World War II, the John Collins Warren Laboratories of Huntington Memorial Hospital, housed at MGH, were headed by Joseph Charles Aub, who had succeeded George Minot in this function in 1928. An oncologist committed to the study of metabolic disease, Aub had reoriented the cancer research agenda of the Huntington, whose medical laboratories had moved to the MGH in 1942. With the approval of the

Harvard Cancer Commission, he had shifted the research program from techniques for producing tumors to the investigation of normal and pathological growth and regeneration.[7] Zamecnik was a medical doctor himself, and when he was able to redefine his research agenda in 1945, he opted for a cancer-related medical starting point that he hoped eventually to trace to the cellular level. In an application to the International Cancer Research Foundation in March 1945, Zamecnik stated, "We would like to attack the problem of protein synthesis in the tumor cell."[8] As Robert Loftfield once mentioned:

Note that we worked in a cancer lab, our director was an oncologist (very distinguished), we were funded by cancer funds, even movie house collections, many had cancer patients and all worked with cancer researchers [not involved in protein research]. We went to AACR (American Association of Cancer Research) meetings and *wanted* to think we were doing something to fight cancer. The Butter Yellow hepatomas and the ascites tumor lines were there because it was a cancer lab.[9]

Zamecnik chose to look at the possible sites of action of tumor-inducing agents, and hence for "the most promising point at which to begin a search for a metabolic distinction between normal and neoplastic tissue."[10]

Because it is a general characteristic of malignant tissue to display deregulated growth and because cell growth is intimately linked to the cell's capacity to make proteins, the regulation of protein metabolism was a likely target of carcinogenic action. Very little was known about the biochemical factors involved in oncogenesis. So the obvious strategy, Zamecnik decided, was not to consider "a single avenue of biochemical study," but rather to start with a "practical" approach and "take advantage of whatever new opportunities become available, in the hope that a definite clue turned up in any corner of the field."[11] Thus, a strategy of "techno-opportunism."

As a student of Bergmann's who had done much work on the specificity of protein-degrading enzymes, Zamecnik considered the then widespread and popular idea of protein synthesis as a simple reversal of proteolysis.[12] Indeed, in the application just mentioned, this hypothesis figured prominently: "The experiments of Bergmann and his group make it appear likely that intracellular proteolytic enzymes may, under proper conditions, be responsible for normal protein synthesis within the cell." On the other hand, Zamecnik was a neighbor of Fritz Lipmann at Harvard. In fact, he was working with him on an enzyme of the bacterium *Clostridium welchii*.[13] As early as 1941, when he had been appointed a research fellow at MGH,[14] Lipmann had speculated that protein synthesis might proceed from single amino acids and that the energy might be provided by activated amino acid intermediates.[15] Theoretically, such an "entirely different mechanism" could also be envisaged.[16] None of the alternatives had compelling experimental support at the time.

Research Policy at MGH

During these first postwar years, the research policy structures of MGH were reshaped. A Hospital Research Council had been in place since 1938. In 1947, the General Executive Committee and the trustees recommended that the proper place of research in the hospital be reconsidered, and upon their request, a Committee on Research and a Scientific Advisory Committee were appointed. The latter included Karl Compton from MIT, Carl Cori from Washington University, Herbert Gasser from the Rockefeller Institute, and Eugene Landis from Harvard, later to be joined by Linus Pauling from the California Institute of Technology. Whereas the research budget for 1935 had been fifty thousand dollars, it amounted to five hundred thousand in 1948, the year in which the Committee on Research assumed its activity with Paul Zamecnik as its executive secretary. In 1955 the budget had been raised to two million dollars. Practically all this money was, and remained, "soft."

The committees were tasked "to promote, facilitate, and guide research affairs at the Massachusetts General Hospital, in the belief that the staff would serve its cause better as partners than as individuals."[17] This was the language of partnership and cooperative individualism that characterized America's early postwar science policy discourse.[18] At MGH, it had a tradition that can be traced back into the 1930s, with a peculiar emphasis on spontaneity and free decision. Already in 1934, the General Executive Committee of the hospital had stated in its annual report,

cooperation in research when truly spontaneous, is bound to be fruitful. The association, however, must spring from natural interest and the curiosity of the workers. When imposed upon them by someone seeking to compel the investigation of a particular subject, there is apt to be sterility. The quality of research is not strained. It flows from the minds and hands of those who have a special gift for doing it. This its patrons should know so that they may invest in investigators not in investigation. That he may make his maximum contributions to scientific knowledge the able investigator should be made secure, provided with the means to live, and the wherewithal to work. Then he should be left to his own devices and allowed to make his own choice of collaborators. In scientific research, perhaps more than in any other field of human endeavor, one thing leads to another. The solving of a problem but opens up a dozen new ones perhaps not even dreamed of before the first was solved. Free choice of problems, and free choice to follow leads disclosed by the solution of a first problem must be the privilege of the researcher. If he strikes a hot scent the means must be found to permit him to follow it with vigor.[19]

Sixteen years later, the new postwar Scientific Advisory Committee recommended in the same vein that the hospital "should hold steadfastly to a policy wherein the direction of research is determined solely by the inspiration of its own investigators."[20]

Joseph Aub had adopted the spirit of this research philosophy, and Zamecnik, backed by the Committee on Research, was closely following his lead in this respect. There was no general formal or deliberate research plan, neither for the MGH research facilities as a whole nor for the Huntington Laboratories in particular. The responsibility for selecting particular problems resided with the individual scientists, and cooperation was practiced on a voluntary basis with no superimposition of preplanned interdisciplinary projects.[21] Zamecnik's scientific trajectory, as I will show, is characteristic of this research philosophy.

Radioactive Amino Acids as Tracers of Protein Metabolism

Within a few years of the end of World War II, radioactive, ^{14}C-labeled amino acids were becoming available for research. Radioactive tracers were an offshoot of cyclotron technology.[22] MIT had a cyclotron at its disposal, whereas Harvard's cyclotron had been shut down in 1942, "since no physical research, either in pure physics or of military significance [was] contemplated."[23] The carbon isotope ^{14}C was produced from nitrogen by bombardment with slow neutrons, and it could be recovered as carbonate from the solution.[24] Robert Loftfield, a Harvard-trained physical organic chemist who worked as a research associate at the Radioactivity Center of MIT, joined the staff of the Huntington Laboratories as a Research Fellow in Medicine in 1948. Since the 1930s, Aub, who is credited with being among the first to use radioactive tracers in the study of metabolism, had had a strong interest in the new isotopes.[25] After the war, Aub ventured a collaboration between his laboratory and his old MIT friend Robley T. Evans, with whom Aub had studied the excretion of radium in 1936.[26] On request, and in the framework of this collaboration, Loftfield worked out a technically improved method for the synthesis of the ^{14}C-labeled amino acids alanine and glycine.[27] The procedure resulted from a careful chemical engineering study during which "every factor was varied including time, temperature, pressures of ammonia and carbon dioxide, and the nature and physical form of the reductant."[28] With practice, skill, and care, Loftfield had managed to obtain radioactive alanine and glycine in small amounts and with an activity sufficient for biochemical purposes.

Zamecnik engaged in a collaboration with Loftfield and Warren Miller, of the physics department at MIT. Miller had been involved in the development of a new method of radioactive gas counting.[29] A prototype of the new instrument had been constructed at MIT. One of these devices was placed at the Huntington, and much technical work was performed to make the counting procedure with this instrument reliable.[30] Evans and Miller were persuaded that gas counting was the only practical approach to the measurement of [14]C. They ruled out solid counting techniques because of the prohibitively high self-absorption of the soft beta rays. As Ivan Frantz, one of the participants, recalled, "this reasoning seemed logical," but "our acceptance of it resulted in considerable unnecessary delay. The internal counters were constructed by the MIT glassblowers, and they tended to be somewhat erratic."[31] After a long series of unconvincing trials, the group switched back to solid counting. The new technique was abandoned.

Zamecnik immediately realized both the potential impact of the tracer technique on his research agenda and the potent source of research money associated with the medical use of atomic energy after World War II. Already in 1948 he suggested to the Committee on Research that it approach the Atomic Energy Commission (AEC): "Dr. Zamecnik mentioned [that] it was possible to obtain large research grants from the Atomic Energy Commission for studies of the medical application of atomic energy. [Mr.] Ketchum recommended that Dr. Zamecnik draw up a list of possible projects which might qualify for Atomic Energy support."[32] During that same year, the Huntington launched a large grant application. A revised proposal—"according to the way the wind is blowing at the moment"—was accepted in 1949, and AEC money continued to support Zamecnik's work during the following decade.[33] The requirements for setting up an experimental system for analyzing malignant growth and the global political situation had entered into a structural coupling.

The first studies on the "incorporation" of radioactive alanine into the proteins of rat liver slices were done in cooperation with another M.D. at MGH, Ivan Frantz. After having served four years in the U.S. Navy, he had been awarded one of the first set of twelve fellowships funded by the American Cancer Society and resumed his scientific career in the laboratories of his former teacher Joseph Aub.[34] Frantz got started studying the splitting of peptides by protein-degrading enzymes, but soon went on to acquaint himself with the technique of incubating sliced livers.[35] This was a remarkable local convergence of technical innovations, institutional collaboration, political opportunities, skills, and craftsmanship from various fields of expertise—organic chemistry, radiation physics, physiological chemistry, and medical laboratory practice. It was a situation that enabled the Huntington group to outline a method for tagging proteins within a very short time and well before labeled amino acids became available on a commercial scale. This local constellation, in turn, determined the choice of the experimental system and, within a few years of tinkering, led to its transformation into a "machine for making the future."[36] In turning that machine on, as Loftfield has noted, it might well have been crucial that none of the initially collaborating persons had a traditional education in biochemistry in the disciplinary sense of the term. They were not restricted by too much knowledge of "what will not work anyhow." It was this element of nonexpertness that enabled them to think differently.[37]

At that time, radioactive amino acids were available only in very limited amounts. Moreover, they posed unusual and new problems for controlling the experimental conditions. One of the biggest concerns of radioactive studies in living animals was keeping control over the specific activity of the injected material. For these reasons, incorporation studies in vivo were hardly a routine perspective to be adopted. Zamecnik and his colleagues, like others, envisaged tissue analysis outside the intact animal.[38] The tissue of choice happened to be rat liver. It was routine in the Huntington to induce liver hepatomas in rats by placing them on a diet containing the chemical Butter Yellow. Rats from the Sprague-Dawley stock had been maintained in the laboratory during the preceding fifteen years. By sorting out the hepatoma-carrying rats after several months, it was possible to compare the livers of sick and healthy animals.[39] The livers could be removed and then cut into slices. They did not lose their metabolizing capacities for at least several hours when kept under appropriate conditions.[40] The procedure allowed experiments to be carried out with fairly small amounts of tissue and correspondingly small amounts of radioactivity.

However, the partially damaged sliced material underwent an enhanced breakdown of its proteins, which caused metabolic distortions of an unknown order. In addition, there was a general slowdown of reactions, with the result that the tissue became progressively inactive. The procedure for counting the labeled samples was laborious, and for unknown reasons the slices exhibited a considerable variance in the "uptake" of tagged alanine. Thus, when the Huntington workers stated that the "use of the tissue slice rather than the whole animal was chosen in order to study protein synthesis under conditions more readily within experimental control than those in the living animal," this was at the beginning more of a projection than experimental reality.[41] Nevertheless, the first results indicated that the arrangement worked "in principle." There was a substantial radioactive signal to be extracted from the protein isolated after the procedure, although it was at times erratic.[42] Moreover, the signal was dependent on oxygen. The utilization of oxygen was compatible with Lipmann's assumption of a linkage between protein synthesis and energy-yielding processes. Last but not least, after degradation of the protein, a large part of the radioactivity could be recovered in the form of recrystallized alanine.

Thus, at least for those involved, the system showed promise. Of course, any critic could have pointed to the table of one of the first papers and asked why there was so little correspondence between the amount of carbon dioxide recovered from the degraded protein sample and the radioactivity assumed to be present in that same carbon dioxide.[43] The experimenters neglected the discrepancy. To account for variable results, they even invoked explanations like cell trauma inflicted by the knife while cutting the livers into slices. They preferred to stick to what appeared to be a significant difference, a yes-or-no signal, irrespective of its actual magnitude. From this difference it seemed possible to penetrate into the details of the system. The oxygen dependence of the slice system seemed the track to be followed.

A Control That Becomes the "Real" Experiment

The results of these first "*tâtonnements*" were sent to the *Journal of Biological Chemistry* in the spring of 1948. Also, in April 1948, a letter to the editor went to the same journal.[44] In this letter, the MGH group communicated,

DNP	normal liver		primary hepatoma	
	O_2 consumption	alanine incorporation	O_2 consumption	alanine incorporation
no	+	+	+ +	+ + +
yes	+	−	+ (−)	−

Fig. 3.1. Dependence on dinitrophenol (DNP) of oxygen consumption and amino acid incorporation in normal and hepatoma liver. +, amino acid incorporation activity; −, no amino acid incorporation activity. Adapted from Frantz, Zamecnik, Reese, and Stephenson 1948, table 1.

in a single table, the combination of two sets of signals. One set was derived from the original, oncological question concerning the difference in activity between normal and tumor cells. The other contained what could be called the first differential answer emerging from the setup of the system itself. The simultaneous presentation of both groups of results is to be read as an expression of the authors' momentary indecision about how to weigh the emerging experimental traces.

The introduction to that letter mentioned a recent finding from the neighboring laboratory of Lipmann.[45] The two groups entertained good contacts at that time. William Loomis, who happened to be a classmate of Frantz, had come to work with Lipmann. He had found that the chemical dinitrophenol (DNP) uncoupled the interlocked processes of cellular respiration and phosphorylation. The substance interfered with the formation of energy-rich phosphate bonds but not with the consumption of oxygen. Because the liver slice system required oxygen, it was tempting to look at what would happen if DNP were added to the slices of normal and hepatoma livers.

A closer look at the schematic presentation shown in Figure 3.1, which is abstracted from a larger table for the sake of clarity, makes it possible to sort out three distinct effects or observations.

The first observation was that normal and hepatoma livers differed both in oxygen consumption and in the incorporation of radioactive alanine. The incorporation of alanine was up to seven times higher in cancer tissue; such a large biochemical difference between normal and malignant tissue had never been observed before. The second observation was that the oxygen consumption in hepatoma slices was depressed by DNP, whereas in normal liver slices the respiration remained unaffected. The third observation was that DNP decoupled the process of oxygen consumption from the incorporation of alanine. In the presence of DNP, respiration continued, but the uptake of amino acids was abolished.

Signal number one fulfilled the original expectations, and it did so to a remarkable degree. Loftfield has emphasized that the sheer magnitude of this "difference"—the news of which brought him from a skiing trip right back into the laboratory—served as a stimulus for further exploration. And it was instrumental in allocating further funds, especially from the American Cancer Society.[46] On the other hand, as work along these lines went on, it took the reassuring appearance of a self-fulfilling prophecy. The difference between normal and malignant tissue remained a difference that did not tell the cancer researchers what to do next. It did not gain experimental significance for "assessing what we do not know."[47] It was a beautiful result with no consequences. Signal number two introduced an additional difference between normal and malignant tissue— that is, an inhibitory effect of DNP on hepatoma tissue respiration, in contrast to the normal tissue, which was not affected. Zamecnik and his colleagues did duly note this oddity, but they did not make any attempt to explain it further. They could well have taken it as a first hint at a cancer-specific metabolic pathway. However, this second signal was overrun by signal number three, which could be interpreted as a linkage of protein synthesis with the energy-yielding process of phosphorylation. It was this latter signal that provoked a shift in the research perspective from the cancer problem to the bioenergetic aspects of amino acid incorporation. It also brought about a definite shift from the conception of protein synthesis as a reversal of proteolysis to the search for energized intermediates in protein synthesis. It promised to serve as a handle for developing the experimental system a step further, away from its hybrid state between in vivo and in vitro and toward a full-blown in vitro system that was no longer dependent on intact cells and, therefore, on an intact respiratory apparatus. This was in 1948. A new option had come into play. As Loftfield put it, it "re-energized" the group,[48] it reoriented their thoughts in terms of endergonic mechanisms, and it prompted a laborious effort toward establishing a cell-free system.

A Landscape of Competing Laboratories

Right from the beginning of the use of radioactive amino acids in the study of protein synthesis, several groups had been involved in the attempts to achieve amino acid incorporation into animal tissue slices. Ahead of Zamecnik's group at MGH, among the first to use tissue slices were Jacklyn Melchior and Harold Tarver, as well as Theodore Winnick, Felix Friedberg, and David Greenberg—all from the University of California Medical School's Biochemistry Division in Berkeley.[49] Chris Anfinsen and his collaborator, Art Solomon, at Harvard Medical School had also been successful in setting up a slice system.[50] Attempts to incorporate amino acids into proteins of tissue homogenates prior to that of the MGH group were made by Melchior and Tarver, by Friedberg, Winnick, and Greenberg, and by Henry Borsook's team at the California Institute of Technology in Pasadena.[51] Initially, they all used different amino acids: sulfur-labeled cysteine and methionine (Tarver), carbon-labeled glycine (Greenberg and Winnick), carbon-labeled glutamic and aspartic acid (Anfinsen and Solomon), and carbon-labeled lysine (Borsook). All these labels were incorporated in vitro, but not all "incorporation" subsequently proved to be related to protein synthesis. Although I will not trace these activities in detail, they indicate that Zamecnik did not work in a void. The community of protein-synthesis workers was small, but it was big enough to constitute a competitive framework of checks and balances.

Flanking Experimental Activities

The risky journey into an unknown experimental landscape, rife with possible pitfalls, had started. But it was flanked by a variety of activities that were not directly coupled to the purpose of establishing an in vitro system related to protein synthesis. Rather, these trials were organized as an experimental exploration of various aspects of protein metabolism that could be approached by the radioactive amino acid tracer technique. The technique served as an institutional organizer for various, mostly cancer-related, activities, and the exploration of its scope in turn served as an "exteriorized" question-generating device. This embedding strategy faithfully translated Zamecnik's hope that "a definite clue [might turn] up in any corner of the field."[52] And it had the added merit of granting the group a continued presence in the cancer research field and access to its funding sources for the years to come.

In those days it was the dream of the small protein synthesis community to have at its disposal a complete set of the naturally occurring amino acids in radioactive form. Ivan Frantz attempted to synthesize them biologically. He grew *Thiobacillus thiooxidans*, an autotrophic sulfur-oxidizing bacterium, in an atmosphere of radioactive CO_2. He isolated the proteins, hydrolyzed them, and separated certain amino acids that could not be synthesized chemically in sufficient quantity and with specific activity high enough "for further studies of tumors in animals."[53] This was a laborious but transient effort. A few years later, all the necessary amino acids were available commercially.

A technique for separating amino acids that could replace the tedious recrystallization procedure and serve for large-scale preparations kept Mary Stephenson busy for quite a time.[54] The technique was based on the use of starch columns pioneered by two workers with whom the MGH group had an intense exchange of information: William Stein and Stanford Moore from Bergmann's lab at Rockefeller.[55] Stephenson had joined the Huntington Laboratories in 1943 as a technician, and she worked with Zamecnik right from the beginning.[56]

Nancy Bucher, too, was a member of the staff of the Huntington Laboratories. A young medical doctor from the Johns Hopkins University, she joined MGH as a research fellow in 1945 and was running a program on regenerating rat liver.[57] Her results indicated that the rate of protein synthesis was enhanced in regenerating liver, a nonmalignant tissue that grows even faster than most neoplastic tissues.[58] Thus, enhanced protein synthesis was not a unique characteristic of malignant cells. As a consequence, it was not necessarily the cause of neoplastic tissue growth. Unfortunately, Bucher's work considerably diminished the significance of Frantz and Loftfield's wonderful difference between normal and malignant protein turnover. Quite understandably, there is no enthusiasm to be found in the following sentence: "An accelerated rate of peptide bond synthesis is evidently not a unique property of the hepatoma, but [at least] does serve to distinguish its protein metabolism from that of resting adult liver tissue."[59] The metabolic behavior of the tumors remained a challenge.

Frantz and Ann Werner studied the equilibrium between dipeptides and amino acids in the presence of peptidases.[60] The use of radioactive glycine and the starch column technique allowed them to separate the tiny amounts of dipeptide formed by these enzymes from large amounts of amino acids. The possibility that protein synthesis might proceed by way of a reversal of proteolysis remained on Huntington's experimental program. Only gradually did it shift toward ruling out the possibility in favor of an endergonic process. Borsook had been promoting the latter view for more than a decade, facing generally what he called an "unsympathetic response."[61]

This experimental puzzle, tied together by the use of radioactive

Fig. 3.2. Fragment of a silkworm cocoon, together with autoradiograph of the same fragment. Reprinted with permission from Zamecnik, Loftfield, Stephenson, and Williams 1949, fig. 2. Copyright © 1949, American Association for the Advancement of Science.

amino acids, culminated in an attempt at the biological synthesis of radioactive silk, a collaborative project with Carroll Williams from the Chemical and Biological Laboratories of Harvard University.[62] The gland of the silkworm produces a protein that contains unusually high amounts of glycine and alanine. Because these were just the two amino acids that Loftfield was able to synthesize in radioactive form, the in vivo synthesis of a complete, radioactive protein seemed within reach. Preliminary experiments indicated that the silkworms did indeed incorporate the label into their cocoons (compare Figure 3.2) and that it was possible to recover a radioactive protein from isolated silk glands. But although the silkworm system looked promising, it was abandoned; it proved technically unfeasible for routine work. It would have required the preparation of hundreds of silk glands to obtain some grams of protein-synthesizing tissue, which then would have had to be homogenized on the way toward a cell-free protein-synthesizing system. In addition, the gland work proved too seasonal, and the homogenates too "gluey."[63] Technical constraints, not experimental failure, precluded pursuing the exotic path of the silkworm. The Huntington Laboratories were a place for such trials: Joseph Aub's growth program included even the study of growing deer antlers.

"A Bridge of Joined Talents"

A close look at Huntington's early postwar years leaves us with the impression of a network of experimental trials, all of them differing slightly in their range and impact, only loosely interlocking, but all, in one way or another, connected with malignant growth and protein metabolism. The network was not restricted to Zamecnik's team alone—collaboration extended to the pursuit by Aub's group of a variety of growth phenomena as well as to the clinic. The technical and institutional environment of the cancer program was not abruptly abandoned. It was pursued further with considerable effort. Although the Huntington researchers worked mostly on their own projects, they shared their expertise, which involved a broad range of professional training.[64] They hoped that "a new research idiom would evolve, consisting of a bridge of joined talents" involving both clinical and nonclinical research staff.

Zamecnik had managed to gather a loosely connected, periodically renewed group of experts from slightly different but overlapping areas of investigation, which was, as Hoagland put it later, absorbed in a "free-wheeling" collaborative effort with no strict assignment of tasks.[65] It was, in the words of Mary Stephenson, a "happy laboratory," and it was operated, to quote Nancy Bucher, as a "democratic place."[66] Zamecnik had brought together different skills and background experiences, and he focused them on the step-by-step development of an experimental system that, as time went on, began to orient the axes of its development through its own output. Robert Loftfield has stressed the balance that existed between the joint efforts of the group and the independence of every collaborator: it was a "tightly connected loose group." In turn, this type of connection favored a scientific climate characterized by a sensitivity to "unprecedented events," and to unexpected effects in one corner of research that might be relevant to work going on in another. Loftfield has spoken in this respect of the "diffused force of the laboratory."[67] Zamecnik had a feeling of how much identity, stability, operational definition, and collaboration a system needed to serve as a basis for the production of unprecedented events. He was attentive to the signals that developed where they had not been expected; he would grope around them and look at them from many angles until they either disappeared as noise or took shape as unexpected epistemic entities.

Cancer Research: A Closure

At a Cold Spring Harbor symposium in 1949, Zamecnik discussed an alternative approach to cancer via protein synthesis.[68] He speculated that carcinogenic agents might alter enzymes. These enzymes would lead to the synthesis of different proteins and/or the synthesis of altered genes, which in turn would yield altered apoenzymes, again producing modified proteins. If so, one might expect that the amino acid composition of proteins from malignant tissue would differ from that of proteins from normal tissue. A feasible method for separating *all* the different amino acids of a protein hydrolysate would be a prerequisite for such an analysis, and starch column chromatography promised to be the method of choice.[69] The results were disappointing. With minor exceptions, the ratio between the different amino acids in normal and malignant tissue was the same. The few pertinent differences that showed up in the liver slice experiments could not be confirmed in vivo. There was "no adequate explanation for the discrepancy between these results and those found in the slice experiments."[70]

It was an impasse. None of the attempts of Zamecnik, Loftfield, and their coworkers to compare normal and malignant tissue had produced experimental differences forceful enough to direct the future experimental program. The slice technique displayed insurmountable limitations. Comparing various sorts of tissue had led to the visualization of differences; but these differences could not be further explored by the technique that had led to their establishment. "The tissue slice is a useful preparation for the study of the overall process [of protein synthesis], but so far it has provided little information concerning mechanisms."[71] Above all, the technique did not allow discrimination between the three mechanistic alternatives for protein synthesis: reversal of proteolysis, utilization of energy-rich phosphate bonds for a de novo synthesis from amino acids, or amino acid exchange into existing proteins. In elucidating these possible mechanisms, one of the promising experimental approaches seemed to follow the line of the Frantz-Loftfield-Miller effect of DNP: it pointed toward a coupling of phosphorylation and protein synthesis. "Of perhaps greater importance," however, remained looking into "the manner of synthesis of several peptide-like compounds."[72] This was the Frantz-Loftfield-Werner approach to studying peptidases. In this confused situation, Zamecnik resorted to poetry: "We dance round in a ring and suppose, but the Secret sits in the middle and knows."[73]

In 1950, on the occasion of a review for *Cancer Research*, Zamecnik contextualized the work done so far.[74] The slice experiments were, in several respects, conflicting and difficult to interpret coherently. In vivo, a balanced mixture of all amino acids including the radioactive one gave optimal results. This brought "into relief the results of *in vitro* slice experiments," in which no such mixture was needed.[75] In the slice experiments, the rate of amino acid incorporation was at least an order of magnitude lower than in the intact organ. So far, there was not the slightest evidence that in such experiments complete proteins were produced at all.

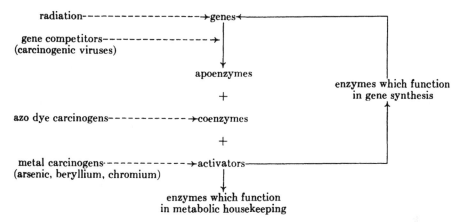

Fig. 3.3. Possible sites of action of carcinogenic agents. Reprinted from Zamecnik 1950, fig. 1.

An evaluation of the few reports on cell-free amino acid incorporation from homogenates was complicated by an increasing evidence that other metabolic processes interfered with what had been taken to be regular protein synthesis.[76] The experiments based on homogenates suffered from an additional slowdown of the amino acid incorporation rate,[77] and there was a dramatic reduction of the time during which the homogenate was active. It occurred to Zamecnik that the "interpretation of what is going on in the homogenate incorporation studies is not as yet clear" and that "the term 'incorporation' covers more than one independent process by which amino acid and protein become closely associated in a chemical bonding."[78] Keeping the choices open and remaining alert to the differences between whole animal experiments, tissue slices, and homogenates therefore appeared to him to be vital. The Huntington group had been at the forefront of those using the slice technique. Others had taken the first steps toward homogenization, and Zamecnik attentively registered the traps in which they got stuck.

In spite of these difficulties, Zamecnik tried to embed his experimental efforts in a theoretical framework. In an attempt to bring the possible modes of action of carcinogenic agents under a "unitary point of view," he conceived a diagram in which "genes," otherwise completely absent from his experimental discourse, entered the scene (compare Figure 3.3). Their products were supposed to either function in "metabolic housekeeping" or be involved in "gene synthesis," thereby closing the feedback loop.

The diagram suggests that, in the words of Zamecnik, "genes (considered here as either nuclear or cytoplasmic nucleic acid-containing materials concerned with hereditary transmission of biochemical characteristics) 'direct' the synthesis of the apoenzyme or protein part of enzymes, which then becomes an active catalyst by combination with a coenzyme, in the presence of the proper ionic milieu."[79] Cancer-inducing agents are inserted in the chart where they are supposed to exert their effects. The scheme suggests a "directivity" from genes to enzymes and visualizes enzymes as concerned with the synthesis of genes in a feedback loop. Yet we have to be cautious and avoid interpreting this diagram as a pathway of biochemical intermediates or even a molecular flow of information between genes and gene products. This is, first and foremost, a *connection* scheme. It is intended to show how carcinogens might affect the basic metabolism of the cell and how their action might become "inherited" through cell division; the effect of these drugs on protein synthesis is only implicitly present in the figure. Because it represents putative entities like genes, apoenzymes, coenzymes, activators, and their regulatory relations in a kind of feedback loop, the biochemical processes too are absent from it. Formulated as a paradox, protein synthesis is present here as an absent process. The formal representation of possible modes of action of enhanced growth—the "second story," as Zamecnik saw it by 1950—appears to have no relation to the underlying mechanisms, or the "first story," which is to say the "mechanisms of protein synthesis itself."[80]

Thus, Figure 3.3 is a fair account of the state of affairs as seen in 1950. The experiments documented that the rate of protein synthesis in tumors was accelerated, but they didn't tell why. As a synopsis, Figure 3.3 connects the whole range of cancer research at Huntington around the turn of the decade. But at the same time, it demonstrates how little this kind of "idle" theoretical conceptualization was able to direct the experimental exploration of the process under scrutiny at that stage.

CHAPTER 4

Establishing an In Vitro System of Protein Synthesis, 1949–52

❖ In analyzing a problem, the biologist is constrained to focus on a fragment of reality, on a piece of the universe which he arbitrarily isolates to define certain of its parameters. In biology, any study thus begins with the choice of a "system."

— François Jacob, *The Statue Within*

In what direction did the experimental system have to be developed to tackle the "first story" of the "mechanisms of protein synthesis itself"? The answer to this question was far from being clear. Did one have to proceed "logically" from intact animals to organ slices to homogenates and finally to model systems? Animals and slices were insufficient substrates for obvious reasons. Homogenates and model systems remained. But something seemed to be refractory there. For the moment, Zamecnik qualified the homogenate "to be a biochemical bog in which much effort is being expended to reach firm ground."[1] With "simpler systems," or model systems, the problem again arose "as to how to interpret the event recorded by the labeling process."[2] Quite obviously, simplicity was not sufficient to confirm the adequacy of the experimental model. The solution had to be found somewhere between the bog and the simple system. When he sent off his review to *Cancer Research*, Zamecnik did not mention that in his lab Elizabeth (Betty) Keller and Philip Siekevitz were heading in that direction. Instead, he inserted an addendum in the galleys, mentioning two reports on enzymatic transpeptidation and transamidation and concluding that protein synthesis might prove a two-step procedure: first, formation of selected peptides via phosphate bond energy and, second, enzymatic transpeptidation with little further expenditure of energy.[3] Hence, he combined the mechanistic alternatives of protein synthesis—reversal of proteolysis, exchange, and utilization of energy-rich phosphate bonds—thus keeping all conceivable theoretical choices open.

How to Go in Vitro?

A research process is a procedure in which, eventually, new things happen to emerge that cannot be predicted by a "theoretical system" and that likewise are not inevitably generated by the "practical system" of experimentation. Thus, the design of experiments is not necessarily determined by theory, and the design of theories is not necessarily constrained by experiments.[4] This mutual nonfitting is exactly what makes the experimental process an explorative endeavor. In the situation just characterized, it was not only unclear which theoretical alternative should be chosen, it was equally uncertain what experimental system would be helpful in making a choice between the conceivable alternatives. The alternatives at hand were themselves far from being stable options. The liver slice system incorporated amino acids. But it did not reveal anything about the underlying mechanism. The slice system had produced one hint—its response to DNP. But the pursuit of this hint was beyond the realm of the existing system—it would have required a physical separation of the processes of oxidation and phosphorylation. On the other hand, the experimental alternatives—cell homogenate and simpler model systems for peptide bond formation—were judged to be poor candidates for showing what went on within the intact cell. The first presented a "biochemical bog" that would be difficult to drain; the second was prone to turn out to be artificial altogether. It might create conditions for a process not usually occurring in the cell. Alternatively, it might be emphasizing a reaction that was a metabolic sidepath not related to protein synthesis.

Zamecnik's strategy, once again, was organized groping. A part of the group continued to work with simple model systems. But other members went on draining the "biochemical bog," and they happened to follow an ingenious procedure: they coupled the in vivo administration of amino acids followed by fractionation to the fractionation of a rat liver cell homogenate with subsequent addition of amino acids.

Simple Model Systems

Frantz and Loftfield went on with exchange reaction studies based on peptides, amino acids, and protein-degrading enzymes.[5] Thinking of protein synthesis in terms of a reversal of proteolysis continued to be an option that, given the present state of knowledge, at least could not be ruled out, at the very least not as a possible explanation of in vitro incorporation. Although Frantz and Loftfield no longer considered it to be a likely mechanism, the "multi-enzyme programme" of protein biosynthesis was still a popular concept. Backed by the powerful analogy to other macromolecular syntheses, at the beginning of the 1950s many leading biochemists, among them Kai Linderstrøm-Lang, Chris Anfinsen, Thomas Work, and Harold Tarver, were not prepared to drop it.[6] Above all, the slower the incorporation rates in the test tube, the less exchange reactions could be neglected.[7] The role of proteolytic enzymes in protein synthesis received renewed interest through the work of Yale biochemist Joseph Fruton and his colleagues, who also went the way of "simple systems."[8]

In liver slices, the protein synthesis machinery discriminated sharply against an uptake of amino acids not belonging to the regular makeup of proteins. Loftfield went on to test proteolytic enzymes from rat liver with respect to their hydrolyzing effect on natural and "nonnatural" dipeptides in a model system. In contrast to the slice system, the enzyme system did not discriminate between natural and nonnatural substrates. All compounds were hydrolyzed with comparable efficiency. Consequently, the same could be assumed for the reverse reaction. Although Loftfield conceded that this line of reasoning was not "conclusive," he took it as further "evidence against the participation of proteolytic enzymes in normal protein synthesis."[9] In addition, Loftfield introduced still another line of argument: after the spectacular work of Frederick Sanger and Hans Tuppy in 1951 on the sequence of insulin, any thought on mechanisms of amino acid assembly had henceforth to take into account the "perfectly definite" sequence specificity as revealed by this protein.[10] It was hard to imagine that such specificity could result from the action of proteolytic enzymes.

Because simple systems as models for protein synthesis ceased to be promising research candidates, the search for alternatives became more urgent. The idea was to set up a protein synthesizing system that worked the other way round, by gradually "simplifying" an utterly complex cell homogenate. Intuitively, Zamecnik followed what Bachelard has described as an epistemology of simplification. Nobody familiar with this procedure will forget that the simple is never a ready-made thing to start from, that it is not the element of science but always the product of a simplification, a degeneration of an elementarily complex situation.[11] One might be tempted to assume that, in the present case, it took a medical doctor to forge such a strategy of draining the biochemical bog. The purity standards of classical protein biochemists, not to mention the aesthetic standards of molecular biologists, might well have guided—and, as a matter of fact, did guide—many researchers in another direction. Indeed, there was a widespread disdain for dirty biochemistry among the members of what Gunther Stent has called the "structuralist" and the "informationalist" schools of early molecular biology.[12] In his autobiography, Mahlon Hoagland has spoken of the "wide gulf" that separated, from the 1950s well into the early 1960s, the experimental culture of biochemists from those who regarded themselves as molecular biologists.[13]

Taking Cells Apart

The line of investigation concerned with setting up a system based on liver homogenates faithfully reflected the practice of shuttling back and forth between the simpler and the less simple. Betty Keller, a Cornell-trained biochemistry Ph.D. who had come to the Huntington as a research fellow in 1949, chose in 1950 to take a coupled in vivo/in vitro approach to the problem. A brief description of her first experiments will help clarify the matter. Living rats, when injected with a comparatively low dose of labeled amino acids, incorporated radioactivity into their liver protein. If, after given time intervals, the animals were killed, the livers removed, the tissue homogenized, and the resulting paste subjected to differential centrifugation through a sucrose solution, the radioactivity could be recovered from the different fractions. It turned out that the bulk of the early label was incorporated in the so-called microsomal fraction sedimenting at 40,000 × g (times gravity) within twenty minutes, from which it later disappeared with a high turnover rate.[14] At this point, a few remarks on the early history of "microsomes" are needed.

An Excursion into the History of Microsomes

Two events marked the beginning of the dissection of the cell's cytoplasm in the test tube toward the end of the 1930s: the characterization of an extraneous epistemic object, that is, a tumor-producing agent embedded in cancer research, and the introduction of a powerful new instrument— the ultracentrifuge.[15]

Disappointed by the results of his biochemical efforts to purify the agent causing the kind of sarcomas in fowl that Peyton Rous had first observed in 1910,[16] Albert Claude turned to the ultracentrifuge in 1936 at James Murphy's Pathology Department of the Rockefeller Institute in New York. News of the high-speed sedimentation of the filterable chicken tumor agent had reached him from England.[17] Murphy encouraged him to take this step, and already the first results were extremely promising. The high-speed pellet derived from infected tissue displayed an enrichment of the agent by a factor of approximately 3,000, as compared to a factor of 30 he had achieved so far with conventional biochemical methods. In parallel experiments, however, Claude pelleted down a fraction from normal embryonic chicken tissue that, in its chemical and physical characteristics, could not be distinguished from the fraction containing the agent, with the exception that it just was not infective.

Two interpretations were possible at this point. On the one hand, the main constituents of the tumor fraction, besides the agent itself, could have represented a "precursor of the chicken tumor principle." The idea of an endogenous rather than a viral origin for the chicken sarcoma had indeed been one of the motives for Murphy's resumption of the chicken agent research toward the end of the 1920s. Alternatively, these constituents simply represented "inert elements existing also in normal cells."[18]

It was the second possibility that first haunted Claude, then intrigued him, and finally, within a few years, led him away from the tumor agent research that had kept him busy for almost a decade. A characteristic displacement of an epistemic object was thus prompted by the incorporation of a new instrument into an existing experimental system. An alternative option had come into play. Claude had introduced differential ultracentrifugation to isolate a submicroscopic principle inducing cancer. Now, the technique promised to become a handle for fractionating the cytoplasm of normal cells. A new cytology emerged on the horizon.

By means of differential centrifugation, Claude managed, within the next decade, to unfold the cytoplasm into a hitherto unprecedented space of representation: a space, that is, for the production, characterization, isolation, and purification of subcellular constituents. For a hundred years or so, cytomorphology had been the domain of observation by the light microscope and the corresponding preparative methods of fixation and staining. Besides the nucleus, which was the most prominent feature of the eukaryotic cell, mitochondria had been visualized for many decades within a basophilic, more or less homogeneous, cytoplasmic ground substance, for which the term *ergastoplasm* had become current. Initially, many traditional cytologists and histologists ridiculed Claude's new approach as resulting in barely more than cellular mayonnaise.

Claude reported about his recent work at a Cold Spring Harbor Symposium on "Genes and Chromosomes" in 1941. The placement of his findings in such a genetic context is remarkable. But it reminds us of the wider scientific network in which the first generation of small cytoplasmic particles found resonance and gained identity: a context of cytoplasmic inheritance and of plasmagenesis, as opposed to chromosomal or nuclear inheritance. Claude, at first, identified his "small particles," which settled at the bottom of the test tube after one hour of centrifugation at 18,000 g, with the cytologically well-characterized mitochondria, or fragments thereof.[19] Their size was just below the power of resolution of the light microscope but still within the realm of a dark-field microscope, where the particles appeared as small reflecting points. Their chemical composition provided reasons for further thought: besides lipids, which comprised about half their mass, they contained a major portion of proteins and, in addition, significant amounts of ribonucleic acid (RNA).

It was precisely the chemical composition of these cytoplasmic granules, and in particular their content in RNA, that immediately caught the attention of Jean Brachet at the Free University of Brussels. In the preceding ten years, Brachet had focused his interest on developing methods of differential histochemical staining, especially for DNA and RNA, and on quantitating these components in different tissues and in different animal species. His overarching goal was an approach to embryogenesis and to nucleocytoplasmic interactions by cytobiochemical techniques.[20] His work during the 1930s on developing sea urchin eggs established the ubiquity of RNA, until then believed to be found only in plant cells and to some extent in the pancreas.[21] By combining RNA digestion with a specific RNA staining procedure based on methyl-green pyronine, Brachet had come to the conclusion that RNA was preferentially located in the nucleolar structures of the nuclei and in the ergastoplasm. Cells actively engaged in protein synthesis appeared especially rich in RNA.[22] "The conclusion to which we are led," he ventured as a result of these efforts in 1942, "is that the pentosenucleic acids might intervene, by a mechanism as yet obscure, in the synthesis of proteins, which perfectly matches the available facts."[23] On the basis of a different technique, Torbjörn Caspersson in Stockholm had essentially come to the same conclusion.[24]

This was the point, as mentioned, when Brachet came across Claude's "small cytoplasmic particle" work. But there was yet another coincidence. One of Brachet's colleagues at Liège, André Gratia, together with André Paillot from the Station de zoologie agricole du Sud-Est de Saint-Genis-Laval (Rhône), had made a fortuitous observation similar to the one that had put Claude on the track of cytoplasmic particle research, although the European group was working on a completely different experimental system related to the French silk industry. Gratia and Paillot were investigating the silkworm virus that causes jaundice disease by means of an air-driven Henriot-Huguenard centrifuge. When comparing viral sediments with sediments from normal tissue homogenates, they found the same fine granules in their healthy control cells, with the only decisive difference that they were not infectious.[25]

Gratia's work had shown the technical feasibility of Henriot's air-driven device for fractionating animal tissue. Emile Henriot happened to be a professor of physics at the University of Brussels, and he had a prototype of his centrifuge in the basement of his institute that served him to measure the velocity of light in different media. Brachet and his colleague, the animal physiologist Raymond Jeener, who had joined forces with some rudimentary equipment, obtained access to Henriot's device. Together with Brachet's doctoral student, the biochemist Hubert Chantrenne, they embarked on a program of isolating what they called "cytoplasmic particles of macromolecular dimensions."[26] Despite the occupation of Belgium by Nazi Germany in May 1940, the group was able to analyze a considerable variety of tissues from different animals with Henriot's tiny, but quickly revolving rotors (one attained 45,000, the other 75,000 rounds per minute). But the working conditions became increasingly difficult. In November 1941, the entire staff of the Free University of Brussels ceased teaching in an act of solidarity with their Jewish and antifascist colleagues, and in July 1942 the university was definitely closed. Jeener and Chantrenne could move to Liège; Brachet was arrested in December 1942 and only released in spring 1943, without being given the possibility to pursue his experimental work further.

In a preliminary and summarizing set of papers published in 1943–45, Brachet, Chantrenne, and Jeener basically came to the following conclusions. In adult cells, the quasi totality of cytoplasmic RNA was located in the macromolecular granules. Moreover, these granules were associated with a host of enzymes serving either hydrolytic or respiratory functions. Reasoning along the lines of the then current ideas about protein synthesis as a reversal of proteolysis, Brachet speculated that the respiratory enzymes would funnel the necessary energy into the process, whereas the hydrolytic enzymes, among them several peptidases, would synthesize, in a reversal of their usual action, peptide bonds. To make the reaction proceed in the direction of synthesis, the RNA would trap the synthesized peptides and so remove them from equilibrium. In favor of these ideas he adduced the additional observation that in specialized cells, such as insulin-producing pancreas cells or hemoglobin-producing blood cells, appreciable amounts of these tissue-specific proteins came down with the cytoplasmic particles.

In summing up this burst of information from a year's effort of high-speed centrifugation, Brachet expressed his doubts about Claude's hypothesis of an identity between the RNA-containing granules and the mitochondria. However, he was no longer in a position to sort the granules out into two or more different fractions, owing, as he later put it laconically, "to war difficulties and lack of equipment."[27]

Such was not the case with Claude. At the time the results of Brachet and his colleagues went into press, Claude had already given up the initial identification of his small particles with mitochondria. Under slightly modified buffering and centrifugation conditions, the resuspended small-particle sediment evidently contained no particulate material of the estimated size of these organelles. In an act of anabaptism, Claude renamed the small particles and, from 1943 on, called them "microsomes."[28] After their short life as mitochondria, Claude had to clear up a confusion, at least with respect to terminology. He did this with the catchy notion of microsomes, which this time, he thought, had the advantage of being "noncommittal" because the term referred only to the particle size—and even this in a rather vague fashion.[29] But Claude retained the idea that these tiny particles might be precursors of larger ones, that they might eventually grow into mitochondria in later stages of the cell's life cycle, and that they might possibly be endowed with the power of self-reproduction. Quite evidently, he wanted to keep the choices open for his particles to become connected, as a sort of "plasmagenes," with the then current, ubiquitous, and quite heated debate surrounding plasma-genetic inheritance, in which he, however, did not participate as an active discussant.

We have come to realize that the ultracentrifuge had been the right tool for the structural dissection of the cytoplasm in the 1940s. In the beginning, however, it caused more trouble than it contributed to the clarification of pending questions. There was a decade still to come of standardization and of hooking up the new space of representation to traditional cytological and biochemical knowledge.

This work connected Claude's name with the new scientific object, the microsomes. In collaboration with a few other biochemists, cytochemists, and enzymologists from Rockefeller—among them Rollin Hotchkiss, George Hogeboom, and Walter Schneider—Claude managed, relatively unhindered by the war, to work out generally applicable conditions for quantitatively separating the mitochondria and other cytoplasmic vesicles from the microsomes and to subject the fractions to what the group called "biochemical mapping," or enzyme mapping.[30] In following this lead, Claude and his Rockefeller collaborators were led to the conclusion that the majority of the respiratory enzymes went with the mitochondria. The major technical breakthrough in this direction came only after years of painstakingly optimizing the sedimentation conditions. Sucrose solutions proved to be the key for success.[31] The enzyme pattern displayed by the microsomes, however, was poor and disappointingly irregular and did not point in any clear-cut direction. As far as their possible function was concerned, Claude still seemed to be at a complete loss when delivering his Harvey Lecture in 1948. He conjectured rather vaguely that they might have a function in some "anaerobic mechanism" or else that they were "an intermediate in the energy transfer for various synthetic reactions."[32] Although he mentioned Brachet's hypothesis of the particles' involvement in the synthesis of proteins, he did not seem to be convinced by Brachet's evidence. At any rate, at the turn of the decade neither Claude nor his colleagues made a published effort to provide further insight into the functional role of the microsomes.

Brachet's group in Brussels could make its way back to regular laboratory work only after the war was over. Hubert Chantrenne took up his work in the summer of 1946. He questioned the size and uniformity of the material, for which the term microsomes had come into general use. Working on mouse liver homogenates and refining the centrifugation conditions with his improvised air-driven Henriot-Huguenard centrifuge, Chantrenne arrived at five different fractions. Their characteristics led him to question the clear-cut dichotomy between mitochondria and microsomes established by the Rockefeller researchers. Although his fractions progressively differed in their chemical constitution with respect to RNA and in their biochemical activity with respect to enzymes, they exhibited qualitatively quite comparable features. Chantrenne came to the conclusion that "it seems that one can partition the granules in as many groups as one wishes, with no experiments or observations indicating that there exist precise demarcation lines between the different groups of particles."[33] The only quite obvious thing was that the smaller the particles were, the richer they became in ribonucleic acid. Thinking along the lines of his earlier prewar observations of free RNA in the cytoplasm of yeast, he now speculated that this continuum might reflect a gradual growth process and that "initially free ribonucleic acid, in the course of development, combines with sedimentable particles."[34] Were the microsomes, as far as their particulate identity was concerned, nothing more than arbitrary cuts into a cytoplasmic continuum? Their boundaries appeared to be due more to centrifugation conditions than to anything of an intrinsic biological significance.

Brachet himself pursued the idea that the microsomes might play a role in tissue differentiation during embryogenesis. The analogy to RNA viruses and the idea of plasmagenes loomed large, not only in the background, but as a quite explicit perspective in his account of the RNA-containing macromolecules. Together with John Rodney Shaver from the University of Pennsylvania, Brachet embarked on a program for testing the possible morphogenetic activity of these granules in the induction of the nervous system by injecting isolated microsomes from various embryonic tissues into cleaving eggs of amphibians. As seducing as this conjecture was, he had to state after a tedious series of trials that with respect to the role of microsomes in neurogenesis "our results have thus far been negative."[35]

During the decade that concerns us here, the 1950s, neither Claude nor Brachet, who can be considered the fathers of the microsomes, stayed at the very forefront of the effort to reveal the mechanistic mystery of the microsome, that is, its biological function. While Brachet's group nonetheless took an active part, Claude had stopped working on microsomes when he returned, in 1949, to Belgium. Brachet appears to have been obsessed with the theme of an involvement of RNA in protein synthesis. His primary interest in morphogenesis, however, made him concentrate on exceedingly complex matters of embryology.

The Microsome-Protein Synthesis Connection

Further experimental support for Brachet's contention of an involvement of microsomes in protein synthesis had to await the pioneering amino acid tracer work of Henry Borsook at Caltech, Tore Hultin in Sweden, and Norman Lee and Robert Williams at Harvard.[36] Betty Keller's experiments in Zamecnik's lab provided considerable additional evidence for "the importance of the microsome in the process of incorporation of amino acids."[37] These experiments reinforced the attempts to localize protein synthesis in a completely fractionated test tube system. If such an in vitro activity was to make sense, however, it had to be related to the in vivo situation. From a coupled in vivo/in vitro approach, Zamecnik and Keller hoped to obtain internal, experimentally convertible referents for "artifact containment." Shuttling back and forth between the simple and the less simple was a way to handle what they perceived as the danger of producing artifacts through homogenization.

From the viewpoint of traditional epistemology, there is agreement that experiments either sustain or refute theories and that theories should provoke experiments. Here, we see another epistemic principle at work. Different experimental practices, not theory on the one hand and practice on the other hand, are brought into mutual resonance. One of the most important procedures for producing resonance in the life sciences is the mutual superposition of in vivo and in vitro approaches. The empiricist/positivist misunderstanding of such stabilization and triangulation procedures lies in the assumption that nature itself is the ultimate instance of resonance.[38] The theorist misunderstanding, in contrast, lies in the assumption that stabilization ultimately occurs on the level of paradigms through which the scientific enterprise is made coherent. Here we see resonance based on establishing internal referents of manipulation, which need not be definitely stable but which are reliable and resonant enough to carry the burden to some next step.[39] For this reason, I prefer the notion of resonance instead of the constructivist metaphor of stabilization.[40]

An Active Homogenate

The first and most important thing to achieve was a homogenate in which metabolic activities did not immediately cease or come to a halt after a few minutes. Such homogenates became available through a technical detail that, though tiny, had revolutionized cytomorphological research toward the end of the 1940s: the resuspension of broken cells in a sucrose solution.[41] As already mentioned, Walter Schneider and George Hogeboom at the Rockefeller Institute had developed the procedure in the course of a cytomorphological and biochemical characterization of mitochondria.[42]

Philip Siekevitz had been working with David Greenberg in Berkeley. After having received his Ph.D. in biochemistry from the U.C. Medical

School's Division of Biochemistry in Berkeley, he joined Zamecnik's group with an NIH fellowship in 1949. Siekevitz adopted the Schneider-Hogeboom recipe for tackling the incorporation of tagged amino acids into a granular fraction of liver cells.[43] His fraction of homogenized rat liver contained mitochondria and microsomes but was freed from nuclei, intact cells, and large cell debris by low-speed centrifugation. Even in the absence of additional respiratory substrates, this granular fraction incorporated a small, but reproducibly detectable, amount of radioactive alanine. When Siekevitz fed the mitochondrial respiratory apparatus with the appropriate substrates,[44] the incorporation activity was enhanced. Most importantly, he also obtained this effect by adding adenosine triphosphate (ATP) to the granular fraction under conditions where respiration was suppressed. Isolated mitochondria did not incorporate alanine unless crude microsomes were added to the mitochondrial fraction. Finally, when Siekevitz incubated a complete homogenate with radioactive alanine and subsequently fractionated it into mitochondria, microsomes, and supernatant, the highest activity went with the microsomes.

Siekevitz's findings resonated nicely with Keller's in vivo turnover studies. Moreover, they confirmed the early hint obtained from the application of DNP: the incorporation of amino acids did not depend on respiration as such but on the presence of ATP. Taken together, this was evidence for both a topological connection of protein synthesis with microsomes and a metabolic connection of protein synthesis with a source of chemical energy. In contrast to the liver slice studies, the first experiments on the "biochemical bog" of the cell homogenate produced differences that did not remain silent but provided operational handles to differentiate the system further. The metabolic handle was the system's energy requirement, and the topological handle was its dependence on a special cell fraction defined as "microsomal fraction." The topological space of cellular components and the metabolic space of energized intermediates in protein synthesis began to intercalate. Protein synthesis began to be represented as a system of cellular fractions and biochemical components that could be added to and withdrawn from a test tube. These elements instigated the game of reconstituting a biological activity.

Differential Fractionation

Within a short time, Siekevitz had managed to provisionally standardize and operationally define a "first generation" of cellular fractions. In July

1951, the assay system was established.[45] The minimal standardization of the system also constituted its differentiality: it became possible to diversify the experimental trials. From the beginning, the few actors in the new field of amino acid incorporation studies had been confronted with, indeed had been haunted by, the possibility that the incorporation activity would turn out to be unrelated, or only indirectly related, to peptide bond formation. The percentage of incorporated radioactivity was low; how could it be distinguished from unspecific adsorption or from other intervening reactions?[46] It simply was not clear in every case what "uptake" or "incorporation" meant. It was not clear whether the same process was being observed in the systems developed by different groups, derived from different tissue (rat liver, guinea pig liver, mouse tissue, chicken tissue), fed with different amino acids (lysine, glycine, cysteine, methionine, and alanine), or tagged with different labels (^{14}C, ^{35}S). Indeed, some early in vitro "protein synthesis" revealed itself to be due to S-S bonding of sulfur-containing amino acids.[47] Siekevitz was well aware of such possible pitfalls. He had worked in Greenberg's lab, first on protein synthesis using labeled glycine in liver slices, then on the metabolism of glycine and serine using again liver slices. Greenberg's paper, in 1947, had been the first report of an active homogenate.[48] Two years later, its protein "synthesis" activity had to be redefined as a conversion of glycine into serine, which in turn was converted into phosphatidylserine and finally appeared with the protein as a lipid contaminant.[49]

Not all these tangles will be traced here in detail. I simply want to stress what seems to me a generalizable statement: the main experimental signal of the system—the incorporation of amino acids—was not a meaningful signal by itself. Whatever its prominence or level above background, it had to be embedded in a network of corollary signals. The elaboration of such a network requires familiarity with the system. This process takes time, which helps to explain why experimenters, once they have established their experimental network, often stick to it in an almost symbiotic fashion. But once the system has become familiar to those who "inhabit" it, its own momentum may take over. It may force researcher and research into a kind of internal exclusion. The more the experimenter learns to manipulate the system, the better the system comes to realize its intrinsic capacities: it starts to manipulate the researcher and to lead him or her in unforeseen directions.

The fractionated cell-free protein synthesis system can serve as an example of this situation. The experimental setup was still far from being handled with virtuosity, but it was already out of the bog to a sufficient degree. Siekevitz homogenized the liver tissue by means of a Potter-Elvehjem blender, and he separated the cytoplasmic fractions, with minor modifications, according to the sucrose centrifugation method of Schneider and Hogeboom.[50] A new mode of representation of the research object—fractional centrifugation of the cell sap—took shape. The technical conditions of the system and the scientific object under scrutiny began to intercalate.

The content of the cells was separated into different fractions (Figure 4.1). The fractions were operationally defined in terms of different centrifugal forces. Some of the fractions contained cellular components that could be identified by microscopic inspection (nuclei, mitochondria); this did not mean, however, that they were defined by them. Rather, centrifugation defined the fractions, and the fractions in turn determined the provisional partition of the scientific object. Techniques and objects became engaged in a mutually deconstructive interaction.

Siekevitz had designed a detailed washing and isolation procedure to assure that the label was indeed incorporated by way of peptide bonds. It started with cold and hot acid precipitation (the latter destroyed the nucleic acids), proceeded through extraction with a warm alcohol-ether-

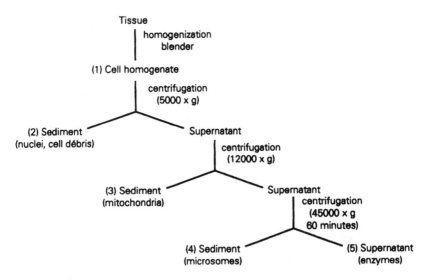

Fig. 4.1. Fractional representation of a rat liver tissue homogenate. Reconstructed from Siekevitz 1952.

TABLE 4.1

Reconstitution of Amino Acid Incorporation from Different Rat Liver Fractions

Fraction	Counts per minute per mg protein	
	Minus energy	Plus energy
Homogenate (1)	1.4	10.8
Nuclei + cells (2)	1.2	2.9
Mitochondria (3)	0.9	1.3
Mixed fraction	1.7	1.1
Microsomes (4)	1.6	1.0
Supernatant (5)	0.1	0.4
Mitos. + Micros.	1.1	4.2
Mitos. + Super.	1.0	6.6
Mitos. + Micros. + Super.	0.8	9.8
All fractions	0.8	10.5

SOURCE: Adapted from Siekevitz 1952, table 1.

chloroform solution (to remove the lipids), and ended with dispersion of the protein in acetone and spreading it on filter paper for counting. Further controls included a ninhydrin treatment of the protein and its hydrolysate as well as a comparison of the radioactivity before and after mild alkali treatment. These procedures reflected the previously encountered difficulties, and all served to make sure that it was peptide bond formation that was measured.[51] But despite their sophistication, the procedures were not simply conditions to be met in order to obtain reproducible results. They "produced" the signals of the system, and they did so at the expense of other possible signals. To focus on the stable peptide bonds inevitably meant destroying other kinds of labile bonding. The rigor of product analysis that was deemed necessary to ensure the identity of the object under investigation precluded access to the conditions of its formation.

There had been earlier attempts at fractionation. As early as 1950, Borsook and his coworkers had fractionated their homogenate into four different components—a 750 × g sediment, a 24,000 × g sediment, a 40,000 × g sediment, and a supernatant. But they failed to differentiate the fractions functionally. All fractions incorporated amino acids.[52] Siekevitz was able to reconstitute activity from different fractions.

The strategy pursued in these experiments proved to be as difficult to realize in practice as it looked simple in theory. The strategy exemplified in Table 4.1 was this: provide incubation conditions such that the complete cell homogenate retains amino acid incorporation activity (1). Take the homogenate apart and determine whether the isolated components carry an activity along with them (2–5). If not, reassemble the isolated components in all possible combinations and observe which components are required to restore the initial activity. Although the strategy worked in principle, virtually everything was left open. The homogenate was active.[53] Neither of the fractions alone displayed substantial activity. However, when different fractions were combined, all sorts of combinations restored a substantial amount of the initial activity. It therefore looked as if different "factors" had to be combined that resided in different cellular fractions and that were separable by centrifugation. What were those factors and fractions?

One of them was the microsomal fraction. This did not come as a surprise. Keller's fractionation assays had already pointed in this direction. With respect to the energy requirements, however, the situation was confusing. The incorporation process required α-ketoglutarate. But in contrast to what Siekevitz and Zamecnik had reported in a preliminary account,[54] the stimulatory effect of adenosine triphosphate (ATP) was gone. DNP, on the other hand, had the same inhibitory effect as it had on liver slices. The emerging picture was that mitochondria, with the help of α-ketoglutarate, produced a "soluble factor," which in turn enabled the microsomal fraction to incorporate alanine. The factor had two rather unusual features: it was extremely heat and acid stable. It seemed unlikely, therefore, that it was an activated alanine or anything like ATP. So the question arose whether "[ATP] is only used for the formation of a factor such as the one described in this paper and it is this factor which, while lacking the high energy phosphate bond, can incorporate amino acids into protein, perhaps with the aid of the ribonucleic acids."[55]

Although Siekevitz's "soluble factor" quickly disappeared from the protein synthesis scene and was never referred to again, it indicated a strategic position: the space between the free amino acids and the completed protein in the emerging protein synthesis display. A third fraction stimulated the activity. It was the supernatant resulting from the last centrifugation step. Nobody paid much attention to it, however. On the whole, the system posed more questions than it resolved. What was the strange mitochondrial-derived factor? Was the stimulatory activity of the supernatant worth being followed at all? What about the nuclear fraction? It also showed some 30 percent activity. The results were both ambiguous and promising.

With respect to quantitative assessment, the assays were, to say the least, at the borderline of resolution. The operational boundaries of the fractions were fuzzy. For a satisfying separation of the microsomes, higher centrifugal forces would be necessary. These forces could not be obtained by an ordinary laboratory centrifuge. The work had all been done with a regular Sorvall laboratory centrifuge attaining approximately 15,000 rpm. Zamecnik and Frantz had requested a high-speed centrifuge as early as 1949, but an ultracentrifuge was installed at the Huntington Laboratories only in 1953.[56]

A New Space of Representation

The fractions obtained so far, although operationally defined in terms of centrifugation velocity and granular contents, biochemically still constituted black boxes. What, then, was the power of the system at that stage? The reconstitution assay linked the construction of a *metabolic* space of reactions to the *topological* space of fractionation. It formed a disciplinary bridge "between the morphologists and the biochemists by providing a means for relating the biochemical events of protein synthesis to recognizable structures."[57] It integrated two tracing methodologies that constituted a new space of representation: radioactive tracing and differential centrifugation. It was the overlap of these two modes of representation that provided a reasonably stable background of reproduction and opened a range of possible orientations.

Soon, the language in which the experimental representation of protein synthesis was captured began to reflect the intricate packing of technical conditions and scientific object as well as the practical power of this package. The laboratory community began to speak about amino acid incorporation, no longer in terms of tissue-specific rates of "uptake," but in terms of centrifugal velocities, sedimentation properties, and precipitation conditions. There were "pH 5 precipitates," "40,000 $\times$ g pellets," and "soluble fractions." This was the cautious wording of experimentalists who like to keep their options open; nevertheless, the new language generated a commitment. There are no noncommittal terms in science. These signifiers constituted a new space of experience and a new kind of practical rationality came along with them, of combinatorial games in the design of experiments, of symmetry considerations with respect to experimental controls, of superpositions, and of matching.

They represented the building blocks of the new experimental system, and they began to determine the dynamics of its differential reproduction.

The initially central, cancer-related medical question of a comparative analysis between normal and malignant tissue had gradually shifted out of focus and become marginalized. The cancer problem had been a hallmark both of the Huntington as an institution and of Zamecnik's professional education, and it had been instrumental in raising funds to finance the research program. The work had been mainly supported by grants from the American Cancer Society, the United States Navy Department, the Atomic Energy Commission, and smaller private funds distributed through the MGH Research Fund.[58] Despite this major change in research perspective, the American Cancer Society saw no actual reason to withdraw its support. In fact, it continued its support throughout the next decade according to a philosophy of flexible distribution. This flexibility with regard to institutional grants is nicely documented by the following sibylline extract of a discussion between a member of the American Cancer Society and the Committee on Research in 1952: "Dr. Lipmann inquired how large a percentage could be used for basic research not directly related to the cancer problem. Mr. Runyon [American Cancer Society] stated that the Society believed both basic and applied research were essential but that unless basic research continued to act as a well for feeding specifically cancer research, then it might not be able to obtain further support. Mr. Spike expressed it in a different fashion, stating that unless Dr. Lipmann would give him a definition of what cancer research was, he could not answer the question."[59]

Step by step, the cancer problem was on its way to being replaced by a deliberately biochemical perspective, that of a search for intermediates of protein synthesis, which could be defined in terms of cellular fractions, which in turn could be manipulated experimentally, and which constituted a particular frame of experimental thinking. To conclude with the words of Robert Loftfield: "In 1950 protein synthesis was important to us because of its possible impact in cancer. [Around] 1953 cancer was important to us because it provided the biological systems to study protein synthesis."[60] Cancer was still present, but it had changed its status. It was no longer a target of immediate analysis; it had become a kind of background screen through which basic research on protein synthesis derived its ultimate sense and justification.[61]

That same year, 1952, Zamecnik went on to spend a term at Caltech with Linus Pauling. He felt that he had to learn more about proteins, and

Caltech was the Mecca of the "protein view of life" at that time.[62] Zamecnik did learn a lot of new things about protein chemistry, but little, as he admitted later, on protein synthesis.[63] Siekevitz left Boston to learn more about oxidative phosphorylation and went to Van Potter at the McArdle Laboratory for Cancer Research in Wisconsin.[64]

CHAPTER 5

Reproduction and Difference

❖❖❖   We may say that when we learn [a] probe, or a tool, and
thus make ourselves aware of these things as we are of our
body, we *interiorize* these things and *make ourselves dwell in
them*.

—Michael Polanyi, *Knowing and Being*

Experimental systems, together with the scientific objects wrapped up in
them, are inherently open, if bottlenecked, arrangements. Their movement is such that it cannot be predicted if they are to retain their character
as research devices. Epistemic things, let alone their eventual transformation into technical objects and vice versa, usually cannot be anticipated
when an experimental arrangement is taking shape. But once a surprising
result has emerged, has proved to be more than of an ephemeral character,
and has been sufficiently stabilized, it becomes more and more difficult,
even for the participants, to avoid the illusion that it is the inevitable
product of a logical inquiry or of a teleology of the experimental process.

Labyrinths

"How does one re-create a thought centered on a tiny fragment of the
universe, on a 'system' one turns over and over to view from every angle?
How, above all, does one recapture the sense of a maze with no way out,
the incessant quest for a solution, without referring to what later proved
to be *the* solution in all its dazzling obviousness?"[1] An experimental system can readily be compared to a labyrinth, whose walls, in the course of
being erected, in one and the same movement, blind and guide the experimenter. In the step-by-step construction of a labyrinth, the existing
walls limit and orient the direction of the walls to be added. A labyrinth
that deserves the name is not planned and thus cannot be conquered by
following a plan. It forces us to move around by means and by virtue of
checking out, of groping, of *tâtonnement*.[2] He who enters a labyrinth and

does not forget to carry a thread along with him, can always get back. But there has not yet been found the thread that would indicate the direction in which to proceed through a labyrinth. There is a striking parallel in this respect between the work of the experimentalist and the way George Kubler sees the work of the artist: "Each artist works on in the dark, guided only by the tunnels and shafts of earlier work, following the vein and hoping for a bonanza, and fearing that the lode may play out tomorrow."[3] The metaphor of the labyrinth matches that of the mine.

Reproduction

Thus, the temporal coherence of an experimental system is granted by recurrence, by repetition, not by anticipation and forestalling. Its future development, on the other hand, if it is not to end in idling, depends upon groping and grasping for differences. Together, this adds up to what can be called *differential reproduction*. The term *reproduction* has many meanings. I would like to briefly indicate in what sense I am *not* using it and which of its connotations are central to my argument. I do not use the term to stress the continuity of an experimental program as against sudden breaks and frequent changes. I do not use it to designate a copying procedure, a process of making replicas from an original. Nor do I use it to indicate that a good experiment should be able to be replicated at will, that is, that its results should be reproducible as that term is commonly understood. Rather, I use the term *reproduction* in a sense somewhat akin to that used in evolutionary contexts. It serves to indicate that experimentation has to be seen as an ongoing and uninterrupted chain of events through which the material conditions for continuing this very experimental process are maintained. Reproducing an experimental system means keeping alive the conditions—objects of inquiry, instrumentation, crafts and skills— through which it remains "productive."

All innovation, in the end and in a very basic sense, is the result of such reproduction. Reproduction, far from being simply a matter of securing appropriate and reproducible boundary conditions for the experiment, characterizes scientific activity as a material process of generating, trans- mitting, accumulating, and changing information. The generation of new phenomena is always and necessarily coupled to the coproduction of already existing ones. Without this coproduction, there would be no basis for comparison; in fact, there would be a sudden dissipation of all knowledge implemented in the experimental process. For this very reason, experimental systems are necessarily localized and situated generators of knowledge. Their reproductive situatedness, not their logicality, marks their cohesion over time, and thus their historicity. To establish a scientific object means that it will have emerged from differential reproduction and that it will be able to be inserted in the reproductive cycle of an experimental system. Epistemic things, therefore, are recursively constituted and thus intrinsically historical things. They derive their significance from their future, which is unpredictable at the real time of their emergence. They are constituted by recurrence. We can state with Heidegger,

The methodology through which individual object-spheres are conquered does not simply amass results. Rather, with the help of its results it adapts itself for a new procedure. Within the complex of machinery that is necessary to physics in order to carry out the smashing of the atom lies hidden the whole of physics up to now. [In] the course of these processes, the methodology of the science becomes circumscribed by means of its results. More and more the methodology adapts itself to the possibilities of procedure opened up through itself. This having-to-adapt-itself to its own results as the ways and means of an advancing methodology is the essence of research's character as ongoing activity.[4]

But let us be cautious. Yes, there is a pressure for connectivity within an experimental setup. First, however, it does not take the form of the usual claims of theoretical consistency and commensurability, and, second, it remains local, that is, restrained to particular setups within a multitude of others that constitute the fractal boundaries of a research field. I will pursue to this problem in Chapter 9.

From another angle, the construction of experimental systems can be described as a "*jeu des possibles.*"[5] The title of François Jacob's essay alludes to the tinkering of evolution as well as to the process of scientific change. As far as scientific research is concerned, we have to conceive of the "possible" in the double sense of the word: it is something that is in the realm of an experimental system, and it is something that is beyond proper control. The possible has a strange and fragile presence. On the one hand, it does not exist in any strong sense of the word, and, on the other hand, "one must already have decided—*il faut déjà avoir décidé*—what is possible."[6] With Jacques Derrida, we could speak of a "game of the difference."[7] I will come back to the implications of this comparison. Here I would only like to draw attention to the point that it is the character of "fall[ing] prey to its own work" that brings the scientific enterprise into propinquity with what Derrida has called "the enterprise of deconstruction."[8] Let me recall, in this context, a remark by nineteenth-century physiologist Claude Bernard in his philosophical notebook: in physiology, he says, "there is a succession of evolving facts which follow each other in time, but which do not necessarily engender each other. It is a chain whose links do not have a relation of cause and effect, neither to the one that follows, nor to the one that precedes."[9] In differential reproduction, as perceived by Bernard and outlined in this chapter, there is no relation of cause and effect, no necessary development, no predetermined direction, but a chance to generate unprecedented events and a possibility to have them retroact on the system and so become concatenated.

Tacit Knowledge

The intricacy of this game of the possible is reflected in the personal experience of those who do research. As Bernard, again, has put it, "one must have felt one's way for a long time—*il faut avoir tâtonné longtemps*—, have been mistaken thousands and thousands of times, in short, have grown old in the practice of experimentation."[10] To feel one's way requires *Erfahrenheit* on the part of the experimenter. "Being experienced," as Fleck uses the expression, is not simply "experience."[11] Experience enables us to judge a particular piece of work or a particular situation. Being experienced enables us to literally embody the judgment in the process of making new experiences, that is, to think with our body. Experience is an intellectual quality. *Erfahrenheit*, that is, acquired intuition, is a form of life. There is a countersense to this expression that is essential to the meaning it is intended to convey. *Erfahrenheit has* to be acquired, and it transcends what *can* be learned. It amounts to what Michael Polanyi has called the "tacit component," the "tacit dimension" of knowledge, or "tacit knowing."[12] It lines up with other attempts to do justice to the "intimacy" of scientific work, to the exuberance of science in action, to what is beyond methodological axiomatization, to "cunning reason," to the fact that plans of intelligent actors do not control action in any strictly definable sense of the word, so that when it comes to in situ action, "you rely not on the plan" but on embodied skills.[13] In Polanyi's vision of personal knowledge, "tacit knowledge" and "explicit knowledge" do not simply coexist in a happy relation of complementarity. He claims that knowing in general, from everyday practical performance to knowledge involved in productive scientific research, is either tacit or rooted in tacit knowledge, thus suggesting that wholly explicit knowledge is simply unthinkable, an illusion produced by analytical philosophy. "Subsidiary awareness" is an integral part of the gestural package of any researcher.[14]

Embodied reasoning does not for these reasons proceed without rules. But "the aim of a skillful performance is achieved by the observance of a set of rules which are not known as such to the person following them."[15] With respect to the type of biochemical reasoning explored in this book, several of such implicit rules can be invoked. Although they can be made explicit, their efficiency rests on being subsidiarily present in the design of an experiment. The first may be called a "symmetry principle." Symmetry considerations govern the arrangement of epistemic and procedural controls within an experimental setup. Usually they take the form of testing all possible combinations of different components within a multi-componential assay. We have seen an instance of this procedure when looking at Siekevitz's in vitro protein synthesis protocol in Chapter 4. A second rule may be called the "homogeneity principle." It refers to the precautions to be taken in the use of preparations of cellular compounds when data sets obtained from different preparations are compared. Never change the batch of material within a single set of experiments, and always repeat the last assay when you try a new batch. The third rule is an "exhaustion principle." It says that the series of similar compounds to be tested within one and the same experimental context should always be complete. Never leave out one of the compounds because you think that it does not work anyhow.

All these rules are learned through action, and their implementation can take widely different forms in different experimental contexts. They constitute, in a certain sense, a materialized and exteriorized system of imagination. They constitute a kind of experimental spider's web: the web must be meshed in such a way that unknown and unexpected prey is likely to be caught. The web must "see" what the spider actually is unable to foresee with its unaided senses. But the web must not become too rigid. In deliberating upon the manner in which a system is to be handled so as to let the unknown intrude and invade it, Max Delbrück has spoken of a "principle of measured sloppiness."[16] "If you are too sloppy, then you never get reproducible results, and then you never can draw any conclusions; but if you are just a little sloppy, then when you see something startling you [nail] it down."[17] Connivance rather than stubborn rigor is at the center of the experimental enterprise. If there is a principle at all that guides the experimental roadmanship, it consists in "being attentive to the answers arising on the margins or even outside the expected discourse."[18]

Difference

We can restate this dynamic, soft, and subtle pattern of embodied and situated knowledge production in more philosophical and general terms and say with Gilles Deleuze, the thinker of the difference: "Difference and repetition have taken the place of the identical and the negative, of identity and contradiction. For difference implies the negative, and allows itself to lead to contradiction, only to the extent that its subordination to the identical is maintained. The primacy of identity, however conceived, defines the world of representation. But modern thought is born of the failure of representation, of the loss of identities, and of the discovery of all the forces that act under the representation of the identical. The modern world is one of simulacra."[19] I will come back to Deleuze's promulgation of the foundering of representation in contemporary thought in Chapter 7. In the present context, my concern is with identity and contradiction. Identity and contradiction have been the key terms of the grand philosophical systems of the eighteenth and nineteenth centuries, from Kant to Frege, from Hegel to Marx. In contrast, difference and repetition are the key terms of a "philosophy of epistemological detail," of a "differential scientific philosophy" in the Bachelardian sense.[20] This shift of attention allows us "to make all these repetitions coexist in a space in which difference is distributed."[21] Where identity is mute, repetition confers tacit knowledge. Where contradictions have to be resolved, differences can coexist and thus enter the game of becoming prominent or marginal, displaced or articulated. Research systems in particular must allow for the emergence and distribution of differences, as we have seen in Chapters 3 and 4. Experimenters are not interested in identities; they proceed in the search for "specific differences."[22] It is not astonishing that observers of laboratory shoptalk have come to recognize that there is a "preference for disagreement" at the bench as an implicit strategy for producing novel, "not previously obvious" features, and that "in practice, it seems, participants prefer a principle of variation over replication."[23] Replication aims at identity, repetition at variation.

Several remarks are in order with respect to the differential reproduction of an experimental system. First, one never knows precisely how the setup differentiates. As soon as one knows exactly what it produces, it is no longer a research system. This makes me hesitate to speak about experimental systems in terms of sheer "systems of production,"[24] an expression borrowed from economic life and laden with connotations such as directivity, efficiency, quantitation of output, and automation. "No method has ever led to an invention."[25] An experimental system that gradually acquires contours, creates resonance between different representations, and conveys manageable meanings to stabilized signals, must create at the same time a space for the emergence of things unheard of. Inasmuch as it is stabilized in a certain respect, it can be destabilized in another. To arrive at new results, the system *must* be destabilized—but without a previously stabilized system there will be no "results." Stabilization and destabilization imply each other. In the language of the laboratory, the result is that unit in which the dynamics of an experimental system is captured. It is not a final statement, it is a piece that fits or does not fit in an ongoing puzzle. To remain productive in an epistemic sense, an experimental arrangement must be sufficiently open to generating unprecedented events by incorporating new techniques, instruments, model compounds, and semiotic devices. At the same time it must be sufficiently closed to prevent a breakdown of its reproductive coherence. It has to be kept at the borderline of its breakdown. In this respect, too, experimental systems in science obey a "rule of order" similar to that of formal sequences in the production of art.[26]

The minds of inventors and scientists, much like those of artists, are not oriented toward recognizing what exists; they "turn more upon future possibilities, whose speculations and combinations obey an altogether different rule of order, described here as a linked progression of experiments composing a formal sequence."[27] The classification into sequences, according to Kubler, "stresses the internal coherence of events, all while it shows the sporadic, unpredictable, and irregular nature of their occurrence."[28]

If research systems become too rigid, they turn into devices for testing, into standardized kits, into procedures for making replicas. They lose their function as machines for making the future. Such transformations, however, happen regularly and are not necessarily dead-end streets. Epistemic things turned into technical objects become integrated as stable subroutines into other, still growing experimental systems and may help to produce unprecedented events in different contexts. The transformation of research objects into stable subroutines of other research arrangements is what confers that special kind of material information storage, or mass action, to the process of experimentation. But by the same token, this process generates a historical burden. With Norton Wise, we could speak of the "resistance" or "resilience" of such a network, whose "structure as a whole puts severe constraints on what it means to know or to explain, or indeed for a thing to exist."[29] Most new epistemic things, therefore, take their first shape from old tools. Yet in the long run and as a rule, technical procedures are completely replaced by subroutines that embody the actual, stabilized knowledge in a subtler way. The historian of science usually looks at a museum of abandoned systems. The liver slice system that provided the starting point for characterizing malignant growth in terms of protein synthesis, described in Chapter 3, was completely replaced within a few years by a fractionated homogenate. There is a life cycle to experimental systems. They are brought into being as research devices, become transformed into kits, and finally are replaced. But there is a symmetrical counterpart to this cycle. Kits can become destabilized and turned into research devices, either by transplantation or by the introduction of new representational techniques.

Différance

To remain a research system, an experimental arrangement must be managed in such a way that it keeps being governed by difference. I use the term *difference* to characterize the specific, displacing dynamics that distinguishes the research process. An experimental system that is organized in a way such that the production of differences becomes the orienting principle of its own reproduction is governed by and at the same time creates that kind of subversive movement Derrida has called *différance*. Pointing to an "irreducible absence of intention" that goes along with all invention, the term conveys a sense for the "event-riddenness" that is at the heart of research experiments.[30] Perhaps the notion discloses its meaning best by way of a language game. It is a word that exhibits its difference to "difference" only in written form and that, when spoken, is made silent by the very act of its pronunciation. With it, Derrida intends to capture a movement he calls "the production of differing/deferring," which is bound to the radical exteriority of inscription, of writing as well as of producing other material traces.[31] *Différance* has become the shibboleth of a literary fashion—deconstruction. But in the eyes of its inventor, it is not just a strategy of literary criticism, as opposed to hermeneutics. If a strategy at all, "one might call this blind tactics, or empirical wandering."[32] Nor is deconstruction a mere intellectual attitude. It describes a movement that pervades all sorts of spaces of experience, their material constitution as well as their discursive framing. "Deconstructions," as Derrida prefers to say in the plural, are specified by "a certain [metonymic] dislocation which repeats itself regularly [in] every 'text,' in the general sense I would like to attach to that name, that is, in experience as such, in social, historical, economic, technical, military 'reality.' "[33] And let me add that these repeated dislocations occur particularly in the experience of scientific experimentation. "Differantial" reproduction in the sense of a permanent dislocation of epistemic entities is precisely what endows a research system with its generative power, and what renders the process genuinely *historial*.[34] This second Derridean neographism hints at the special temporal character of the *différance*. "Historial *différance*" is an operator, not a parameter, of a system.

The attentive examination of experimental systems and the networks they constitute will help us to gather indispensable information for a future "differential typology of forms of iteration."[35] The epistemology and history of biology can contribute to such a typology precisely because the life sciences are so prolific in creating these specific forms of iteration we call experimental systems. We have to pay much more attention to the synchronic as well as diachronic dimensions of these forms of transfer, their spacing, and their power of dissemination. The characterization of such transfer patterns has been initiated by Isabelle Stengers under the notions of "operations of propagation" and "operations of passage."[36]

This book is about the iterative enforcement of a local research system and its subsequent dissemination. I have chosen to narrate the history of the rat liver in vitro system of protein biosynthesis in sufficient detail so that the experimental moves at the microlevel become visible, the fortunate choices as well as the abortive trials. As I mentioned in Chapter 3, the Huntington Memorial Hospital, under the directorship of Joseph Aub, had embarked upon a program of cancer research based on the study of growth phenomena. Zamecnik's group at MGH had spent the first years of work looking at growth deregulation in malignant tissue within the framework of a biomedically oriented research environment. Protein synthesis was assumed to be a possible target of the neoplastic behavior of cancer cells. But between 1947 and 1952, through a series of differential, dislocating experimental events, the emphasis shifted. The perspective on cancer became deferred, and although it was not completely and abruptly abandoned, it glided into, was subverted by, and became reoriented as an inquiry into the conditions of cell-free incorporation of amino acids into proteins of normal cellular tissue. A first in vitro system was in place in 1952. I will now outline its differential reproduction between 1952 and 1955.

CHAPTER 6

Defining Fractions, 1952–55

❖❖   Biologists abhor abstraction. Let there appear a ghost, a
creature of pure reason, and they will immediately form a
visual representation of it. [To] make them accessible to
experimentation, the questions had to be changed, broken
down into parts.
                                    —François Jacob, *The Statue Within*

The rat liver homogenate was a big step toward an experimental, operational representation of protein synthesis. At the same time, it produced
major problems. The activity of the system was, to put it mildly, at the
borderline of resolution. The whole edifice rested on a figure of ten
counts per minute. The radioactivity taken up into protein was at least
one order of magnitude below the values obtained in the previous slice
experiments. The fractionation pattern suggested that virtually everything in the cytosol, including the mitochondria, was needed to obtain
that faint signal showing up as the integral of reconstitution. For the most
part, the fractional partition did not sort out well-characterized particles
or molecules of the cytoplasm. In this respect, it was a representation
without clearly delineated objects, a "system without referent." This expression exactly translates the laboratory situation in which the Huntington researchers were engaged. It was impossible to determine the
fractions by the macromolecular constituents of the biochemical process
under scrutiny; instead, the centrifugation technique determined the sutures of what could be taken apart and reconstituted. Here we see in
action what Freud has described as the precarious interplay of deriving
concepts from and imposing them upon the material of experience (see
Chapter 1). To use the words of Bachelard, there was an increasing tension between the "space of ordinary intuition" with its objects of experience and the "functional space" in which the phenomenon of protein
synthesis came to be represented.[1]
    Why not substitute a metabolically highly active bacterial extract for
the liver homogenate? As early as 1951, Zamecnik and Mary Stephenson,

in a joint effort with David Novelli from Lipmann's laboratory, had made an attempt to break up *Escherichia coli* cells.[2] In fact, they got something that amounted to an amino acid incorporation, but, unfortunately, they did not manage to clear the system sufficiently of intact bacterial cells. Consequently, these intact cells, as a contaminant, could also account for the incorporation effect. Stephenson remembers doing those endless bacterial counts under the microscope: "It was just a matter of a few thousand bacteria that would completely foul up the system."[3] Finally, they gave up and put the work on the shelf.[4]

During that time, in June 1951, microbial chemist Ernest Gale from Cambridge University paid a visit to the MGH laboratories.[5] Over the years following the war, he had established an in vivo protein synthesis system based on the gram-positive bacterium *Staphylococcus aureus*. Gale became interested in Zamecnik's in vitro studies, and shortly thereafter, in 1953, he and his coworker Joan Folkes announced an in vitro bacterial incorporation system based on sonicated *Staphylococci*.[6] Needless to say, the MGH group judged Gale's system to be delicate. It only worked in the presence of, albeit disrupted, cell walls, and the microscopic distinction between disrupted and undisrupted cells was ambiguous and difficult. In turn, Gale adopted a skeptical attitude toward what he disparagingly called "incorporation studies" as a measure and surrogate "of protein synthesis in living cells," as long as such experiments were not accompanied by a net increase of protein substance. And this was not the case, either at MGH or elsewhere.[7]

Gentle Homogenization: Another Difference

Indeed, the activity problem was the specter that lurked behind the whole enterprise. In this precarious situation, help came from a neighbor—an example of collaboration *hors de programme* resulting from the loose coupling of different research activities at the Huntington Laboratories. Nancy Bucher had earned her M.D. from the Johns Hopkins Medical School in 1943 and had been a clinical and research fellow in medicine at MGH since 1945. In 1952, she was working on a method for isolating intact liver cells.[8] She minced rat livers in the presence of glass beads, but the mechanical disintegration of the tissue appeared to make the cells leaky. Ivan Frantz recalls: "Nancy cast about for a good biochemical process on which to try her cells. Cholesterol synthesis looked like a good prospect. It had been demonstrated *in vitro* only in tissue slices. She came to me for instruction in the incubation techniques and methods of assessing the results, which, incidentally, I had learned from Gordon Gould."[9] Indeed, when suspended in Bucher's "witch's brew,"[10] the cells formed cholesterol from ^{14}C-labeled acetate. But the preparations contained broken cells and cell debris as well. Bucher repeated the experiment and included a control: she centrifuged the cells and tested the supernatant. To her surprise, "the debris worked better than the cells."[11] An in vitro system for the formation of cholesterol resulted.[12] Henceforth, Bucher used a Potter-Elvehjem homogenizer with a loose-fitting pestle for gentle disruption of her cells. With this little trick, she also set off a new era in the study of protein synthesis: Zamecnik tried loose homogenization in his own cell-free system, and the procedure proved an elegant means for enhancing the incorporation activity. The incorporation was increased by at least a factor of ten. It was still far from net protein synthesis, but counting the radioactive signals now became reliable.

In the course of the years, the homogenization mixture gained considerably in complexity. It grew by apposition. Components that had been found to stabilize the system were kept and carried along throughout the subsequent stages. Sucrose, for example, was present in all the experiments for the next ten years.[13] Experimental practices create local customs. By traffic, trade, and trust, such customs can sweep through a whole community, even if they are by no means the only ones that work.

Small Molecules and Big Machines: ATP and Ultracentrifugation Intertwined

The experiments of Philip Siekevitz had been severely limited by the lack of a centrifuge yielding more than 45,000 × g. Under these conditions, the microsomal material, which contained the bulk of the cell's ribonucleoprotein, did not sediment completely, and a quantitative separation of the microsomes from the postmitochondrial supernatant was beyond the realm of the technique. Siekevitz had tried acid precipitation as an alternative.[14] But this procedure brought down additional material that, when centrifuged, remained in the supernatant. The result was either a microsome-containing supernatant or a supernatant lacking all acid precipitable material. Neither helped to clarify the protein synthesis partition.

A refrigerated preparative ultracentrifuge was installed at the Huntington Laboratories in 1953, upon request of Lipmann, to be used jointly by the laboratories of Lipmann and Aub.[15] Access to an ultracentrifuge changed the situation. In the beginning, the instrument entered the experimental scene of protein synthesis silently and as a supplement. "In some experiments," we read in a paper of 1954, "the 5,000 × g supernatant fluid was separated [in] a Spinco preparative centrifuge."[16] But within a year, the instrument helped to reorganize the whole fractionation procedure. Preparative high-speed centrifuges had been available commercially for more than a decade. Claude and others had been using a first generation of such machines for the identification and structural characterization of cytoplasmic particles since the early 1940s.[17] Before metabolically active homogenates became available, however, the instrument was of little help in the elaboration of functional in vitro protein synthesis systems. With the advent of such homogenates, it gained relevance. It is not the instruments per se that guide and shape experimental systems. It is the experimental system with its many parameters that defines whether a particular technical procedure or instrument, if available, makes sense in the context of the ensemble.

Around the same time, a refinement in the low-speed range of centrifugation proved to be of equal importance. Already in 1951, Siekevitz and Zamecnik had reported a stimulatory effect of ATP on a mixed fraction of mitochondria and microsomes.[18] But the stimulatory effect had vanished in the more elaborate assay system of 1952. During the following year, Betty Keller managed to remove the cell debris together with the mitochondria in a single preliminary centrifugation step. The resulting homogenate became dependent on ATP, and protein synthesis became independent of mitochondria and the aerobic energy-converting mechanism associated with them. Given Lipmann's long-standing claims about phosphorylated intermediates in peptide synthesis this did not altogether come as a surprise.[19] Nevertheless, the complete displacement of the mitochondria from the system was a major event in draining the biochemical bog. From now on, the experiments could be performed without the complicated and messy respiratory machinery as a source of biochemical energy and, therefore, without the tedious aerobic incubation procedure. The cellular "energy fraction" could be replaced by a commercially available biochemical substance: ATP, a mono-nucleotide.

Zamecnik realized the potential impact of this finding and immediately transmitted a grant application to the American Cancer Society on "purines and pyrimidines as sites for activation and transfer of metabolic intermediates."[20] The grant application was approved by the Committee on Research in October 1953. To my knowledge, this proposal was the first in which Zamecnik explicitly envisaged the possibility of amino acid activation by nucleotides and in which he used the concept of "transfer" for this reaction. A topological picture of protein synthesis emerged, comprising sites of synthesis and shuttles for amino acid transport. By the same token, Zamecnik felt encouraged to propose to the general director of MGH, Dean Clark, that the institutional relationships between the MGH and MIT be strengthened for the sake of a "freer flow of talent and knowledge" in the hope of "translating medical disease into molecular terms."[21]

Another achievement was directly due to the ultracentrifuge. Using high-speed centrifugation at 105,000 × g, Keller sorted out a "microsome-rich sediment" from a "105,000 × g supernatant fraction." Neither the microsomal sediment nor the supernatant was active in amino acid uptake. But the combination of both fractions plus ATP and an ATP-regenerating system yielded activity.[22] Thus, it would appear, Keller and Zamecnik concluded, "that the 105,000 × g supernatant fraction of liver contains some protein or proteins essential for the utilization of ATP for the incorporation reaction."[23] Besides the microsomes, the supernatant, too, began to catch experimenters' attention. It was no longer just carried along for the sake of completing the record. It surfaced from the obscurity in which it had been buried since Siekevitz's pivotal first efforts (described in Chapter 4).

The Differentials of Replication

This redistribution of interest, from mitochondria plus microsomes in 1952 (the former now replaced by an ATP-regenerating system) to microsomes plus a high-speed supernatant in 1954, looks minor, but it deeply changed the scene. It led to a factor quite different from the "soluble factor" that Siekevitz had described in 1952. The latter disappeared from experimental discourse together with the bulky mitochondrial fraction. The new factor emerged from the postmicrosomal supernatant.

The coherence of an experimental system does not, as we can see, depend on the explicit resolution of contradictions. As long as its differential replication goes on, the appearance of a new trait related to the

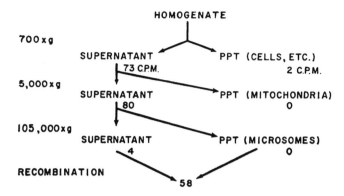

Fig. 6.1. Reconstitution diagram of an active homogenate. Fractionation before incubation. The numbers in the figure represent radioactive counts indicating the incorporation of ^{14}C amino acids by the respective fractions per milligram of protein. PPT precipitate. Reprinted from Zamecnik and Keller 1954, fig. 2.

epistemic object under scrutiny need not eliminate earlier traits. They may, however, decrease in prominence, be reduced to marginality, dissolve into background noise, or simply be forgotten. In the present case, what had been a finding well worth highlighting in 1952 was now reinterpreted as an obstacle that had to be overcome before reliable results could be obtained. Siekevitz's factor turned into a price that had been paid for going in vitro. In retrospect, the experimenters invoked the small extent of incorporation in these studies that had "made it difficult to explore the relationship of this process to energy-yielding mechanisms."[24]

Another Subversion

The new fractionation pattern created another subversion comparable to that which had replaced the cancer-related medical perspective of the slice experiments described in Chapter 3.

Figure 6.1 visualizes the experimental space as a fractional space exclusively. Paradoxically, it subverts the energy requirement of the system by making it invisible. It is absent from the picture. Where did the energy requirement interfere biochemically? The identification of a primary energy source—ATP—had been pushed toward molecular resolution; unfortunately, this resolution was far ahead of that of all the other components participating in the process. Localizing the function of ATP in the metabolic chain of reactions had not become easier.

Fractional Dynamics

The next experimental steps did not follow from the energy aspect of the system but from its fractionation characteristics. One of the components was "a microsome-rich fraction into the proteins of which the amino acids are bound by a linkage as stable as the peptide linkages of the protein." The other was "a soluble, heat-labile, non-dialyzable fraction which facilitates the incorporation of amino acids into the microsome protein."[25] Two years earlier, this fraction had been perceived as simply stimulating the mitochondrial activity. In fact, if we take a close look at Table 4.1 in Chapter 4, we see that the mitochondria were the central component to which everything else was added. Fractionation had been introduced as a technique to come to grips with the energy aspect of the protein synthetic machinery; now the fractionation pattern in turn began to capture and define the possible components of that machinery. This transition is reflected in Figure 6.1, where the energy aspect of the reaction pattern was literally removed to a footnote stating that the energy-regenerating compounds were added to *all* the fractions. The focus shifted and drifted in an unfolding space of fractional representation. As Mary Stephenson put it, "it was a matter of one thing giving the next obvious thing to do."[26] But dislocation did not mean disappearance in this game of differentiating options and deferring actions. The energy aspect was not lost from sight, and it was to reappear in a most surprising way. The whole episode illustrates the microdynamics of experimentation. It shows how the potentials of an experimental system are brought into play, and how its movements gain momentum by continued deconstruction through a permanent process of resignification.

Other issues remained open as well. Despite claims to the contrary based on similar experiments,[27] the MGH incorporation system still did not respond to the addition of a full complement of nonradioactive amino acids to the radioactive one. Were there enough endogenous amino acids present in the various fractions? This was the standard explanation for what remained an oddity.[28] In this situation, it became of major concern to the small protein synthesis industry to produce a complete radioactive protein. Without being able to demonstrate this, the question whether the observed incorporation of radioactivity into proteins could be taken as a model of cellular protein synthesis remained open for debate. But what was the mechanism of cellular protein synthesis? It was *the* epistemic thing: the thing unknown. Ernest Gale, in ruminating on the shortcomings of in vitro incorporation, even decided to "[abandon] the use of isotopes and incorporation studies as means of studying protein synthesis *per se*."[29] As an alternative to this option, Zamecnik's group resorted for many years to the rigorous demonstration that the products of their in vitro system did contain the radioactive amino acids in α-peptide linkage. "The responsibility for demonstrating α-peptide bonding remains inherent in every new experimental system."[30] Whose responsibility? In this transfigured claim, it is the investigations themselves that appear as the ultimate participants in the dialogue.

Models for Models for Models

"No method of isolating and washing radioactive protein can be accepted as a 'standard procedure,' or considered to be free from artifact until it has been proved so for the system under consideration."[31] To translate this tautologous form of circular postulate: there is no absolute referent for the construction of an in vitro system that would permit us to judge whether it is a "model" for the "real" in vivo situation. Neither models nor reals are givens; models do not stand for absolute referents. Models are simply the privileged objects of manipulation. They become privileged, not by the things themselves they pretend to model, but through comparison with other model systems. This situation has been described as the "experimenter's regress," and Harry Collins claims that there is only a "sociological resolution" to the problem of experimental circularity.[32] But as a rule, scientists do not follow Collins's postulate. They do not resort to social instances, either implicitly or explicitly, to stabilize their facts. They redouble and mutually recalibrate their model representations within the ongoing differential replication of their experimental systems.[33]

The only thing fairly undisputed about protein synthesis within the cell was that as a result of the process, the completed proteins contained their amino acids in an α-peptide linkage. Admittedly, this was a rather poor reference for comparison, which was of little help for answering questions about mechanisms in the formation of such linkages. Consequently, the only feasible, though circumstantial, way was to compare what both systems, the intact organism and the fractionated homogenate, delivered as their products. The nature of the linkage could be investigated by comparing the conditions of its *destruction* in both instances. The result of this analytical procedure served as indirect evidence for the process of *formation* of the linkage.

Reinforcements

Robert Loftfield tried to find a more direct approach.[34] He wanted to retrieve a particular, well-characterized radioactive protein from rat liver. Loftfield induced rats to produce the protein ferritin and, after twenty hours, analyzed the distribution of the product between the liver microsomes and the nonparticulate supernatant fraction. The pattern he obtained was, as he claimed, "entirely similar" to the radioactivity pattern of a parallel assay performed in Zamecnik's in vitro system. Thus, he deemed it conceivable that the latter system, too, was "capable of synthesizing a native, authentic, isolable protein."[35] This kind of experiment did not produce new insights into the mechanism of protein synthesis. But it produced another comparative argument, based on considerations of similarity, for a close relationship between what went on in the animal and in the test tube. The ambiguity of such comparisons, however, is elucidated by a case where it failed to work. Harold Tarver had observed that ethionine, an analogue of the amino acid methionine, inhibited the incorporation of glycine and of methionine in vivo.[36] Consequently, Zamecnik and Keller added ethionine to their in vitro system, but they failed to detect any inhibition. Could this finding be taken as an argument against the test-tube system? Clearly, their answer was no, although the negative result did not lend any comparative support for the question at issue.

RNA: The Unanswered Question

The efforts described in the last section were characterized by a well-defined problem for which no clear-cut experiment could be designed. In the case that follows, we have a clear-cut experiment for which there was no well-defined explanation. The addition of ribonuclease to the in vitro system completely abolished the amino acid incorporation. This observation neither supported nor refuted the reliability of the system. Its epistemological status was quite different. The ribonuclease test hinted at an involvement of RNA in protein synthesis. The involvement of microsomes in amino acid incorporation had lent new urgency to an old speculation.[37] Did the RNA moiety of the microsomal sediment play an active role in protein synthesis? Other workers had come up with similar observations—Gale in Cambridge on fragmented staphylococcal cells, Alfred Mirsky at the Rockefeller Institute on fractionated rat liver cells.[38]

The microsomes appeared to contain the bulk of the cytoplasmic RNA. The problem was finding a place for the microsomal RNA in the metabolic pathway from amino acids to proteins. Until then, there was no role conceivable for RNA in terms of a metabolic intermediate in this pathway. Although Zamecnik and Keller vaguely envisioned a "relationship of the ribose nucleic acid of the microsome fraction to the incorporation of the amino acids into the microsome protein," they considered such a relationship "by no means conclusive."[39] When Zamecnik wrote down this conjecture early in 1954, RNA had not yet entered the experimental game of differential reproduction. Although it was in the minds of the MGH group, RNA was not yet an object of inquiry. Instead, the group took the RNA-to-protein ratio of the nonmicrosomal fractions as a measure of the *contamination* of the different fractions with microsomes.[40] Loftfield recalls, "I know we acknowledged the problem, but dismissed it—we expected leakage—broken microsomes, incomplete sedimentation."[41]

Purifications: The Soluble Fraction

For the time being, the work continued along other lines. Betty Keller made various attempts to free the soluble fraction further from small molecules, amino acids, and nucleotides. Chromatographic ion exchange on Dowex retained nucleotides. Alternatively, she precipitated the active "protein components" of the soluble fraction by adjusting the pH to approximately 5. The material recovered from a subsequent low-speed centrifugation, and resuspension was able to replace the soluble fraction. A system composed of microsomes and this "purified" soluble fraction required the nucleoside diphosphate GDP, in addition to ATP, for optimal performance. Did a GDP or GTP derivative also "[function] in the formation of peptide bonds"?[42] This was the first hint that another nucleotide might be involved in protein synthesis, and it instigated a long-term search for its function. Zamecnik had obtained GDP from Rao Sanadi of Wisconsin, before it became available from Sigma Company.[43] The experimenters' rationale behind the search had been simple, and it was successful: if ATP plays a role in the energy turnover of the system, try whether the remaining three ribonucleotides are also effective. If they are ineffective at one stage of fractionation, do not leave them aside, but try them again throughout the subsequent stages of fractionation. Eventually, a signal will show up and produce a difference that can be pursued.

Although the active principle of the 105,000 × g supernatant was now characterized as a protein component that could be acid precipitated, and although GTP seemed to play a role in the energy regime of the process, the whole system still remained silent with respect to the functional connections underlying the reconstitution of its activity. Fractional representation and functional representation did not map neatly onto each other. This *asynchronicity* of representations was, at one and the same time, an embarrassment and one of the main driving forces of the differential machinery. Centrifugal, physical analysis was a prerequisite for biochemical analysis, but they neither implied nor complemented each other without friction. The two modes of representation employed different tools working at different analytical levels.

The Microsomes

In parallel efforts, the material sedimenting in the ultracentrifuge at 105,000 × g was subjected to further fractionation. To obtain active, "purified" microsomes became one of the major concerns in developing the cell-free assay system.[44] Another Harvard M.D., John Littlefield, who had joined the laboratory of Zamecnik as a research fellow in 1954, pursued this task over the next three years.

For purification, Littlefield took advantage of the detergent sodium deoxycholate. In the context of their work on oxidative enzymes in rat liver, Cornelius Strittmatter and Eric Ball, neighbors from the Department of Biological Chemistry at Harvard Medical School, had observed incidentally that deoxycholate decreased the opacity of a microsome suspension.[45] Obviously, it solubilized the protein-lipid aggregates of this fraction. Upon treatment with deoxycholate, Littlefield was able to recover a loosely packed sediment consisting of protein and virtually all the microsomal RNA. But Littlefield also encountered unexpected, by no means trivial, difficulties inherent in his technique. What he recovered from the detergent-insoluble sediment in terms of RNA-rich "ribonucleoprotein" largely depended, in its RNA/protein composition, upon the concentration of the solubilizer. At moderate concentrations, the RNA-to-protein content increased from roughly 10 to approximately 50 percent.[46] Thus, the representation, or definition, of the particle again was a matter of the preparative operations performed on it, and because the solubilization procedure brought all subsequent incorporation activity in the test tube to a halt, there was no correlate to the preparative, operational definition in terms of biochemical function. In this situation, alternative criteria had to be introduced to derive a "robust" particle through a new round of triangulation and calibration, which involved an increasing scientific community of cytologists, biochemists, and cancer researchers.

Resonance and Stabilization: Electron Microscopy, Analytical Ultracentrifugation, Solubilization

One of the criteria that became widely used operated on size and shape. With it, the quest for microsomal function joined another line of research: that of an ongoing comparative in situ and in vitro inspection of the cytoplasmic ultrastructure by means of electron microscopy. We can take the seminal work of Albert Claude at Rockefeller on mitochondria as its starting point.[47] Through a series of further studies, electron microscopy led to the characterization of what Keith Porter came to term the "endoplasmic reticulum," a membrane structure extending throughout the cytoplasm.[48] A little later, George Palade, using an ensemble of advanced specimen preparation techniques, in turn correlated the microsomes with fragments of the endoplasmic reticulum to which small, electron-dense particles appeared to be attached.[49] Philip Siekevitz, after having spent two years with Van Potter in Wisconsin, had joined Palade in 1954. He added his biochemical expertise on protein synthesis to the structural work at the Rockefeller Institute. Palade and Siekevitz aimed at a correlation of what they called the "cytochemical concepts" of microsomal particles in vitro with the "morphological concepts" derived from careful electron microscopic in situ inspections.[50] It was in the course of these studies that the microsomal fraction, as stated, came to be identified with fragments of the endoplasmic reticulum. The in situ visualization of the reticulum with its electron-dense particles produced the kind of resonance with the in vitro biochemical work that is characteristic of what scientists in the process of shaping their epistemic objects like to call an "independent evidence."

Postmicrosomal fractionation became fashionable. In their episodic electron microscope studies, the group at MGH collaborated with Jerome Gross of Harvard Medical School. In contrast to the crude microsome

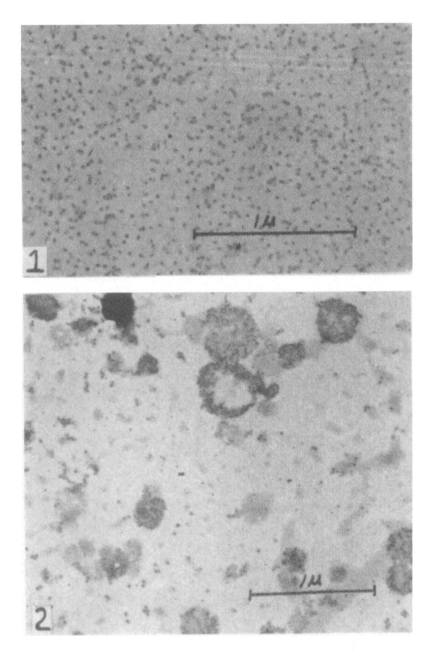

Fig. 6.2. Top: Electron micrograph of deoxycholate-treated microsomes, un-stained, unshadowed. Magnification, 45,900×. Bottom: Electron micrograph of whole microsome fraction, unstained, unshadowed. Magnification, 35,200×. Reprinted from Littlefield, Keller, Gross, and Zamecnik 1955a, pl. 1.

fraction, which contained chunks of irregularly shaped, large granular material, the deoxycholate particles appeared relatively homogeneous when inspected without further treatment under the electron microscope (Figure 6.2).[51] Yet the use of this technique brought with it serious operational problems of another order: that of specimen preparation for inspection. Because electron microscopy is based on the physical interaction of an electron beam with the object to be visualized, biological specimens were prone both to destruction by the beam and/or to deformation by the addition of electron-dense heavy metal solutions used to stain and to fix the material. Due to preparation differences, Littlefield and Gross's particles measured between 19 and 33 nanometers, which in itself was a quite considerable variation, whereas Palade and Siekevitz's osmium-treated particles were only 10 to 15 nanometers (nm) in diameter.[52] Were the particles homogeneous and small, or were they heterogeneous and larger? The problem could obviously not be solved within the representational space of electron microscopy alone.

A further technique of representation brought into play was analytical ultracentrifugation. To obtain the sedimentation pattern and sedimentation coefficient of his particles, Littlefield engaged in a cooperative effort with Karl Schmidt, another staff member of Huntington's. Mary Petermann's analytical ultracentrifugation of mouse spleen and rat liver homogenates had hinted at the occurrence of several discrete particles of different size.[53] Just like Zamecnik, Petermann, working at the Sloan-Kettering Institute in New York, had entered the field from cancer research. She had begun by searching for differences in the microsomal fraction between normal and malignant tissue. But the difference she found was not a distinction between normal cells and tumor cells. The difference that reoriented her research program pertained to the constitution of the microsomal fraction itself, irrespective of the kind of tissue it was derived from. Littlefield's particles formed a major "47S peak" in the optical record. This peak resembled the main macromolecular component of Petermann and her coworkers Mary Hamilton and Nancy Mizen. A broader peak running ahead of the 47S particle disappeared upon treatment of the material with 0.5 percent of the detergent. But there was also a smaller peak running behind the 47S particle that did not disappear under the same treatment (Figure 6.3). Was the ribonucleoprotein portion of the microsomal fraction itself heterogeneous, after all? Again, the question could not be answered within the framework of this representational technique alone.

Still another mode of recovery of ribonucleoprotein particles was based on the change of their biochemical composition as a function of the solubilizer concentration. Upon raising the level of detergent, the non-solubilized microsomal protein decreased more or less monotonically, whereas for RNA, a defined boundary appeared. Below 0.5 percent deoxycholate, virtually all RNA of the fraction remained in the unsoluble material. Beyond this threshold, the RNA was gradually lost to completion (Figure 6.4). This biphasic behavior by the RNA with respect to the solubilizer could be taken to indicate an edge at which a qualitative change in the cohesion of the particle occurred.

For none of these representational techniques did there exist an a priori referent that could have served as an external guide in shaping the scientific object under preparation. Its shape was not derived from comparing a "model" particle with a "real" particle; it gradually gained its contours from a correlation and superposition of different representations derived from distinct biophysical and biochemical techniques. In paraphrasing adequation theories of truth, Latour has called such a procedure, in his wonderful kitchen Latin, "*adaequatio laboratorii et laboratorii.*"[54] Because the material no longer incorporated amino acids after the various isolation procedures, there was no functional reference point for comparison either. The experimental representations partially mapped upon each other, and they partially interfered with and thus eliminated each other. The deoxycholate particle, so prominent in these studies, entered the field of in vitro protein synthesis around 1953. Around 1956 it became obsolescent and disappeared from the scene because no ways were found to render it functionally active. Nevertheless, it had a transitory function quite paradigmatic for the production of epistemic things. It was a tenta-

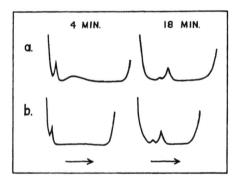

Fig. 6.3. Ultracentrifugal analysis of rat liver microsomes without (a) and with (b) deoxycholate at 37,020 rpm. Reprinted from Littlefield, Keller, Gross, and Zamecnik 1955a, fig. 2.

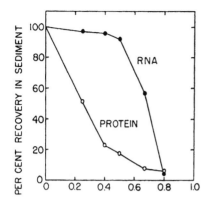

Fig. 6.4. Effect of sodium deoxycholate concentration on recovery of RNA and protein from the microsome fraction. The X-axis represents the percentage of solubilizer. Reprinted from Littlefield, Keller, Gross, and Zamecnik 1955a, fig. 1.

tive invention. It was a step on the laborious way of bringing the fractional representation of the cell sap into resonance with some functional features of protein synthesis. Ribonucleoprotein particles that were active in vitro became available only some years later in the course of a process that involved the recomposition of a suitable ionic milieu, the exchange of the solubilizing material, and the switch to another cellular source for the particles.[55]

A Dynamic Turn

Some evidence concerning the function of the particles came from fractionation following kinetic studies on living rats.[56] After a pulse of radioactive leucine, Betty Keller was able to follow the distribution of labeled protein in the deoxycholate-soluble and in the deoxycholate-insoluble material. This distribution displayed an intriguing pattern. The RNP-particles took up the radioactivity very rapidly and then approached a steady state value. In contrast, the deoxycholate-soluble protein was labeled much more slowly, and it continued to accumulate counts. The turnover of the isotope on the RNP-particles could thus be calculated. Only a tiny amount of the amino acids appeared to turn over rapidly. What happened in the course of this rapid process? Was it "an essential step in protein synthesis," or was it a "non-essential equilibrium reaction"?[57] There was no chance to answer the question by continuing these in vivo studies. It looked as if the deoxycholate-insoluble particles were the locus in which proteins were assembled before they were transferred to the cytoplasm. A year earlier, Zamecnik had interpreted similar results in favor of an alternative cytoplasmic nonmicrosomal mode of protein synthesis. According to this hypothesis, the microsomes would be responsible for tissue-specific proteins, the cytoplasm for housekeeping proteins.[58] The new dynamic turn was made possible by, and was the consequence of, a differentiated *fractional representation* of the microsomal components, combined with a kinetic approach that had been refined over several years.

A Complex Space of Representation

The experimental space of representing protein synthesis had become a rather complex *dispositif.* It involved the preparative ultracentrifuge, the analytical ultracentrifuge, the electron microscope, and a kinetic tracing method, in addition to the whole set of already routinized procedures for RNA and protein determination and radioactive counting. Despite this sophistication, all the biological materials still had to be freshly prepared for each experiment. To the despair of the experimenters, none of the components could be frozen and stored. Unpredictable deviations from one preparation to the other resulted.

The overall fragmentation pattern of the rat liver cell sap now appeared as in Figure 6.5. The fractional representation had attained considerable resolution. But it did not resonate with, indeed it was far

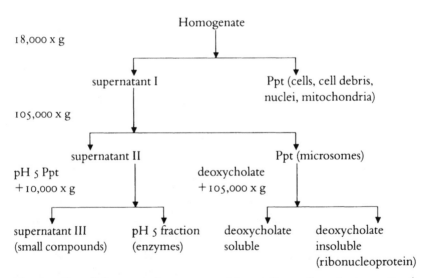

Fig. 6.5. Overall fragmentation pattern of the rat liver cell sap by 1955. Ppt ↕ precipitate. Author.

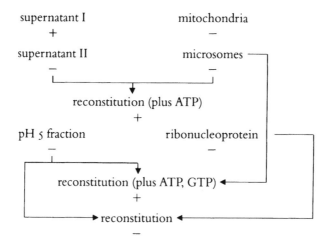

Fig. 6.6. Functional reconstitution of protein synthesis in the test tube as seen in 1955. +, incorporation activity; −, no incorporation activity. Author.

ahead of, the attempts at functional reconstitution (see Figure 6.6).

All attempts at getting amino acid incorporation from purified ribonucleoprotein particles had failed so far. On a mechanistic level, the nonresonance was even more pronounced. The system still did not respond to the addition of a full set of amino acids. Therefore, the relation between "incorporation" and net protein synthesis remained unclear. The energy suppliers ATP and GTP were not assigned to any of the components of the fractions. A conceivable mechanism for coupling the soluble and microsomal fraction was lacking. This bundle of inconsistencies and unknowns, however, was held together by the system's *différance*: It had the capacity to produce distinctions. It was a generator of experimental events.

Experimental reasoning in a particular research system consists in breaking down the questions into parts.[59] Their representation, that is, the manner of their "translation into visible effects" and recordable traces, may be triggered through outside challenges or from inside the system.[60] Such translations may be realized suddenly or may develop slowly. It does not matter whether they are initiated by bold assumptions or technical purposes; by educational predilections or instrumental facilities; or by opportunistic grant considerations, chance, or odd signals of the system itself. In any case, the instances that create the differentials on which the scientific events depend finally have to be implemented in the system. They have to become instances of its differential reproduction.

CHAPTER 7

Spaces of Representation

❖❖ Thus everything depended on the representation we
formed of an invisible process and on the manner of its
translation into visible effects.
                                  —François Jacob, *The Statue Within*

Given its formal dynamics of differential reproduction, how does an experimental system deploy its epistemic power? If we follow Jacob, everything depends "on the representation we [form] of an invisible process
and on the manner of its translation into visible effects." This formulation
is ambivalent. Is representation just a manner of rendering the invisible
visible, something hidden but ready to be disclosed? Is it a hide-and-seek
game? Or is it a manner of translation, which literally means converting
signs into other signs, traces into other traces, concatenating transformations? Is it a tracing game?

Whatever escape we may seek, when it comes to the heart of what the
sciences are about, we touch on representation. The sciences, so the story
goes, aim at a specific, in-the-limit, "true" representation of the world.
This was the grand project leading to enlightenment: science's duty is to
represent the world as it is in order to make its domination possible. Two
accompanying metanarratives of how this project might be realized developed in the seventeenth and eighteenth centuries: empiricism and rationalism. Empiricism claimed that true representation comes through
the undisturbed, if aided, senses and saw itself as based on observation. But
then, how to dominate? Rationalism claimed that representation comes
through actualized concepts and thus saw itself as based on intervention.
But then, how to represent? Since Kant's critical attempt to avoid the
conventionalist pitfalls of empiricism as well as the constructivist pitfalls of
rationalism by grounding the possibility of experience on some transcen-
dental conditions of reasoning, philosophy has become uneasy with both
solutions. This uneasiness has persisted until today, as has the distinction
between empiricism and rationalism, observation and experiment, and
induction and deduction, under various labels.

The Meanings of Representation

"So what are we talking about when we speak of representation?"[1] Intuitively, we connect "representation" to the existence of something that is represented. "In sum, representation of an object involves producing another object which is intentionally related to the first by a certain coding convention which determines what counts as similar in the right way."[2] Upon closer inspection, however, the term reveals itself to be polysemic. If we speak about the representation of something given, the common sense of the notion is plain: we speak about a representation "of." If, however, we claim that we have seen the actor Bruno Ganz yesterday evening representing Hamlet, we speak of a representation "as." In this case, representation takes on a double meaning: that of vicarship and that of embodiment. Every play is governed by this tension, this "paradoxical trick of consciousness, an ability to see something as 'there' and 'not there' at the same time."[3] If, finally, a chemist tells us he or she has produced or represented a particular substance in his or her laboratory, the meaning of "representation of" is gone, and instantiation in the sense of the production of a particular substance has taken over. In this latter case, we deal with the realization of a thing. There is a continuum from vicarship to embodiment to realization.

These three connotations of commonplace *representation* have their equivalent in scientific representation. Roughly speaking and without unnecessarily stretching the parallel, we are, in the first case, accustomed to speaking of analogies, of hypothetical, more or less arbitrary constructs (symbols in C. S. Peirce's sense). In the second case, we may speak of models or simulations (Peirce's icons). In the third, we deal with an experimental realization (comparable to an index in Peirce's semiotic system, i.e., a trace).[4] Jacob argues in a similar fashion, although casting the triplet in a questionable hierarchical order, when he claims that experimental biology proceeds from "analogies" to "models" to "concrete models."[5] It appears historically contingent and a matter of case-by-case evaluation as to which of them figures prominently within a given scientific context. Note, in addition, that I am speaking here of the function of representation on the level of scientific practice itself, as it gets enacted in the materialities of the laboratory. The transition to the presumably pure semiological realm and hence to the problem of how the results of this work are graphically and linguistically recorded is certainly not an abrupt one. It proceeds continually and through any number of intermediates. We might easily recognize here again the three modes of representation, and we could follow their trajectories. I am not concerned here in the first place with the relation between theory and reality, between concept and object as such. With that, we would locate ourselves on another level—a level that a long tradition of analytical philosophy has judged to be the *Kampfplatz* of philosophy of science. I am concerned with describing the process of making science as a process in which traces are generated, displaced, and superposed, in the sense of the connotations of representation just mentioned. I contend that if the perspective of a clear-cut dichotomy between theory and reality, between concept and object, is being adopted precociously, this process tends to disappear from sight.

An Inextricable Interconnection

In recent years, the issue of representation in scientific practice has received increasing attention, not only from philosophers, but also from historians and sociologists of science and culture.[6] Let me stress that representation, as viewed in the previous section, not only tends to be "overrated," but should be put "*sous rature*."[7] First to be put under erasure must be the traditional meaning of *representation* as referentiality, its association with what has been called the copy theory of representation. I claim with Nelson Goodman that "the copy theory of representation [is] stopped at the start by inability to specify what is to be copied."[8] There is no such thing as a simple representation of a scientific object in the sense of an adequation or approximation of something out there, either conceptually or materially.[9] Upon closer inspection, any representation "of" turns out to be always already a representation "as." As Michael Lynch and Steve Woolgar note, "representations and objects are inextricably interconnected; [objects] can only be 'known' through representation."[10] Basically, my argument is that anything represented, any referent, as soon as we try to get hold of it, and, concomitantly, as soon as we try to shift it from subsidiary to focal awareness, is itself turned into a representation. As a result, the term loses its referential meaning. Engaging in the production of epistemic things means engaging in the potentially endless production of traces, where the place of the referent is always already occupied by another trace. To use a terminology familiar from linguistics, there is a permanent gliding replacement of any presumed "signified" by another "signifier."[11] Science, viewed from a semiotic perspective, does not escape the constitutive texture of the inner workings of any symbol system: metaphoricity and metonymicity. Its activity consists in producing, in a space of representation, material metaphors and metonymies. From semiotics, we have learned that symbols take their meaning, not from the things symbolized, but from their relation to other symbols. At this point, it is more important to stress that the sciences share this structure with other symbolic/material worlds than to follow the worn tradition of specifying what makes them distinct.

Second, there is no representation without a chain of representations, with the immediate caveat that this is, as Claude Bernard already put it (although he did not make available the statement to the public), "a chain whose links do not have a relation of cause and effect, neither to the one that follows, nor to the one that precedes."[12] In the final analysis, then, the activity of scientific representation is to be conceived as a process without "referent" and without assignable "origins."[13] We deal with a "strange, transversal object, an operator of alignment whose power of veridiction resides in that it permits us to pass from a preceding transformation to that which follows."[14] As paradoxical as it may sound, this is precisely the condition of the often touted objectivity of science and of its peculiar historicity as well. If we accept this statement, any possibility of a deterministic referential account of science, be it based on nature or on society, is excluded.

Graphemes in Spaces of Representation

A scientific object investigated in the framework of an experimental system such as the in vitro system of protein synthesis is, as we have seen, articulated from material traces, or graphemes, within particular spaces of representation. Spaces of representation are coordinates of signification. They are opened as well as limited through the technicalities of the system. They disrupt the immediacy of presence of a phenomenon by rendering it as a mark; they are forms of iteration in a differential typology whose most obvious and prolific form, according to Derrida, is "writing."[15] Graphemes and spaces of representation do not exist independently. They mutually engender each other. There is no space of representation prior to an articulation of graphemes. And outside such a space, a particular graphematic trace remains without assignable meaning. Graphemes are what logicians abhor: elements without simplicity, that from which the simple arises through a process of degeneration. "Even before being determined as human (with all the distinctive characteristics that have always been attributed to man and the entire system of significations that they imply) or nonhuman, the *grammè*—or the *grapheme*—would thus name the element. An element without simplicity. An element, whether it is understood as the medium or as the irreducible atom, of the arche-synthesis in general."[16] We recognize, in this passage from Derrida's *Of Grammatology*, the fundamental form of Latour's conception of networks of human/nonhuman actants, although Latour certainly would not like to see the empire of his "hybrids" qualified as Derridean.[17]

Articulations of graphemes, or systems of signification, within the limits of an experimental setting, constitute the objects of a science. They channel the noise produced by the research arrangement and translate it into further traces, graphemes, "inscriptions," or "marks."[18] Graphemes, in the first place, are material articulations of significant units. Of course, they can be, and usually are, transformed into elements of more formalized spaces of representation. But this is not what makes them work as such. What makes them work is their oscillation between "density" and "articulation," a distinction that Nelson Goodman introduced in order to be able to conceive of hybrids of all sorts between continuous and discontinuous marks, from pictures to texts, and of all sorts of diagrammatic transitions on the basis of a "grammar of difference." Together with the transverse continuum between analogies, models and realized traces, this grammar of representative differences forms a system of coordinates in which resemblance has disappeared as a criterion of evaluation and which leaves room for all sorts of hybrids, these warrantors of "innovation, choice and unprecedented shifts."[19] Latour, too, takes representation as a particular kind of activity, as a process of inscription that results in a particular category of things, called "immutable mobiles."[20] They are characterized, not by what they depict, but by how they work. Immutable mobiles fix transient events (make them durable), and in doing so, allow them to be moved in space and time (make them available in many places). This is their power. What is significant about representation qua inscription is that things can be re-presented outside their original and local context and inserted into other contexts. It is this kind of re-presentation that matters. Inscriptions are thus not mere abstractions. They are durable and mobile purifications, which in turn are able to retroact on other graphematic articulations—and, what is most important, not only on those from which they have originated.

Graphematic articulations are the material forms of the epistemic things under investigation. Their materiality renders them recalcitrant to the theoretical forms of coherence we would like to impose on them. Their recalcitrance perpetuates the game of the "machine for making the future." Whether the traces that are produced in an experiment will prove "significant,"[21] depends on their capacity to become reinserted into the experimental context and to produce further traces. No experimental work can escape this recursive action, this iterative process of detaching an inscription from its transient referent and turning the referent itself into an inscription. The significance of an experimental event is its further signification, which, necessarily, comes ex post. If there is anything special about *scientific* inscriptions, it is their specific form of differential iteration, the complexity of which resists any shortcut description and which, therefore, I have tried to assess through the case study in Chapters 3, 4, 6, 8, 10, 12, and 13.

Experimental representation, then, may be taken to be equivalent to bringing epistemic things into existence. In their transiently stable forms, they may act as embodiments of concepts, as "reified theorems," to use Bachelard's expression.[22] But once reified, they are no longer interesting per se for those who do research. They remain interesting only as tools, as technical objects for constructing novel research arrangements. Jacob again: "Already, the results we had obtained no longer interested me. The only thing that mattered was what we were going to do with this tool."[23]

Uncertainty, fuzziness, and fugacity are at the heart of the experimental process of producing graphemes. The "epigraphy of matter" proceeds through groping.[24] "The abortive trials, the failed experiments, the false starts, the misguided attempts" prevail.[25] In the end, it is only the "*Verfahrungsart*" that counts. Goethe's essay on "The Experiment" closes with the following words: "The method [*die Verfahrungsart*] itself will fix the bounds to which they [imagination and wit] must return."[26] In this return, stratagems are at work that Derrida has characterized as "wandering in the tracing of *différance*."[27] Bernard, somewhat more prosaically, has described this process as "*tâtonnement*."[28] In it, there is no overarching telos, no ultimate perspective, no vanishing point at which the movement of research could come to a rest. This state of affairs is intimately related to the nature of the means by which the experimental game of producing graphemes is realized. The epistemic techniques through which it engenders its inscriptions lead, again and again, to unprecedented excesses that cannot be anticipated but appear only in the making. Through engagement in this game, the purpose of representation is continually subverted. As Goodman has put it, to represent is not to "mirror," but to "take and make."[29] All representation is production/reproduction, and the "eventuation" of epistemic things is distinguished by lack of reference to prototypes. Representation *is* "eventuation" (it is about intervention, invention, and the creation of events). Yet the ruse of this dialectic of fact and artifact consists precisely in that it functions by permanently deconstructing its constructivist aspect: the New does not enter through the obvious door but through some fissure in the walls.

Models

Let me try to approach the quandary of representation from a slightly different angle and to relate it to the story of this book. I have posited that, with an experimental system, a space of representation is established for engendering things that otherwise cannot be grasped as objects of epistemic action. Biochemical representations, in particular, create an extracellular space for reactions assumed to take place within cells. Traditional wisdom has it that such a representation constitutes a model of what is going on "out there in nature." Thus, biochemical "in vitro systems" would be models for "in vivo processes." But what goes on within the cells? There is no other way to learn about this than to provide a model for it. In other words, nature itself only becomes real, in a scientific and technical sense, as a model. Of course, there are *in vivo* experiments. But insofar as they are parts of a research arrangement, they are model systems too. The very necessity of representation in terms of modeling implies that unmediated evidence is excluded. The process of modeling is one of shuttling back and forth between different spaces of representation. Scientific objects come into existence by comparing, displacing, marginalizing, hybridizing, and grafting different representations with, from, against, and upon each other. It might be argued that this view amounts to standard conventionalism, if not relativism or even anarchism. I disagree. I would like to strongly assert the claim that it is not conventions by which scientists agree, for whatever reasons, that something is such and such. It is *convenience* in the etymological sense of the word: being drawn together in space and time, in a concurrent and recurrent gesture of involvement and involution.

The biochemist speaks of "model substances," of "model reactions," of "model systems," even of "model organisms." The concept of model, as it is used here, differs from its use in mathematical logic, where "models" are taken to be semantic interpretations of syntactic chains of signs.[30] The laboratory use of the term *model* is more instructive. The bench work language of the scientific practitioner translates with much more appropriateness what his work is actually about than what a particular philosophy of science declares him to be doing. The language games of scientific practice suggest that model things are substances, reactions, systems, or organisms particularly well suited for the production of inscriptions. The power of these "material generalities" resides in their ability to be disseminated through the capillaries of an experimental culture. More precisely, they create anastomoses through their spreading. Models are, to use laboratory language again, "ideal" objects of research in two respects: First, they are particularly well suited for experimental manipulation. This is the practical meaning of ideal. Second, they are idealized objects in the sense that they are, to a certain extent and in some respects, standardized, reduced, purified, isolated, contracted, and monofunctionalized entities that can be transported and subjected to local modifications. Models embody questions "accessible to laboratory experimentation."[31] But through this very embodiment, they also tend to change their "nature." Whether a bacterium, for instance, is considered as a model of genetic replication, of the production or function of antibiotics, or of the cause of certain diseases will materially change its structure with respect to its life as an epistemic thing.[32]

Nature as such is not a referent for the experiment; it is rather a danger. It counteracts the scientific enterprise. It is a constant threat of intrusion. If cells are fractionated, unfractionated cells have to be excluded from the space of representation. If one works with an in vitro system, every whole cell in it tends to behave as an artifact, a "whole cell artifact," as Zamecnik once aptly called it.[33] One must not contaminate an in vitro experiment with "nature." We end up with the uncomfortable situation that something we are used to calling natural, by remaining so, turns into something artificial. Consequently, we have to conclude that the reference point of an experimentally controlled system turns out to be another experimentally controlled system. The reference point of a model is another model. The basic rule of the scientific tracing-game reads: what engenders the traces can only be assessed by means of further traces. It is impossible to go behind a signifying chain, or "battery of signifiers." The quoted term is Lacan's, who uses it to indicate that the production of significance does not take place between the essence and the phenomenon but has a collateral constitution.[34] It is not the measuring devices that simply produce inscriptions *of* an object. The scientific object itself is cast into a traceable configuration. Temporally and spatially, epistemic things *are* and behave *as* inscriptions. The very movement of signification, the very making of scientific sense and non-sense, as the result of a structural necessity, turns presumptive referents, time and again, into signifiers.

As a result, the concept of model itself, much like that of representation, comes under erasure. The practice of modeling subverts the privileged position we grant to the model with respect to what is going to be modeled. Within the continuum between epistemic things and technical things, what we usually call a model occupies a middle position. As a rule, qua epistemic things, models are already sufficiently established to be regarded as promising areas of research and therefore to function as research attractors. On the other hand, they are not yet sufficiently standardized to serve as unproblematic subroutines in the differential reproduction of other experimental systems. Thus, an experimental model system has always something of the character of a supplement in the sense Derrida confers on the notion.[35] It stands for an entity that draws its effectiveness from its own absence. The supplement presents itself as a simple addition, but it has the potential to redirect the differential movement of the whole system. The subversion of an experimental system by a supplement shows both aspects characteristic of that movement: it tends to change the identity of the system's components, and it fails to do so by its very presence as a supplement. This is because a supplement, by definition, tends to be supplanted by the next. Likewise, a model is a model only in the perspective and by virtue of an imaginary reality at which it fails to arrive.

Traces

If epistemic things can be viewed as inscriptions, traces, and graphematic articulations, they can duly be addressed as grammatograms and, hence, generalized forms of "writing." The French notion of *écriture* is only inadequately grasped by its translation into writing. *Ecriture* is the "writing" and the "written," and it is the "how to be written" as well. It covers the graphematic space *and* the things from which it is built. According to Derrida, "to write is to produce a mark that will constitute a sort of machine which is productive in turn, and which my future disappearance will not, in principle, hinder in its functioning."[36] If epistemic things do not intrinsically display recordable marks that transform them into machines that become themselves productive, tracers are introduced into them: radioactive markers, fluorescent dyes, pigments, anything that gives inscriptions.

It is thus unnecessary to distinguish between machines that "transform matter between one state and another" and apparatuses or "inscription devices" that "transform pieces of matter into written documents."[37] What is a centrifugal fractionation? It transforms matter—it separates molecules—and it produces an inscription, say, a pattern of bands in the centrifugation tube. The argument has to be radicalized. The whole experimental arrangement—including both types of machines—has to be taken as a graphematic articulation. Written tables, printed curves, and diagrams are further transformations of a graphematic disposition of pieces of matter, a disposition that is embodied in the design of the experiment itself. Its elements—"elements without simplicity," "archisynthetic elements"[38]—are "big spots," "heavy peaks," "little shoulders," "yellow soups." Fractions, centrifugal pellets, and supernatants *are* a partition of the cytoplasm. They are handled as inscriptions. It is not simply the measuring devices that produce the inscriptions. The scientific object itself is shaped and manipulated "as" a traceable conformation. Temporally and spatially, the object *is* a bundle of inscriptions. It displays only what can be handled in this way.

What does the production of a chromatogram amount to? of a centrifugal partition? of an array of tubes to which pieces of filter paper are aligned, to which in turn counts per minute of radioactive decay are correlated? These are all material signs, entities of signification. The arrangement of these graphemes composes the experimental writing. What are graphemes made of? A polyacrylamide gel in a biochemical laboratory is, on the one hand, an analytical device for the separation of macromolecules; at the same time, it is a graphematic display of compounds visualized as stained, fluorescent, absorbent, or radioactive spots. It is a graphematic articulation. You cannot compare such an inscription, for example, a microsomal pellet, to microsomes tout court. You can only compare it to other representations to determine whether they match, whether they reinforce, displace, or marginalize each other. You can only compare it to other graphematic traces from other spaces of representation. Their matching gives us that sense of "reality" we ascribe to the scientific object under investigation. The "scientific real" is a world of traces. Both the concept of model as well as the concept of representation thus face the limits of their conventional meaning of imaging and delegation, of vicarship and correspondence. Upon close inspection, they assume the sense of a production of traces. There is a German word for the chemical process of production, characterization, isolation, and purification of a substance: *Darstellung*. This is what representation is about.

Scientific activity notoriously consists in undercutting the opposition between representation and referent, between "model" and "nature." Representation is not the condition of the possibility of getting knowledge about things. It is the condition of the possibility for things to become epistemic things. The scientific object is re-*presented* in being produced; and it is *re*-presented in the sense of a repetition, of an iterative act. Any production of representations is also, always already, a reproduction. I have developed this argument in Chapter 5. There is no external referent for this activity. Nature can be subjected to experimentation only insofar as it is already representation, insofar as it is an element, however marginal, of the game. Instead of conceiving of the epistemic activity of representing or modeling as an asymmetric relation, we should consider it as being symmetric: both terms of the relation are representations or models of each other. Instead of upholding the classical, biased relation between the signifier and the signified, we should displace it by an inter-conversion of signifiers: the very movement of signification turns the signified itself into a signifier, incessantly. The question of how the energy gets funneled into the process of protein synthesis, for example, successively took the form of its "oxygen dependence," of "inhibition by DNP," of "stimulation by a mitochondrial sediment," of a "mitochondrial factor," of an "ATP-regenerating system," just to mention a few of the translations we have encountered in the preceding chapters.

The production of inscriptions is not for that reason arbitrary. Although scientists are "always already within the theater of representation," they meet with "constraints."[39] For representations, in the long run, do count in scientific practice as far as they can be made coherent and resonant, as far as they take on consistency. Fortunately, on the other hand, this process is not deterministic. It is not just dictated by the technical conditions and the instruments involved in the endeavor. Producing traces is always a game of representation / depresentation. Every grapheme is the suppression of another one. Enhancing one trace inevitably means suppressing another one. In an ongoing research endeavor scientists usually do not know which of the possible traces should be depressed and which should be made more prominent. So, at least for shorter spans of time, they have to conduct the game of representation / depresentation in a reversible manner. Epistemic things must be allowed to oscillate between different significations.

A Pragmatogony of the Real

In conclusion, things come to look differently from the viewpoint of a pragmatogony focusing on representation as rooted in and emerging from scientific practice. Ian Hacking, too, has sketched such a pragmatogony in the "Break" of his *Representing and Intervening*, which in turn certainly shares elements with Edmund Husserl's comments on the origin of geometry.[40] "*Human beings are representers*," says Hacking. "Not *homo faber*, I say, but *homo depictor*. People make representations."[41] But Hacking explicitly does not want to take representations as mental representations or as visual images in the first place. He wants to take them as physical objects that owe their character of "likeness" to the process of their own replication. As Jean Baudrillard once put it: "We are in a logic of simulation which has nothing to do with a logic of facts and an order of reasons. Simulation is characterised by a *precession of the model*. [Facts] no longer have any trajectory of their own, they arise at the intersection of the models." With that, Baudrillard concludes: "The very definition of the real becomes: *that of which it is possible to give an equivalent reproduction*. [At] the limit of this process of reproductibility, the real is not only what can be reproduced, but *that which is always already reproduced*. The hyperreal."[42] Thus, the concept of reality, as a second order concept, can only take shape against the background of such first order representations as a reflection on the status of the replica. "The real [is] an attribute of representations."[43] The concept of reality only makes sense within a context of replication, and it only becomes a problem when *alternative* systems of representation come into play. But what holds for the concept of reality holds also for the concept of representation. "The *problem* of representation arises as a function of analytic efforts to assign stable sense and value to structures of practical action that, in the interest of analysis, have first been dissociated from the particular occasions in which they are used."[44]

To bring alternative spaces of representation into existence is what scientific activity is about, and this is why the question of reality as an attribute of alternative representations, and the question of representation as an attribute of its alternative uses, will continue to stay at the center stage of the scientific enterprise.

CHAPTER 8

The Activation of Amino Acids, 1954–56

❖❖ Of that life of worry and agitation there lingers most often
only a cold, sad story, a sequence of results carefully
organized to make logical what was scarcely so at the
time. [What] guides the mind, then, is not logic. It is
instinct, intuition.

—François Jacob, *The Statue Within*

What agitation had the protein synthesis system to offer? What twist led
to the creation of an experimental situation in which instinct and intuition would enter the realm of intermediates of protein synthesis? We will
see in this chapter how, with respect to the energy requirement of peptide
bond formation, a solution of dazzling obviousness emerged out of the
maze of the rat liver cell homogenate and how it tended, after the event,
to render logical what had led to its production.

Crafts and Grafts

Three components had been singled out as necessary for protein synthesis
in vitro: first, ATP and GTP as energy suppliers; second, a soluble protein
fraction mainly consisting of enzymes; and third, a particle composed of
proteins and RNA. It looked as if this particle would be the "site of initial
incorporation of free amino acids into protein" and as if RNA was an
essential part of it.[1] But the fractional representation that had guided the
work over the past three years had revealed few if any hints concerning the
mechanistic interaction of these different components.

Lipmann's suggestion was still pending: amino acids, to become substrates for peptide bond formation, might have to become activated, and
the phosphate bond energy of ATP might be involved in this process. But
how? At that time, Lipmann's coworkers Werner Maas and David Novelli
were working on the mechanism of pantothenic acid synthesis,[2] and

Lipmann thought of the "translation," as he said, of this mechanism into a "model for polypeptide synthesis" (see Figure 8.1).[3]

According to the model, an enzymatic "template" was phosphory-lated from ATP. In a second step, the phosphates were attacked by the car-boxyls of specific amino acids. Once aligned in this way, the C-terminus of one amino acid reacted with the N-terminus of the adjacent one, thus yielding a polypeptide chain. As Lipmann later recollected, he "naively thought then that protein synthesis could be more or less solved if one understood the mechanism of amino acid activation." It took him "a long time to realize that, in contrast to most other biosyntheses in the making of a protein, this was just a premise."[4] Even the establishment of this premise did not follow Nobel laureate Lipmann's prescription.

It was clear that whoever tackled the problem would have to be skilled in the handling of high-energy compounds and their biochemistry. But quite obviously, this was not enough. Tackling the problem also depended on a fractionated in vitro protein synthesis system. And it depended on the availability of radioactive ATP, which was necessary to trace the con-versions involved in the reaction. All these prerequisites were given at MGH. Zamecnik and Lipmann were close neighbors, and they met reg-ularly at sessions of the hospital's Committee on Research. Yet they made no attempt at a joint effort. As Zamecnik diplomatically put it, Lipmann "kept friendship and science in different boxes."[5] What follows is an example of what we might call "inadvertent collaboration."

Mahlon Hoagland took his degree as a medical doctor from Harvard Medical School in 1948 and that same year became a postdoctoral fellow at the Huntington Laboratories. Here he began to work on phosphatases, on phosphate turnover, and on the effect of beryllium on growth.[6] Dur-ing that time, he came into contact with Zamecnik and his protein syn-thesis group. Upon Zamecnik's suggestion, Hoagland decided to com-plement his medical education and to "retool" in protein chemistry and biochemical energetics.[7] Like his mentor, he spent a year with Kai Linderstrøm-Lang in Copenhagen, then another with Fritz Lipmann at MGH, on the sixth floor, before he returned to the fifth and joined Zamecnik by the end of 1953.[8] For a year, he participated in one of Lipmann's big enterprises, the synthesis of coenzyme A and the activation by ATP of acetate. Together with Maas and Novelli, he became immersed in handling ATP and its cleavage products.[9] At that time, Lipmann recalls, "protein synthesis was much on our mind, but the preoccupation with coenzyme A and acetate activation was still strong; there was some kind of

PROTEIN SYNTHESIS.   MODEL CYCLE

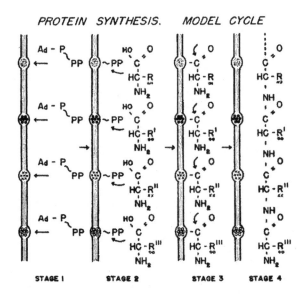

Fig. 8.1. A model cycle of protein synthesis. Ad-P~PP, Adenosinetriphosphate; R, specific amino acid side group. Reprinted from Lipmann 1954, fig. 2, by permission of the Johns Hopkins University Press.

a block that prevented us from taking an active part. Werner Maas was in our laboratory and so was Mahlon Hoagland, and of course Dave Novelli. But they were sidetracked by some facets of acetylation and of CoA synthesis."[10] What later appeared to Hoagland as one of those "vagaries of fortune in science," was that he realized that he could use the "phosphate-ATP-exchange" as a tool in Zamecnik's project on "purines and pyrimidines as sites for activation and transfer of metabolic intermediates."[11] The technology of these isotopic exchange studies went back to the end of the 1940s, and Maas had exploited it for analyzing a reaction that resulted in a peptide bond.[12] Hoagland grafted the technique onto the fractionated protein synthesis system, and by December 1954, less than a year after he had joined Zamecnik, he submitted a note on an "Enzymic mechanism for amino acid activation in animal tissue" to *Biochimica et Biophysica Acta*.[13] The paper comprised two pages, but it was a landmark.

Symmetry Considerations, Model Reactions

There was a symmetry consideration at the beginning. If there was anything like amino acid activation in the test tube system of protein synthesis, it could occur in the supernatant as well as on the microsomes. And, indeed, in both the 100,000 × g supernatant and in the microsomal fraction radioactive pyrophosphate (^{32}PP) made its way into ATP. But in contrast to the microsomal fraction, the incorporation in the supernatant was stimulated by a complement of amino acids. This was the differential signal Hoagland needed. It hinted at an amino acid-dependent turnover of ATP in the soluble fraction. It was a very crude signal at the beginning: the phosphate-exchange background was considerable, and the stimulation was low. Nothing was settled, therefore, especially since the effect of a single amino acid was below the level of detection.[14]

Another observation intrigued Hoagland. He was familiar with the effects of potassium fluoride (KF) from his earlier work on phosphatases, and he knew that KF prevented the hydrolysis of pyrophosphate. Consequently, he added KF to the assays. To his surprise, no accumulation of pyrophosphate was measurable. It looked as if there was a perfect equilibrium exchange and that *if* amino acids were intermediates in the reaction they would stick to their activating enzymes. There seemed to be no cellular acceptors of the hypothetically activated amino acids in the dialyzed soluble fraction. Of course, the microsomes were assumed to act as acceptors. But adding microsomes to the reaction mixture would have spoiled the signal. Hoagland's trick was to introduce a model substance that substituted for the putative acceptor. Hydroxylamine (NH_2OH) was available from the shelf. The rest was chemistry. The compound had to be salt-free, and it worked only in the absence of KF. Although this complicated matters considerably, the model reaction, when properly conducted, produced two results. First, the complete system including NH_2OH yielded measurable amounts of hydroxamates, indicating the presence of activated amino acids as a donor for the reaction. Second, the formation of hydroxamate depended on the addition of amino acids, and it could be correlated to the decomposition of ATP and a corresponding formation of inorganic phosphate (P_i). Taken together, the data were the result of an intricately constructed network of controls. In fact, it was this network that made the data meaningful.

As Table 8.1 shows, the system displayed a huge intrinsic loss of ATP (second column) and a corresponding formation of inorganic phosphate (third column). The addition of amino acids did not change these parameters (third and fourth rows). Given these figures alone, there would have been no signal at all. The differential signal was only created when NH_2OH was included (second row). For unknown reasons, the addition of NH_2OH resulted in a depression of the endogenous ATP hydrolyzing

TABLE 8.1

*ATP Loss and Hydroxamate Appearance in the Presence of Amino Acids
and Hydroxylamine*

Addition	Hydroxamate formed	ATP lost	P, formed
—	0	2.31	4.64
NH$_2$OH	0.34	1.39	2.78
AA	0	2.25	4.51
AA + NH$_2$OH	0.69	2.25	4.51
$\triangle$ due to AA alone	0	0	0
$\triangle$ due to AA in presence of NH$_2$OH	0.35	0.86	1.73

SOURCE: Hoagland 1955a, table 2.
NOTE: Numbers are given in μMol per milliliter.

activity of the mixture. Only when the amino acids were introduced under these conditions did their stimulating effect on the decomposition of ATP become visible (compare rows 2 and 4). The difference hinting at an amino acid activation was entirely hidden within the endogenous background of the system. Without a symmetrical and exhaustive arrangement of controls, Hoagland would not have been able to measure anything. It was this exteriorized "thinking machinery" of controls that provided what could be taken as evidence for amino acid activation. Only from a sophisticated juxtaposition of the measurable quantities, from the placing of the different experiments into a "landscape of controls" did the experimental traces become significant with respect to each other. No single one of them signified anything separately. Even when arranged in a way that rendered them significant, one of the parameters, "hydroxamate formed," failed to correlate with the other, "ATP lost," as can be seen in the last row of the table. Twice as much ATP was used up in comparison to the hydroxamate that had been formed. Because this feature did not fit the picture, the construction of the experimental web of traces not only made sense of the data, it also indicated what to try next.

Although there was still no activated amino acid—nobody had yet isolated such a compound—this signal shifted the experimental efforts from general energetical considerations toward the search for a specific metabolic intermediate. It provided a first outline of peptide synthesis involving enzymes and intermediates in a reaction chain.

Toward a Biochemical Representation

Let us stay for a moment with this representation. It consisted of two reactions in the test tube that taken alone, would both have abolished protein synthesis. The first, that is, pyrophosphate exchange into ATP, was the reverse of the reaction that could be supposed to occur in protein synthesis, namely, the release of pyrophosphate from ATP. However, it was the only feasible way to achieve a measurable signal: radioactive ATP could be identified by adsorption onto charcoal. The second reaction that was crucial in the test-tube visualization of amino acid activation rested on a model compound. In the apparent absence of an amino acid acceptor in the soluble enzyme fraction, the chemical NH_2OH served as a device for catching the activated amino acids and thus for pushing the overall reaction from its educts to its products. But as a substitute it blocked all subsequent reactions. Both reactions together served as an experimental model in a peculiar sense: as an epistemic object constituted by substitution and reversion. They resulted in a "tentative formulation" of a reaction sequence (see Figure 8.2).[15] Part of it rested on experimental evidence; part of it was interpolated.

As I mentioned, Hoagland had not identified the activated compound itself, $AMP\text{-}AA_1$. He linked both test reactions together by resorting to another system: the already mentioned reaction sequence of pantothenic acid synthesis.[16] His biochemical evidence was fragmentary, but he "tentatively" closed the gaps by referring to an experimental analogy that Lipmann had exploited with less luck a year before. We also note that NH_2OH was supposed to take the place of the "natural intracellular counterpart" of the amino acid. The presumed acceptor was expected to be the amino group of "peptide chains in the microsomes where arrangement of amino acid sequence and condensation of peptide chains would occur" without further intermediates.[17]

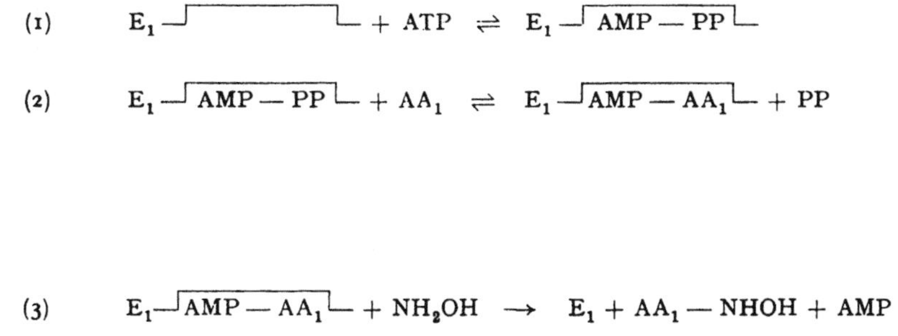

(1) $E_1 \overline{\quad\quad\quad\quad} + ATP \rightleftharpoons E_1 \underline{}^{\overline{}} AMP - PP \underline{}$

(2) $E_1 \underline{}^{\overline{}} AMP - PP \underline{} + AA_1 \rightleftharpoons E_1 \underline{}^{\overline{}} AMP - AA_1 \underline{} + PP$

(3) $E_1 \underline{}^{\overline{}} AMP - AA_1 \underline{} + NH_2OH \rightarrow E_1 + AA_1 - NHOH + AMP$

Fig. 8.2. Reaction scheme of amino acid activation. Reprinted from Hoagland 1955a, p. 289.

Stabilizations

Hoagland's mechanistic partial model of protein synthesis resulted in a remarkable turn in the fractional representation of the system. It translated into a set of amino acid activating enzymes, what until then had been the "soluble fraction," or the "105,000 × g supernatant," or the "pH 5 precipitate." The "heat-labile soluble factor(s)" were reconfigured as a class of enzymes with a specific function, despite the fact that they still were a cytoplasmic mix of completely undefined composition. A new, integrated picture emerged, which had to be consolidated by new experiments.

Fine-tuning his system cost Hoagland half a year.[18] The resulting changes were small but significant. He replaced the supernatant by an acid precipitate, the pH 5 enzymes. Their activity varied from day to day, but they were considerably more active than the reference soluble proteins. The precipitate reduced the background of the hydroxamate reaction. With it, the amino acid-related hydroxamate increment quantitatively corresponded to the amount of phosphate released from ATP.[19] The effect of individual amino acids on both exchange and hydroxamate formation scattered considerably, but it was cumulative, with no evidence of competition, and GTP could not replace ATP in the activation reaction. Using ammonium sulfate precipitation of the soluble protein fraction, Hoagland partially purified the enzymatic activities specific for methionine, leucine, tryptophane, and alanine. Biochemically, this was a huge challenge, and a "cumbersome" one,[20] because a distinct enzyme for every amino acid had to be expected among the pH 5 enzymes. With respect to the purification of the tryptophane-activating enzyme, Earl Davie from Lipmann's lab took the lead.[21] Hoagland recalls: "After discovering amino acid activation, I had jubilantly dashed upstairs and told Fritz Lipmann of my findings. Before I knew what hit me, he put a young associate to work on the obvious next step of isolating and purifying one of the many amino acid-activating enzymes presumed to be in the soluble fraction of the cell."[22]

Other workers came up with similar observations shortly after Hoagland's preliminary note had been published. One of them was David Novelli, who had just left Lipmann and moved to the Department of Microbiology at Case Western Reserve University in Cleveland. By September 1955, John DeMoss and Novelli had established an amino acid-dependent PP/ATP exchange reaction with microbial extracts.[23] Paul

Berg's organism of choice was yeast. Berg had started to work on the activation of acetate at Washington University School of Medicine in St. Louis, where he was an American Cancer Society Fellow from 1952 to 1954 and a scholar in cancer research from 1954 through 1959. Berg, like Zamecnik, was financially backed by the American cancer connection in the early 1950s, although he did biochemical work hardly connected to cancer. His first experiments on the PP/ATP exchange reaction date from November 1953. By the end of March 1955, we find him engaged in a search for a methionine-activating enzyme in his yeast extracts,[24] and in April 1955, in preparing "salt free NH_2OH according to Hoagland." Using amino acid hydrolysates from yeast as a reference, Berg performed the decisive experiment on methionine activation in May 1955.[25] By the end of April 1955, however, he had already inserted a preliminary remark in the communication made to the *Journal of the American Chemical Society* on the adenylation of acetate.[26] Within half a year, the "quite general character" of the carboxyl-activation mechanism appeared to be "established."[27] Its stabilization involved the refinement of the assay system as well as the embedding of the finding in a broader field of related reaction systems based on other organisms and studied by other colleagues, who all "were hot on the same trail."[28]

To obtain more direct evidence for the formation of an aminoacyl adenylate and on the nature of the bond involved in the reaction, Hoagland and Zamecnik finally engaged in a collaboration with Lipmann's laboratory and with Paul Boyer and Melvin Stulberg from the University of Minnesota.[29] Lipmann contributed a purified enzyme, and Boyer carried out the necessary measurements of isotopic oxygen from L-tryptophane-^{18}O. It was work "after the event," a piece of corroboration, after the claims had been marked out between the two competitors at MGH, and it involved a third party on which both depended. But this exchange of materials and the combination of experimental subroutines including biochemical, enzymological, and biophysical resources gave the model as well as its proponents additional credibility.

With respect to the further, still completely unknown sequence of reactions, the following statement by Hoagland, Keller, and Zamecnik reads like a tentative announcement, vague enough to be overlooked, and precise enough to be read as a prophetic conjecture in retrospect: "The enzyme-bound amino acid~AMP compound would then react with a natural cellular acceptor: either another nucleotide carrier or the nucleic acid of the microsome. The next step would be polypeptide condensation, which appears to occur in the ribonucleoprotein particles of the microsome fraction."[30] To this announcement I shall return in Chapter 10.

A Context of Corroboration

There was a broader context of corroboration as well as of debate around amino acid activation. As I mentioned in Chapters 4 and 6, there had been an ongoing discussion at the beginning of the decade as to whether the proteins of a cell were made exclusively from free amino acids or whether they would result from peptide fragment reshuffling. By 1955, the work on bacterial enzyme induction by Jacques Monod and his colleagues as well as Sol Spiegelman and his coworkers had strengthened the view that new protein within the bacterial cell was made from the pool of free amino acids.[31] With respect to higher organisms, however, there were conflicting reports. Work in different laboratories on rabbit muscle, on goat milk, and on the synthesis of amylase could be interpreted in favor of a synthesis from free amino acids.[32] Experiments on rat placenta and tissue pointed in the opposite direction.[33] The difficulties were connected with the choice of the respective systems. In lower organisms, radioactive pulse/chase experiments in combination with the induction of specific enzymes yielded reliable in vivo signals. With respect to higher animals, the situation was much more complicated because of protein breakdown, amino acid transport phenomena, and the lack of enzyme induction mechanisms comparable to those of bacteria. These shortcomings were one of the main reasons why workers in animal protein synthesis aimed at the construction of in vitro systems much earlier than workers in bacterial protein synthesis were forced to do.

At MGH, Loftfield and Anne Harris combined the techniques of ferritin induction in vivo in rat liver with the chromatographic determination of the intracellular specific activity of tagged amino acids.[34] Loftfield's experiments could be interpreted in favor of the free amino acid hypothesis, but the results could not definitively settle the debate. Thus, one of the most basic questions of protein synthesis still remained at issue. Another question Loftfield tried to approach by his procedure was the speed of peptide bond formation. Widely different estimates for the synthesis of mammalian proteins existed. They ranged from less than two seconds to up to 100 minutes.[35] It was clearly desirable to get more reliable data under conditions in which the production of a well-defined protein could be assessed. This would also be vital for judging the quality of

MGH's in vitro system. From his kinetics, Loftfield calculated the time needed for the synthesis of ferritin as between two and six minutes.[36] This was much quicker than what had been observed so far in the test tube. Hence, again, these experiments could not settle the worry that had provoked them: the suspicion that the test-tube incorporation process might be artificial.

The Question of the Microsomes

The structure and composition of the microsomes remained at issue, too. Based on the most advanced techniques of electron microscopy and analytical ultracentrifugation, their analysis had become intricately connected with the endeavor to partition the black box of protein synthesis. Meanwhile, Philip Siekevitz, whose work in Zamecnik's lab I described in Chapter 4, had joined George Palade, Rockefeller Institute's pioneer in the electron microscopy of microsomes.[37]

Let me recapitulate very briefly. During the 1940s, with the centrifugation studies of cell homogenates by Albert Claude, a particulate fraction of the cytosol came to be referred to first as "submicroscopic particles," then as "microsomes."[38] They were rich in proteins, in phospholipids, and in RNA. But despite Caspersson's and Brachet's hypothesis of a connection between cytoplasmic RNA and protein synthesis, it was not until the beginning of the 1950s and in a context quite different from their original characterization that the purely operational definition of these granules in terms of fractionation, optical inspection, and chemical composition became experimentally linked to protein synthesis in vivo and in vitro.[39] It took another decade for this linkage to take the form of a definition of these particles in terms of protein synthesis.

Microsomes had gradually become an integral part of the subcellular morphology through application of comparative electron microscopy of intact cells and fractionated material. This work both provoked and was stimulated by the introduction of new specimen-embedding techniques, and of microtomes with which craftsmen were able to cut sections 20 to 50 nm thick.[40] The in situ distinction of an "endoplasmic reticulum" with "small dense particles" attached to it and the identification of the microsomal fraction with ruptured reticulum had interesting methodological implications and important consequences. Palade and Siekevitz described the methodological aspect as follows: The small electron-dense particles, once identified in situ, could be used as a kind of "label," as an internal tracer for "following the fate of the endoplasmic reticulum through the various steps of the homogenisation and fractionation procedure."[41] Thus, they served as objects of investigation and as research tools at the same time. As a consequence, attempts were made to separate the particles from the rest of the microsomal fraction. Postmicrosomal fractions were obtained by various treatments (compare Chapter 6). Besides electron microscopy, the calibration of these "macromolecules" involved velocity sedimentation and electrophoretic mobility studies.[42] In turn, this battery of technical attacks led to a cytochemical reconfiguration that came to be reflected in a new name: "ribonucleoprotein particles." These ribonucleoprotein particles were tentatively identified as the cytoplasmic sites of protein synthesis around 1955.[43] Henceforth, the granules became a synonym for cytoplasmic RNA, although the postmicrosomal supernatant invariably also contained RNA—approximately 10 percent of the cell's total RNA.[44] The impact of this shift of RNA into the focus of attention will be discussed in Chapter 10.

Within this field of epistemic transformations, preparation procedures played a prominent role, and the terminology faithfully reflected their operational character. The different means and modes of representation were mutually interacting: choice of material, tools of inspection, physical separation, chemical treatment. Following a lingering trajectory, they eventually resulted in operational concepts that could be linked either to the subcellular morphology or to biological functions, without necessarily leading to a merger of both aspects. For instance, in the unbroken cell, Palade and Siekevitz were able to distinguish between membrane-bound and free particles. But they were unable to prepare a homogenate that would retain this distinction. This was especially disappointing in that the distinction had led to far-reaching speculations about the differential function of these two sorts of granules: the membrane-bound particles were supposed to be responsible for tissue-specific protein production, whereas the free particles would maintain the general protein metabolism. In contrast, deoxycholate particles could be prepared in a fairly routinized manner. But functionally, they were no longer active. The different representations only partially overlapped, and there was no way to circumvent the uncertainty inherent in any new trial.

Plant Systems

Mary Stephenson pursued a different line of research between 1954 and 1956. With the help of Ivan Frantz, Zamecnik had finally convinced his former technician to complete her graduate studies. For reasons of generalization, comparison, and complementation, she tried to establish a cell-free protein synthesis system based on tobacco leaves. One of the unsettled issues of the rat liver system was its persisting nonresponsiveness to a complete mixture of amino acids. Another missing link was the recovery of a single, specific protein from the homogenate. To solve the latter problem had been one of the main purposes of the tobacco leaf enterprise.[45] Because tobacco leaves produced large quantities of tobacco mosaic virus (TMV) upon infection, Stephenson's idea was to synthesize TMV protein in vitro from leaf homogenates. Leaf disks (comparable to liver slices) incorporated radioactive amino acids. But the standard homogenate fractions showed no differential radioactivity pattern. Surprisingly, however, Stephenson's *chloroplast* fractions displayed some activity. She found that incorporation of amino acids into chloroplasts was stimulated by light and oxygen, and abolished by heating or hydroxylamine. A whole arsenal of other metabolic inhibitors had no effect.[46] From the impressive list of compounds "without effect" little corroboration resulted. No effect of DNP and no effect of ribonuclease? This was, to say the least, surprising and confusing in view of earlier results. A single attempt, finally, to recover radioactive TMV protein from a virus-infected homogenate gave "no activity in the virus." At the end of two years of work, Stephenson drew the conclusion "that the intact cell or a less disrupted cell preparation is necessary for incorporation of labeled amino acids into virus protein."[47] "It wasn't a great thesis," she recalls, "because I could not achieve cell-free synthesis of viral protein, it wasn't a great project, but—it was a project. And it was interesting."[48]

Subversion by Repetition, and Choices

Why this insistence on a failed attempt? It tells us a story about the heuristics of experimentation. The attempt to simply reproduce the rat liver system on the basis of plant tissue by application of a strategy of "repetition" (see Chapter 5) did not in fact lead to a repetition. Instead, the whole research direction tended to be subverted by these experiments. Because the chloroplasts produced the largest incorporation signal, by a concatenation of small experimental events they became the main target of investigation in Stephenson's plant system. That protein synthesis occurred in a membrane-bounded, cellular organelle, namely, the chloroplasts, was a conclusion at which only Norair Sissakian from the Bach Institute of Biochemistry in Moscow had arrived in the course of the same year, 1955.[49] For a while, Stephenson even thought about a possible coupling of photosynthesis and protein synthesis. Something quite different from what she had been looking for showed up in her experiments. But although the chloroplast signal was intriguing, it did not appear to provide an immediate clue to the further elucidation of mechanisms. Because mechanisms underlying the synthesis of proteins were the urgent topic, a choice had to be made. The choice was against the chloroplasts. After earning her Ph.D. in biochemistry from Radcliffe College, Stephenson returned to the rat liver system. She had learned another lesson: a plant incorporation system equivalent to the mammalian system would not result from the simple application of an established centrifugation protocol. It would imply reshaping the whole fractionation pathway. And TMV? Producing the viral protein in vitro was a bold idea. But the metabolism of the virus was still a black box, and Stephenson's attempt was disconnected from contemporary virus research in general and TMV work in particular. Had she succeeded, it would have been an event comparable to the reconstitution of TMV from its components in the test tube,[50] although her motivation for doing the work had been quite different; she wanted to show that in vitro systems were "good" protein-synthesizing systems.

Most of the daily laboratory work surrounding and grounding the emergence of major, unprecedented events, the bulk of all that *tâtonnement*, leads to lists of "no effect." It diverts, it produces things that remain in the laboratory folders. Nevertheless, it is an activity as necessary for exploring an experimental space as the breakthroughs. In construing such a space the information about what does not work is as important as the information about what does. The entries "no result" are constitutively interwoven into the implicit narrative that a laboratory tells, but, for the most part, they are lost for and have no place in the stories that are told in public.

The Magic G-Nucleotide Pursued

Betty Keller, meanwhile, was pursuing another question: the involvement of G-nucleotides in protein synthesis.[51] Her interest in guanosine diphosphate (GDP) and guanosine triphosphate (GTP) had arisen from an observation of Rao Sanadi at the University of Wisconsin, Madison, who had identified GTP as a cofactor in the phosphorylation of ADP.[52] Sanadi had sent GDP and his purified transphosphorylase. Zamecnik wrote back,

"GDP seems to help our system but we have been unable to arrange conditions so as to get a constant effect."[53] A prerequisite for testing the nucleotide requirements of the system was to deplete the fractions from low molecular weight compounds. The soluble fraction could be acid precipitated. The microsome fraction was more difficult to handle. Deoxycholate inactivated the particles, and washing the microsomes through ion resins did not work either. Finally, a very gentle procedure opened the window for further analysis. Keller sedimented the microsomes from a diluted supernatant through sucrose. Sucrose-washed microsomes gave only a faint incorporation signal in the presence of ATP, but further supplementation with GTP restored the system to its full vigor.

What was the role of GDP/GTP? Did it take on the activated amino acid and act "as a donor of the amino acyl group in the elongation of the polypeptide chain?"[54] Was it the cellular counterpart of the NH_2OH model acceptor in Hoagland's amino acid activation assay? In any case, the effect of GDP/GTP was bound to the presence of intact microsomal RNA,[55] although its RNA-requirement remained obscure. The G-nucleotide became another of those walls in the experimental labyrinth, which, at one and the same time, prevented and guided vision.

Cancer Cells Reengaged

All this functional work depended on microsomes. As soon as the endoplasmic reticulum fragments were stripped off the particles, they became inactive. Was the reticulum essential for protein synthesis? In search of active, reticulum-free particles, John Littlefield reactivated a resource from the abandoned arsenal of cancer research.[56] This reengagement, however, placed the cancer tissue in a quite different context: as a source of microsomes rather than as an object of inquiry in its own right. The Scientific Advisory Committee of MGH appreciated this broad conception of cancer research when it stated at its meeting in December 1955, "the Scientific Advisory Committee strongly urges that the spirit in which the cancer research program has been conducted in the Massachusetts General Hospital be not only continued but further strengthened. This implies the expectation that much of the research will necessarily not be tied too intimately to objectives directly identified as cancer, or even as yet objectively identifiable as cancer-related, except in the minds of the investigators."[57]

An Ehrlich mouse ascites tumor had been introduced at Huntington in 1950, and it had been maintained in laboratory mice since then. Ascites cells were relatively deficient in endoplasmic reticulum.[58] The idea was that ribonucleoprotein particles might be derived more easily from their microsomes. To isolate ascites particles, Littlefield and Keller reactivated another technique dating back to the end of the 1940s.[59] Instead of deoxycholate for washing the microsomes, they used sodium chloride. The resulting particles had an RNA content comparable to the particles obtained with deoxycholate. Roughly 20 percent of the cytoplasmic RNA remained in the soluble fraction, and another 5 to 10 percent was solubilized from the microsomes by salt. The MGH workers carefully recorded the fate of cytoplasmic RNA throughout the purification procedure. But they did not account for it in terms of a functional distribution. They accounted for it in terms of loss from a single source: the microsomes. Yet although the RNA-to-protein ratio of deoxycholate- and NaCl-particles was virtually the same, contrary to any expectation their sedimentation coefficients were dramatically different. Littlefield and Keller had no explanation for this behavior. Some of the peaks derived from analytical ultracentrifugation were comparable to those identified by Petermann and her associates; others were not.[60] The sedimentation pattern was unstable and simply confusing.

Despite all this confusion, for the first time, the salt-washed particles were active in the amino acid incorporation reaction. In this respect, they were "good" particles, good enough to live with in spite of the other discrepancies. The ascites homogenate also had the advantage of a low ATP-degrading activity. Consequently, ATP without an energy regenerating system could be used. The pH 5 enzymes, the microsomes from liver and the particles from ascites mice, respectively, could even be interchanged. The membranous component of the microsomes was thus dispensable for protein synthesis. With that, the fractionated system had undergone considerable further purification.

The changes just described were small, but they resulted in another shift of perspective. Half a decade earlier, tumor tissue analysis had been dropped because it had led to an experimental impasse. Tumor cells were now reintroduced for a quite different purpose: they yielded active RNP particles. But their reintroduction also led to a by-product that again opened a new experimental perspective: the possibility of active hybrid systems composed of fractions from different cell types. In turn, hybrid systems allowed researchers to come back to unresolved questions related to the behavior of neoplastic tissue. The day-to-day laboratory bench work, the game of dissolution and formation of new attractors, remains truly unpredictable.

The Beginning of a Story

The resolution of the rat liver incorporation system, and in particular the activation of amino acids, began to attract the attention of a larger scientific community. The MGH system now became a reference for protein synthesis research. As a consequence, the history of the differential reproduction of the rat liver system began to be transformed into a story. In review presentations, its proponents now constructed a straight line from an early "prophecy" by Lipmann on amino acid activation, through the step-by-step establishment of the rat liver system, to the "reasonable" consideration that "either the microsomes or the amino acids themselves were being activated by adenosine triphosphate."[61] According to this story, the system was purified until ATP alone no longer sustained protein synthesis. Hence, another nucleotide had to be involved, and it was found to be GTP. What had been a central part of the enterprise, to find out what the incorporation of amino acids amounted to, became a story of avoiding pitfalls in identifying peptide bond formation. Now it looked as if, from the beginning, there had been a checklist of what had to be avoided in order to really deal with "true" protein synthesis. The empirical groping through the "biochemical bog" began to assume the form of logical alternatives to be explored and decided by clearly delineated crucial experiments. The heuristic principle of controlled impurity, which oriented the experimental process, became re-envisioned as a logic of purification.

The Actual State of Representation

So far, the representation of the system had been fractional. With Hoagland's amino acid activation, the fractional design began to be superimposed on an energy design. The emerging representation was a mosaic of centrifugal fractions, of energy transfer mechanisms, and of a possible sequence of intermediates on the way from free amino acids to proteins. It contained topological, energetical, and biochemical elements.

In contrast to earlier representations,[62] the energy requirement was now assigned a place in the fractional display. It became located within a special fraction (the soluble fraction) and connected to a special activation reaction. The activated amino acids were thought to be transferred to the

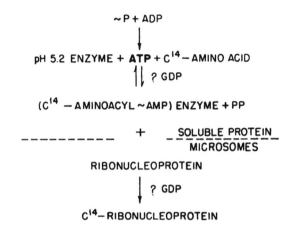

Fig. 8.3. Schematic summary of the mechanism of amino acid incorporation into protein as viewed in 1955. Reprinted from Zamecnik, Keller, Littlefield, Hoagland, and Loftfield 1956, fig. 5, by permission of The Wistar Institute, Philadelphia.

RNP particles, either directly or via another nucleotide carrier, and finally incorporated in the protein moiety of the ribonucleoprotein. Figure 8.3 reflects Zamecnik's awareness of "the participation of ribonucleic acid (RNA) in protein synthesis."[63] It appeared as a black box that also attracted the attention of other biochemists and, gradually, of molecular geneticists who were using bacteria and viruses as experimental models.[64] Might it be possible that the particle's RNA played a role in the "sequentialization of activated amino acids"?[65]

Template Models

Ideas of molecular fit in terms of "templates" had been amply discussed among biochemists during the preceding years. At the end of the 1940s, Felix Haurowitz still favored proteins as templates for autocatalytic protein synthesis.[66] A few years later, Alexander Dounce proposed phosphorylation of an RNA template, which then would allow the formation of a covalent RNA-amino acid intermediate (involving the amino-moiety of the amino acid), which in turn would lead to a polypeptide chain.[67] As we have already seen, Lipmann's ideas were different in that he proposed pyrophosphorylation instead of simple phosphorylation, that his RNA-amino acid intermediate involved the carboxyl-moiety and not the amino-moiety of the amino acid, and that he cautiously left his template unspecified.[68] Based on a suggestion of Hubert Chantrenne, Victor Koningsberger's model also involved a linkage between the amino acid's carboxyl group and the phosphate of the template as its starting point, as did, for instance, Alexander Todd's.[69] All these RNA template models assumed covalent intermediates, and none could account for protein sequence specificity in terms of the chemistry on which the models relied. To account for sequence specificity, enzymes had to be postulated that would mediate between the different amino acids and the sequence neighborhood of the nucleotide to which they became attached. In 1954, George Gamow, inspired by Watson and Crick's double-helical model of DNA, favored a geometrical fit between DNA and amino acids.[70] As Lily Kay shows, he set off a discussion among molecular biologists centering on formal, combinatorial considerations about a nucleotide "code" that ran for some years independently of and parallel to the developments I have described in this book.[71]

Zamecnik and his colleagues envisaged "sequentialization" (see Figure 8.4) as a process in which "activated aminoacyl nucleotide compounds line up along a ribonucleoprotein template, with their side-chain R groups determining the sequence by their ability to fit into particular spaces occurring on the ribonucleoprotein surface."[72] Thus, they spoke of

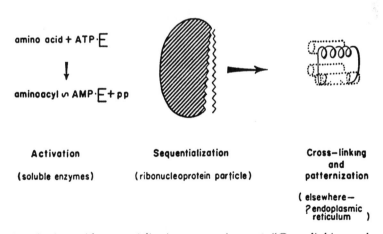

amino acid + ATP·E

$\downarrow$

aminoacyl ⌒ AMP·E + pp

Activation

(soluble enzymes)

Sequentialization

(ribonucleoprotein particle)

Cross—linking
and
patternization

(elsewhere—
?endoplasmic
reticulum)

Fig. 8.4. Amino acid sequentialization as seen in 1956. "Cross-linking and patternization" refers to the three-dimensional folding of protein. Reprinted from Zamecnik, Keller, Hoagland, Littlefield, and Loftfield 1956, fig. 2, by permission of J. & A. Churchill Ltd., London.

"ribonucleoprotein" when they used the notions of "template" and "se-quentialization" for the first time in public in 1955. They did not make any attempt at a more detailed definition. In contrast to Dounce and Koningsberger, the MGH group assumed a complex template formed from proteins and RNA that would recognize the specific side chains of the different amino acids in a lock-and-key manner. On the whole, these ideas in no way directly interlocked with the experiments Zamecnik and the other members of the group were reporting. The presence of these concepts was confined to the discussion section of review papers. They were not yet connected to epistemic things manipulable in the experi-mental space of rat liver protein synthesis. But the role of RNA in gene expression and in protein synthesis had become a topic of serious concern.

As I will outline in Chapter 10, a negotiation started between mo-lecular geneticists and biochemists at several meetings in 1955 and early 1956. Molecular biologists still "were often contemptuous of the bio-chemists, who focused on the dissection of cellular machinery." The bio-chemists, in turn, "tended to view the molecular biologists as party crashers, whose novel use of physics and genetics was glamorously over-shadowing the more traditional biochemical approaches."[73] The problem, however, was not so much one of maintaining or establishing disciplinary and corporate identities: the contenders spoke two different languages. They belonged to different experimental worlds. Everything depended on what would show up as a possible common object of desire and manipulation.

CHAPTER 9

Conjunctures, Hybrids, Bifurcations, Experimental Cultures

❖ [That] moment when the facts combine to indicate a new and unforeseen direction.

—François Jacob, *The Statue Within*

If experimentation has "a life of its own,"[1] experimental systems do not for that reason live alone. As a rule, experimental systems come in populations of multiple variants inhabiting overlapping areas of investigation, and the development of any of them can lead to points of conjuncture. Zamecnik spoke of a "juncture" when reviewing, in 1959, the events I described in the last chapter.[2] The term *juncture* indicates the emergence of an extraordinary constellation. The notion should not be confounded with that of an "anomaly" or with that of a "paradigm shift" in the sense Thomas Kuhn has attributed to these expressions.[3] It designates neither an irritating irregularity within an established and accepted conceptual frame nor the replacement of an encompassing theory by a new one; it rather points to unforeseen directions opened by the experimental process.

Conjunctures

Junctures, or *conjunctures*, as I prefer to say, come along with unprecedented events and may lead to major rearrangements and recombinations between given partial spaces of representation of an experimental system. Likewise, I prefer the notion of "unprecedented events" to the often used notion of "discovery."[4] The latter is part of a positivistic lexicon that I have sought to avoid throughout this book. With respect to another term frequently invoked in similar contexts, serendipity,[5] I find Royston Roberts's clarification quite useful. He says, "I have coined the term *pseudoserendipity* to describe accidental discoveries of ways to achieve an end sought for, in contrast to the meaning of (true) *serendipity*, which describes accidental discoveries of things not sought for."[6] Unprecedented events are about things and concatenations not sought for. They come as a surprise but nevertheless do not just so happen. They are made to happen through the inner workings of the experimental machinery for making the future. And yet they may commit experimenters to completely changing the direction of their research activities. Conjunctures can take different forms. I will describe one of them, but it remains the business of historical case studies to elaborate these forms in detail.

The emergence of a soluble, small RNA molecule in the cell-free protein synthesis system to be described in the next chapter perfectly matches the criteria of a major conjuncture. The molecule came along as an unprecedented event. It first appeared as a compound that had *not* been looked for. Subsequently, it changed the character of the whole system from a representation of intermediates in the metabolic pathway of protein synthesis to the representation of what came to be viewed as genetic information transfer from DNA to RNA to protein. Put in more general terms, it generated a crystallization point and locus for a juncture between classical biochemistry and molecular biology and, consequently, for a shift of the protein synthesis system toward molecular biology. How are such events brought about? Because they cannot be foreseen, there is no logical prescription, no algorithm to make them happen. Very broadly speaking, their production depends on the open structure of the investigative process. What looks like background noise with respect to a particular epistemic thing may take on significance with respect to the ongoing transformation of that very thing. As I outlined in Chapter 5, listening to noise and transforming it into a signal depends as much on acquired intuition, on "following one's instinct without knowing exactly where it will lead,"[7] as it does on the design of the experimental process itself. The experimental tracing-game has its heuristic rules, which do not simply serve for error and artifact discrimination.[8] They organize the experimental groping. They serve as generators of surprise, as exteriorized imagination devices.

We are familiar with the concept of conjuncture from economics, where it designates a fluctuation in the production volume corresponding to a changing constellation of economic factors. The use of the concept for characterizing investigative processes needs justification. Such justification leads back to the Latin meaning of the word: "a situation deriving from the connection of different phenomena."[9] Here, I use the expression for the emergence of unpredictable constellations in the development of experimental systems resulting from a connection of phenomena that do not derive from an expected relation of cause and effect but that, once set in place, can enter into a kind of structural coupling.

Experimental systems, as introduced in Chapter 2, designate those material formations, or dispositions, of epistemic practice within which a scientist or a group of scientists generate the epistemic products that they characterize as the "results" of their craft. What is a result? A—positive— result is a finding that, in principle, can be reintegrated as a component of the system and can thus enlarge or change the setup. Following Bachelard, I have characterized such retroactive couplings, appropriations, or annexations with the notion of recurrence. Let me repeat that there is no logical necessity for such recurrence. Despite their integrative tendency, or reproductive coherence, experimental systems are not mechanical constructs, and they exhibit fringed contours. They are not closed in any appreciable sense, not even operationally closed. On the contrary, they operate, insofar as they are research systems, at the border of their breakdown. They derive their dynamics not from the vantage point of a prospective closure but rather from their quasi-ecological character as niches. They are defined with respect to, and in distinction from, other neighboring experimental systems. If it is their articulation that conveys microsignificance to the traces within an experimental system, it is the articulation of experimental systems within a field of experimentation that conveys macrosignificance to the individual systems. What, then, is the relation between unprecedented events, conjunctures, and recurrence? We may say that unprecedented events, through recurrence, lead to conjunctures.

Hybrids

There is another kind of events that derive from the fringed, or even fractal, contours of experimental systems: events that produce linkages between mutually independent systems. Interfaces can be created between two or more experimental arrangements. The fusion, for example, of François Jacob's bacterial conjugation and phage replication system with Jacques Monod's system of induced enzyme synthesis led to the emergence of messenger RNA, to be described in Chapter 13, and to a pathbreaking model of genetic regulation.[10] Such coincidences connect particular experimental systems to integrated setups. From a hybridization of different, originally unconnected experimental systems, research arrangements with totally unexpected qualities can result. My use of the notion of "hybrid systems" draws on and at the same time differs from Latour's. He characterizes *hybrids* as amalgamations of social and natural constituents and scientific enterprises as exemplars of such hybrids.[11] What I would like to retain as a connotation of the term, here, is its meaning of bringing together things whose articulation, amalgamation, or even blending was not assumed to lie in the nature of the things so brought together. Such hybridizations are a prerequisite of what Lindley Darden has described as "interfield connections" eventually leading to "interfield theories."[12]

Bifurcations

A third type of event is complementary to hybridization. It can lead to the *bifurcation* of an experimental system and thus to offspring systems. Such offspring systems tend to form ensembles that construe the experimental space, not of a single, localized group of scientists, but of a circumscribed scientific community. I will present examples of the bifurcation of an experimental system in the next chapter. The pursuit of soluble RNA led to several offspring systems still connected with but different in perspective from the in vitro protein synthesis setup. Generally speaking, bifurcations of an experimental system occur when it has reached a certain complexity that allows researchers to pursue slightly diverging epistemic tracks but which are sufficiently different to enable them to arrive at significantly different results. Typically, clusters of such bifurcated systems remain linked for a while by sharing one or more of their material constituents so that scientists may take advantage of each other's achievements or each other's services. But this must not necessarily remain so. Bifurcated systems can become completely disconnected from the maternal system or integrated into other ensembles.

To assess these different modes of experimental articulation of epistemic objects and procedures, we need to define three operations that I would like to call operations of "insemination," operations of "grafting," and operations of "dissemination," respectively. In drawing upon the distinctions Isabelle Stengers has used to describe the organization of an ensemble of phenomena in a particular field of research and across previously established disciplinary boundaries, we could also speak of "operations of propagation," by way of tempering and hooking, and "operations of passage," respectively.[13]

Experimental Cultures

So regarded, the concepts of conjuncture, hybridization, and bifurcation permit us to envisage an ensemble of experimental systems and their intricate interactions. They permit us to conceive of an articulated experimental network of objects and practices whose coherence, just as in the case of individual experimental systems, is a tinkered and patched-up coherence with a collateral constitution. In other words, what holds it together is not its vertical relation to a hidden referent but its horizontal concatenation, both in time and in space. Its cohesion is due to, and reaches exactly as far as, the circulation and the exchange of epistemic entities, model compounds, technical subroutines, and tacit knowledge throughout the network. Conjunctures, hybridizations, and bifurcations basically describe types of shifts, linkage, and descent through which the dynamics of reorientation, of fusion, and of proliferation of particular experimental systems is made possible. The consideration of these processes permits a transition from the microdynamics of localized and situated experimental systems to the macrodynamics of broader fields of experimentation.

Conjunctures, hybridizations, and bifurcations, then, constitute ensembles of experimental systems. Within such ensembles, the systems interact and remain connected through the exchange of sufficiently stabilized procedures and epistemically attractive objects. But they can also finally become isolated. From evolutionary biology, we are familiar with the processes of species change, of speciation, and of hybridization. As tempting as the analogy with biological processes may be, precaution is needed. Of course, experimental systems are not organisms. Yet both kinds of entities share something fundamental that forms the basis for conceptualizations that may be quite different in detail but compatible within a general, broadly conceived ecohistorical perspective. As coherent arrangements of practices in space and time, experimental systems are entities that transmit, and in doing so change and accumulate, information. In developing an argument from the last chapter of his *Structure of Scientific Revolutions*, Thomas Kuhn has spoken in this respect of "process[es] of proliferation" and of the "dy[ing] off" of scientific specialties. And he has compared them with a "variety of niches within which the practitioners of these various specialties practice their trade." Those niches interact, but "they do not sum to a single coherent whole."[14]

Thus, ensembles of experimental systems can be seen as clusters of materials and practices that evolve through drift (conjunctures), fusion (hybrids), and divergence (bifurcations). We may go yet a step further and look for linkages between ensembles of experimental systems. With that, we reach the level of what might be called an *experimental culture*. Experimental cultures, then, are clusters of ensembles of experimental systems. They share a certain material style of research. Instead of using Fleck's notion of "thought style" in this context, I prefer to speak of styles of experimental reasoning.[15] Experimental cultures are characterized by specific forms of what Hacking calls the "laboratory style."[16] The concept of experimental culture does not merge and is not coextensive with the classical notion of a discipline. On the contrary, experimental cultures constantly tend to shift, to displace, and to subvert the contours and confines of established disciplines with their educational curricula and institutionalized structures of communication. Experimental cultures, not disciplines, determine how far at a given time materially mediated scientific cooperation, competition, and epistemic negotiation reaches. They determine the possible circulation channels of epistemic things. They circumscribe the fluctuating boundaries of those spontaneously arising informal communities of researchers that regulate the flow of information below the level of institutionalized scientific organizations and corporations. What I call experimental culture is thus in the first place an epistemological and not a sociological concept.

Toward an Epistemology of Experimentation

Epistemic things, experimental systems, ensembles thereof, and experimental cultures are thus the conceptual entities with which I try to prepare the ground for a history and epistemology of experimentation that dissolve the traditional hierarchy between context of justification and context of discovery and free the experiment from its subsidiary role in rationalistic accounts of theory development and theory change. My narrative, however, does not focus on the history of scientific institutions and disciplines. It is an attempt to understand the epistemic dynamics of the empirical sciences in terms of the peculiar structure of practices from which these sciences spring and in which they dwell. Experimental cultures are not homogeneous spaces. They are as patched up and tinkered with as the experimental systems they are composed of. But they are held together by a specific kind of glue: material, not institutionally formalized interaction; epistemic, not theoretical compatibility in its narrow and constricted sense.

The biological content of evolutionary and ecological concepts is of no particular value for the historian of science and should, therefore, be dropped from his or her account. What matters, however, is their potential for conceptualizing multiply interacting systems as reproducing entities that embody and change knowledge over time. If experiments are said to have a life of their own, both components of the statement have to be explained: what it means for a system of practices to possess a "life" and what it means for such a system to have its "own" life. If we are to take Hacking's notion seriously, we have to explore in what specific sense scientific practice engenders bits of knowledge. What I would like to retain from the evolutionary metaphor is the following, to speak again with Kuhn: "Scientific development is like Darwinian evolution, a process driven from behind rather than pulled toward some fixed goal to which it grows ever closer."[17]

Not only the history, but also the historiography of the sciences is a process "driven from behind" (as is the production of a book like this). The concepts of historical narration are shaped and reshaped from immersion in an "epistemological laboratory." As Eduard Dijksterhuis once remarked, "the History of Science forms not only the memory of science, but also its epistemological laboratory."[18] This laboratory is constituted by the historical development of the sciences in *all* their trifles and details. Historians of science, then, have to conceive of themselves as being engaged in an experimental exploration of experimental reasoning. In his *Introduction to the Study of Experimental Medicine*, Bernard has clearly sensed the necessity of such an epistemology of detail: "In scientific investigation, minutiae of method are of the highest importance. The happy choice of an animal, an instrument constructed in some special way, one reagent used instead of another, may often suffice to solve the most abstract and lofty questions." And, he continues, "one must be brought up in laboratories and live in them, to appreciate the full importance of all the details of procedure in investigation, which are so often neglected or despised." Finally, he compares the life sciences with a "superb and dazzlingly lighted hall which may be reached only by passing through a long and ghastly kitchen."[19] In a similar vein, Bachelard has argued for a "*distributed* philosophy" in order to assess the sciences in the making.[20] Such epistemology must resist the temptation of the homogeneous and the hegemonic, of grand narratives, be they affirmative or critical of their object of inquiry. Instead, such epistemology will stress the heterogeneous and the regional, that which resists being classified and consequently tends to be lost between the lines.

For a long time, historians of science have preferred to stay in those "dazzlingly lighted" halls and to enjoy the imaginary desire of historical narration for coherence, integrity, totality, and closure,[21] that is, for a life beyond the "ghastly kitchen" full of recipes, annals, and chronicles. This predilection, however, has come under criticism. Under the heading of "science as practice and culture" we are currently witnessing a lively debate concerning the contingent, contaminated, local, and situated making of science. In this context, the notion of culture refers to the multiple, epistemic, technical, institutional, and social resources involved in giving shape to experimental practice. The notion of practice refers to the ensemble of activities sustained by these resources.[22] "Culture," within the self-perception of the modern world, is tightly interwoven with the basic issue of innovation; "experimental culture," therefore, is but another expression for a specific ensemble of conditions that allow the generation of unprecedented events. To characterize science as practice and as culture does not amount, as far as I apprehend it, to determining the social influences hindering or furthering the sciences. It does not amount to a critique of ideologies of science in the traditional sense. Rather, it amounts to characterizing the sciences themselves as cultural systems that shape our societies and all the while trying to find out what makes the sciences different and confers on them their peculiar drive, not privileging them with respect to other cultural systems.

How shall we describe this endeavor? As early as his inaugural address to the Collège de France in 1970, Michel Foucault traced a series of concepts characteristic for such a history of science:

The fundamental notions now imposed upon us are no longer those of consciousness and continuity (with their correlative problems of liberty and causality), nor are they those of sign and structure. They are notions, rather, of events and of series, with the group of notions linked to these [including regularity, the aleatoric, discontinuity, dependence, and transformation]; it is around such an ensemble that this analysis of discourse I am thinking of is articulated, certainly not upon those traditional themes which the philosophers of the past took for "living" history, but on the effective work of historians.[23]

Foucault has characterized his version of such an endeavor as archaeology and, with respect to the history of science, has spoken of an "archaeology of knowledge."[24] Of knowledge, not of science: systems of knowledge are not to be equated with the sciences conceived as systems of theoretical propositions. The archaeologist digs out the material sediments, the dispositions and depositions in which those theories are embodied.

In speaking of experimental cultures—not in terms of truth and causality but in terms of things and *epistemata*—I seek to gain access to these sediments in two respects. First, I would like to accord the experimental and instrumental life forms of the sciences the same attention that we have come to attribute to their theoretical dynamics. Second, I would like to point at the plurality of terrains, a plurality that is hidden by our deliberate speaking of "The Science" in the singular. But then, I would like to go a step further still. I would like to assume that the development of a particular horizon of scientific problems and concepts necessarily remains concealed without revisiting its experimental texture: its texture, I say consciously, before its context.

All experimentation is technically implemented. For that—and with all the necessary reservations that have led me to draw a functional distinction between the technical and the epistemic aspects of experimental systems—I do not hesitate to follow Heidegger's claim that the dynamics of the modern sciences are a consequence of a historically unique technocultural conquest, and not the other way round, namely, that all modern machinery and machination would rely instead on the conceptual force of the sciences qua systematized theories. The epistemic things that ground the experimental sciences emerge from the deposit of the technical and its potential for tinkering. Whence it follows that time and again they lend themselves to becoming reincorporated in that deposit. This constellation is what Heidegger has called the "enframing" or the "stand" (*das Gestell*) of contemporary science and technics.[25] Let me recapitulate: "The essence of what we today call science is research. In what does the essence of research consist? In the fact that knowing establishes itself as a procedure within some realm of what is, in nature or in history."[26]

A Note on Protein Synthesis Before the Language of Information

In the next chapter, I will deal with a particular historical conjuncture through which biochemical research on protein synthesis became connected with the discourse of molecular genetics.[27] I will take a close look at how precisely this conjuncture took place at the level of experimentation, and I will show how the implementation of the discourse of "code" and "information," at the point of this particular juncture, initially took the form of a supplement. Although Zamecnik's soluble RNA opened the field for viewing protein synthesis as a process of translation, the language of code and information appears to have been of no heuristic value for the emergence of what we today call transfer RNA. This language, however, became of paramount importance for choosing between the options that were opened by the in vitro system of protein synthesis after the advent of transfer RNA. The contenders in the discourse of protein synthesis came to know differently what they had done only when they entered the next round of events. Which means that they were not the subjects, in the classical interpretation of rationalist philosophy, of the process; rather, they became subjected to its consequences. It is to such peculiar situations that Lacan alludes when he says that in modern science the subject is sutured to its object, an object refractory and protractory at one and the same time. I will show that, with respect to this particular conjuncture, a process full of frictions was set off between biochemical refraction and molecular biological protraction. On the level of experimentation itself, it did not lead to a sudden replacement of biochemical reasoning by reasoning in terms of genetic information transfer. Rather it led, via supplementation, to a kind of blending between discourses that reflected the hybrid state into which the experimental system developed in coping with Francis Crick's adaptor hypothesis. The movement of supplementation I will trace with respect to the emergence of soluble RNA appears to be a general characteristic of the formation of new discursive regularities at the interface between biochemistry and molecular genetics in the second half of the 1950s. As far as I can see, this situation did not change throughout the following decade. We historians should not try to make our reconstructions cleaner than the experimental and rhetorical patchwork that presents and absents itself wherever trans-disciplinary conjunctures come into operation and start to challenge and redefine accepted spaces and boundaries of negotiation. Languages, scientific ones not exempted, do not describe the world, they inscribe themselves into practices—whence their power, their seductive force, and the cross-fertilizing hubbub to which they give rise. Science does not work in spite of the fact that there are different languages on different operational levels, it works *because* there are many of them, thus creating the possibility for differential contexts, for unexpected hybridizations, and for all sorts of effects of interference and intercalation without which the thing we call research would not exist.

CHAPTER 10

Toward Molecular Biology:
The Emergence of Soluble RNA,
1955–58

✤✤✤ There is in research a unique moment: when one suddenly
sees that an experiment is going to overturn the landscape.
[When] the change taking place is due more to a feeling,
to a premonition, than to the chilly logic of facts. Where
the dream of novelty suddenly takes on consistency
without being fully assured of becoming reality.

—François Jacob, *The Statue Within*

In a 1979 discussion, Zamecnik responded as follows to a question from
historian Robert Olby:

When we found that the ribosome was an important feature in protein synthesis,
I talked to Paul Doty, who was the high priest of the RNA field in the Boston
area, and said, "How do you think the sequencing step of protein synthesis
occurs, and what is the relationship to DNA?" He replied, "I don't know, but
there's a young fellow over here named Watson who just happens to be visiting
me and I'll ask him if he'll go over and see you. He has a model of the double
helix which he and his colleague Crick have recently described." I had not heard
about that model, in 1954.

Zamecnik continued:

Watson came over and visited me, and I said, "It seems to me that we have
evidence going back to Brachet and Caspersson that RNA plays a role in protein
synthesis. At least they showed that organs such as the pancreas, in which protein
synthesis for export was high, had a high concentration of what they called
cytoplasmic nucleic acid." Then Watson showed me his model. Here we have a
model of DNA; now where does RNA fit in? If Brachet and Caspersson have a
point and the DNA is a generator of the code for protein synthesis, how is its
message transmitted? We looked at the model together and I said, "Could RNA
possibly fit into a groove here?" Watson just startled, shrugged his shoulders,
lifted his hands, and went off to look at birds.[1]

There are some interesting features to this 1979 rendering of a 1954 conversation. First, the news of the DNA double helix had obviously no immediate impact on Zamecnik's thinking about protein synthesis in 1954. Second, the possibility of experimentally exploring protein synthesis in vitro had obviously no direct impact on Watson's thinking about how genetic information might get transferred to proteins. Third, 25 years later, the language of message and code had acquired such universal currency that Zamecnik inadvertently recast the episode of 1954 in these terms—although they were completely absent from his vernacular for speaking about protein synthesis at that time. In this chapter, I try to recapture the experimental context of protein synthesis as it evolved between 1955 and 1958. I retrace the shifts of epistemic entities, metaphors, and rhetoric associated with one of its major events—the emergence of soluble RNA—which began to tie the system of protein synthesis to molecular genetics.

RNA Synthesis Revisited

Something clearly was going on with RNA in relation to protein synthesis. As I mentioned at the end of Chapter 8, by 1955, the RNA of microsomes was generally seen as directing the sequential assembly of the amino acids within that particle.[2] Workers in the field assumed RNA to function as a template that, in one way or another, conveyed specificity to the protein in the making. This, however, was not the actual point of a growing discussion. Over the past few years, many laboratories had gathered indirect evidence for a coupling of the synthesis of proteins, not to RNA as such, but to the de novo *synthesis* of RNA. In 1952, Jacques Monod had observed that the induction of the enzyme β-galactosidase in the bacterium *Escherichia coli* depended on uracil, the base characteristic for RNA, and had postulated an "organizer" directing the induced synthesis of the enzyme.[3] Other hints came from studies on enzyme formation in yeast, on protein synthesis in cells depleted of their walls and in enucleated cells, on in vivo and in vitro protein synthesis systems, and on phage replication.[4] In the course of experiments in 1953, Alfred Hershey, at Cold Spring Harbor, had noticed that after phage infection, a small amount of RNA was made very rapidly in the bacteria. According to Hoagland, Hershey "did not see what to do with the finding, so he published it and forgot about it."[5] Be this as it may, the RNA problem loomed large in the minds of all those concerned with protein synthesis around the middle of the decade. For Ernest Gale in Cambridge, there was no doubt in 1955 that, "in induceable systems at any rate, protein synthesis is accompanied by, if not dependent upon, RNA synthesis."[6]

Zamecnik certainly also remembered the wartime RNA turnover measurements in rat liver of his former Huntington colleague Waldo Cohn, when he wondered, in 1954, "whether the same cell-free conditions which we had found a year previously to make cell-free protein synthesis possible, might also serve for synthesis of RNA."[7] He had asked his new coworker John Littlefield to undertake this study. Littlefield recalls: "I agreed that I should do a few experiments to look for the possibility of RNA synthesis in the protein synthesis system. [The] few experiments I did on RNA synthesis were unfortunately not especially promising, while the work on the ribosomes [microsomes at that time] was going well, so I dropped the work on the former and concentrated on the latter."[8] During a meeting at Oak Ridge, Tennessee, in the spring of 1955, Erwin Chargaff had asked, "Dr. Zamecnik, does your soluble 100,000 × g supernatant fraction contain RNA? Also, would we not visualize it as playing a role, together with the microsomal particles, in what you consider to be evidence of protein synthesis?" Zamecnik had responded cautiously: "One cannot say that the RNA present in the 100,000 × g supernatant protein fraction does not play a special role in the incorporation process." Then he pointed to the low concentration of RNA in this fraction and the possibility of a residual microsomal contamination. But, "your point remains, Dr. Chargaff," he said, "that there may be some critical nucleic acids in this 100,000 × g supernatant fraction."[9] Zamecnik and his coworkers had repeatedly noticed that the enzyme supernatant contained small amounts of RNA. They had taken this RNA as a measure of a residual *contamination* of the soluble fraction with broken microsomal RNA.[10] It was an impurity that could not readily be eliminated. In any case, it was not considered to be of functional relevance. So, Zamecnik answered as he did and forgot about Chargaff's suggestion.[11]

In the course of that same meeting, Sol Spiegelman from the University of Illinois, Urbana, drawing on his enzyme induction studies in yeast, argued that the presumed template function of RNA certainly could not be that of a "passive mold" and that the template had to be seen as an active, short-lived material. He concluded that "either protein synthesis is mandatorily coupled to the synthesis of the new RNA molecules, or the destruction of an RNA molecule is mandatorily coupled to its function as a protein-synthesizing machine." Richard Roberts of the Carnegie Institution in Washington had done some experiments on bacteria that suggested that the amino acids might be trapped on a matrix prior to becoming assembled into protein: this would reflect a Gamow-type of template. And Walter Vincent of the State University of New York, Syracuse, remarked, "I have in progress some studies on the nucleolus of the starfish oocyte that strongly suggest that this organelle contains two classes of RNA, [one] a soluble, metabolically very active, fraction. [The] other is a less active, tightly bounded fraction. [One] exciting implication of the active, or labile, form would be that it is involved in the transfer of nuclear 'information' to the synthetic centers of the cytoplasm."[12] The concept of "information transfer" was on its way to entering the world of biochemistry with its consolidated and familiar talk of enzyme specificity, metabolic circuitry, and macromolecular synthesis in terms of growth.

Also in 1955, Marianne Grunberg-Manago, who was working as a postdoc from Paris in Severo Ochoa's lab at the New York University College of Medicine, eventually hit upon an enzyme that she termed "oligonucleotide phosphorylase." It was able to assemble RNA from nucleotides. She published a short note on the enzyme in June, and Zamecnik was aware of her finding by the end of 1955.[13] Enzymatic RNA synthesis in the test tube became an experimental prospect.

Talk about RNA also pervaded the CIBA Foundation's "Symposium on Ionizing Radiations and Cell Metabolism," held in London in March 1956. The discussion of Zamecnik's contribution was again dominated by RNA. Thomas Work from the National Institute for Medical Research in London had separated two ribonucleoprotein peaks, but his radioactive label stuck to only one of them. Nobody had any idea how to deal with the finding. George Popjak of the Hammersmith Hospital in London suggested the existence of further intermediates subsequent to amino acid activation. But nobody had as yet observed such a compound. Jean Brachet from Brussels asked for the effects of ribonuclease. Zamecnik's answer was that it seemed to break down the particle. Maybe, Zamecnik conjectured, this was a way to find where the radioactivity was attached. In fact he had made some attempts to decompose the particle without success. Barbara Holmes of Cambridge wanted to know whether there was any RNA synthesis going on in the system. Zamecnik dropped the remark that it looked as if labeled ATP would become incorporated into RNA. "We have not done definitive experiments in our laboratory," he said, "but I suspect, from the few we have carried out, that RNA is being synthesized, since ^{14}C-labeled ATP makes its way into RNA during that same time [of protein synthesis]."[14]

DISPLACEMENT I: From a Contaminant of the Soluble Fraction to an Intermediate in Protein Synthesis

What was Zamecnik's RNA synthesis remark about? Late in 1955, Zamecnik himself had begun to look for an RNA synthesis activity in his fractionated system after Littlefield had been heading in another direction. Who did what at MGH had always been and continued to be a matter of personal decision. Zamecnik felt encouraged by experiments communicated to him by Van Potter from Wisconsin, who was in the process of studying the incorporation of nucleotides into RNA under conditions similar to those of the protein synthesis system.[15] Hoagland remembers the situation as follows: "In ruminating upon the activation reaction, Zamecnik wondered if the enzymes involved might also participate in the synthesis of RNA. Could adenyl-amino acids be, in a sense, double-barreled—donating their amino acids to the protein polymerizing machinery or, alternatively, their adenyl portion to a nucleotide polymerizing system?"[16]

Looking for India, Finding America

The assay that opened the game was straightforward: Zamecnik added radio-labeled ATP to a mixture of his enzyme supernatant and the microsomal fraction. To his surprise, the nucleotide indeed seemed to become incorporated into an RNA component of the system. But cautiously he asked himself, "are the counts in the 'hot TCA' extract in RNA to begin with, or not?" And he noted: "Would it be worthwhile to repeat this type of expt, with greater care in washing procedure?"[17]

There was another, even more intriguing, feature. Hoagland's experiments had shown that the pH 5 enzyme fraction was able to activate amino acids. The question was whether aminoacyl-AMP would act as a donor in the observed attachment of ATP to RNA. If so, the radioactive amino acid would be released again. So, in a parallel experiment, Zamecnik incubated nonradioactive ATP and ^{14}C-labeled leucine instead of nonradioactive leucine and ^{14}C-labeled ATP together with the fractions. This was more than just a specificity control that eventually "became the real experiment."[18] Later commemorations of both Zamecnik and Hoag-

Expt to repeat the essential conditions of 10/31/55, but to
1) improve the washing procedure; and 2) introduce various
controls and variants.

	0°				0°	0°			
	↓				↓	↓			
	1	2	3	4	5	6	7	8	9	10

	1	2	3	4	5	6	7	8	9	10	
0.05 ml l-leucine-C14 ✓ 8K .002M [80,000 cpm, 0.1μM]	X	X	X								
0.3 ml microsomes ✓ (4 vol.)	✓	✓	✓	✓	✓	✓	✓	✓	✓	✓	3.6 cc
0.2 ml pH 5.2 ppt ✓	✓	✓	✓	✓	✓	✓	✓	✓	✓	✓	2.4 cc
0.1 ml Mg Na2 ATP ✓ 0.01M [1μM]	X	X		X	X	X	ⓞ	ⓞ	ⓞ	ⓞ	
0.05 ml GDP ✓ .005M [.25μM]	X	X		X	X	X	X	X	X		X
0.1 ml PEP 0.1M ✓ [10μM]	X	X		X	X	X	X	X			X
0.05 ml pyr. kinase ✓ 1:20 dilution	X	X		X	X	X	X	X			
0.1 ml K-orotate-C14 ✓ 0.012M [1.6μc, 1.2μM]			X	X	X						
0.1 ml Mg Na ATP-C14 ✓ 0.012M [0.59μc, 1.2μM]						X	X	X	X		
0.05 ml 0.005M UTP ✓ [.25μM]				X	X					X	
0.05 ml 0.005M CTP ✓ [.25μM]				X	X					X	
0.1 ml 0.01M ribose-5-P ✓ [8μM]				X	X						
Isotonic KCl ✓	.85	.85	.55	.9	1.1	1.1	.8	.8	.6	.9	
	.25	.25	.55	.2	-	-	.3	.3	.5	.2	
cpm/aliquot Hot TCA fraction azul	28	73	22	34	27	23	56	74	21	62	Counter: Nuclear micromil
Total cpm hot TCA " (x25)	700	1800	550	850	700	575	1400	1850	525	1550	
cpm/mg protein	4	140	2	8	5	25	8	4	2.5	7	" "
Total cpm in protein	47	175	25	76	43	28	87	40	26	80	

Incubate 10' 37° 95%N2-5%CO2; or else 0° for 10'. Stop reactions
by plunging test tubes (they better be lustroid) into powdered
CO2 snow. keep in CO2 ice box overnight & then work up.

Fig. 10.1. Protocol of an early assay (November 1955) that indicated amino acid
incorporation into RNA. Reprinted from Zamecnik, laboratory notebooks.

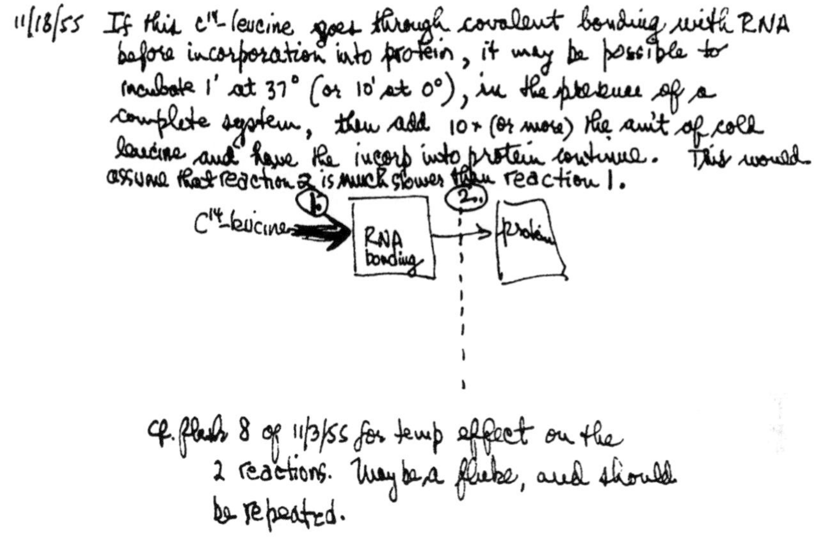

Fig. 10.2. First sketch of amino acid bonding to RNA, dating from November 1955. Reprinted from Zamecnik, laboratory notebooks.

land should not be taken as indicating that there was nothing but chance involved in the action.[19] Zamecnik followed a symmetry principle. He would not have thought of adding *amino acids* to an *RNA* "synthesizing" system had he not been "ruminating" upon the adenylation reaction. Nevertheless, the result was totally unexpected.

The assay of November 3, 1955, whose protocol is reproduced in Figure 10.1, suggested the contrary of what Zamecnik had thought would happen. Radioactive leucine became attached to the RNA as well. On November 10, 1955, the counts were processed and attention became focused on experiment no. 3 (third column in Figure 10.1). One-third of the batch, that is, 182 cpm of the 550 cpm that had been hot-acid precipitated, was subjected to a ninhydrin reaction. In a note, Zamecnik concluded, "the above indicates that only 44 cpm out of a calculated 182 cpm of radioactivity is released by the ninhydrin-CO_2 procedure. If true, this leucine-C^{14} is tightly bonded to the RNA." And he commented, "it appears unlikely that the radioactivity in 1N [the nucleic acid fraction] is due to contamination from protein (at least as a sole explanation) because the specific activity of 1N is higher than 1 [the protein fraction]."[20] During the same day, a preparative assay was performed to obtain RNA, "which has C^{14}-ATP and C^{14}-leucine associated with it, so that partial degradation and paper electrophoresis may be carried out at a future date."[21] On the back of his protocol, Zamecnik summarized that "if this C^{14}-leucine goes through covalent bonding with RNA before incorporation into protein," a two-step procedure could be assumed, where "reaction 2 is much slower than reaction 1." He drafted an experiment to find out how this could be tested and added the sketch shown in Figure 10.2.

Diversions

With the technical assistance of Meredith Hannon and Marion Horton, Zamecnik had completed these assays late in 1955.[22] During the following months, he was absorbed by clinical duties. In January and February 1956, he had to work on the board, and in March he departed for the CIBA Foundation meeting in London. When he came back he was appointed as Joseph Aub's successor as Collis P. Huntington Professor of Oncologic Medicine and director of the John Collins Warren Laboratories of the Huntington Memorial Hospital of Harvard University. Zamecnik now had to take care of all the laboratories, in which some 60 scientists were working. Also, he was waiting for Liza Hecht, a student from Potter's laboratory in Wisconsin, who was supposed to continue with him as a postdoctoral fellow. She was experienced in working with nucleic acids and, Zamecnik said, "I trusted her more than I trusted ourselves."[23] With RNA, Zamecnik entered a new field, and, as in the case of his earlier moves, he did not want to explore this field without the simultaneous input of someone with expertise, this time in RNA biochemistry. But Potter wanted to keep Hecht until the end of the year, and so her arrival was delayed until the beginning of 1957.[24]

From November 1955 to July 1956, some 25 preparations and tests including pH 5 fractions derived from all sorts of tissue show up in the laboratory notebooks. These assays do not seem to have a direct bearing on the newly established RNA-amino acid connection.[25] It was work related to the ongoing characterization of the soluble enzyme and the microsome fractions, respectively. During June Zamecnik tested the ability of the rat liver system to incorporate the radioactive amino acid tryptophane in the presence of washed microsomes and the purified tryptophane-activating enzyme of Earl Davie and Victor Koningsberger.[26] The latter had come to work with Lipmann after completion of his thesis with Theo Overbeek in Utrecht.[27] Much to the distress, if not anger, of Hoagland and Zamecnik, Lipmann had reoriented virtually his whole lab toward protein synthesis. Lipmann had become a competitor.

The exchange continued, but mostly on corroborating matters. The tryptophane experiments failed to work and therefore remained unpublished.[28] But they were instructive precisely in that they failed. Their failure indicated that some essential component, maybe an RNA, was lacking in this overpurified system. Things began to intercalate.

Reengagements

Meanwhile, Mary Stephenson had finished her thesis.[29] She joined work on the labeling of RNA with ATP in July 1956. The experiment from July 12, 1956, is another example of the effects of a surplus of controls by which, as Stephenson recalls, the work was pushed forward: "I think we [did] extremely well-controlled experiments. [I] think it was just a matter of one thing giving the next obvious thing to do."[30] The experiment was "to test effect of (1) pyrophosphate on incorp. of C^{14}-ATP into RNA; (2) amino acid mixture on same."[31] As additional controls, Stephenson included radioactive leucine instead of ATP, and in one assay of each set she omitted the soluble fraction from the mixture. Neither pyrophosphate nor the amino acid mixture, which had been the explicit target of the experiment, showed an appreciable effect. Instead, it turned out that the incorporation of ATP (and leucine) into RNA completely depended on the soluble fraction—a result deriving from the "surplus"-control. "Both systems seem to require pH 5 enzymes," Stephenson tentatively concluded. When she pursued this requirement,[32] it became apparent that, contrary to the implicit assumption at the beginning, it was not the microsomal RNA that picked up ATP and leucine but the RNA "contaminant" of the pH 5 enzyme solution! Accordingly, the RNA was termed *soluble RNA* (S-RNA).[33]

Hoagland, too, at that time, had joined the common venture of mapping soluble RNA. He recalls: "It was not until June that Paul told me of the five-month-old findings, in his usual low-key, half-puzzled, half-deprecating way. [Paul] was happy to have me pursue the lead."[34] In the following months, Hoagland made sure that the labeling of S-RNA with leucine was stimulated by ATP and that it was additive when different amino acids were given simultaneously. Moreover, he found that the label, once attached, was nondialysable, nonexchangeable, acid-stable, and alkali-labile, in remarkable contrast to the alkali-stable complexes that Joseph Potter and Alexander Dounce just had isolated as possible intermediates in protein synthesis.[35] Most importantly, the amino acid could be transferred from S-RNA to the microsomes, and this latter reaction was enhanced in the presence of GTP. Taken together, a new scientific object was in the process of emerging, and, quite characteristically, it first took the shape of a list of what it did and of what it did not do, changing its contours with every new entry.[36]

Quantitative measurements were complicated by the fact that the RNA was extracted from the precipitate by means of a hot salt solution.[37] The procedure fortunately worked because the acid pH of the solution had not been readjusted. The happy impact of this sloppy neglect became apparent when Stephenson tested Liza Hecht's "improved" nucleic acid extraction method. The yield of RNA was indeed improved at neutral pH, but the radioactivity was lost. The method deemed to be the appropriate one would have abolished the new signal altogether.

In terms of stoichiometry, the situation was confusing. Incorporation figures did not tell very much, as long as nobody knew what size the RNA was. Zamecnik calculated that roughly 5 percent of the AMP-residues in the RNA fraction were replaced, but he hesitated to draw any conclusion. "We don't know whether the C^{14}-ATP incorporation would continue at this rate or not, nor indeed whether it is RNA and whether the C^{14}-ATP is in the fabric of the 'RNA.' "[38] With respect to leucine, a first calculation indicated that about 1 in 300 nucleotide residues acquired the label. Whence, "maybe only the terminal nucleotide in a chain is involved: either at a terminal $5'$ or $3'$ or cyclic $2'3'$ residue."[39] Soon, however, higher values showed up. This meant, as a laboratory note points out, that "leucine could all be terminal, but 20 amino acids could not be. Around ⅓rd of the mononucleotide residues of the S-RNA would have to be occupied."[40] Could soluble RNA carry several amino acids at a time? Nobody knew in January 1957 when Hoagland, Zamecnik, and Stephenson sent their "Preliminary Note" to *Biochimica et Biophysica Acta*.[41] Despite all these uncertainties, the experiments suggested that the initial activation of amino acids was followed "by a transacylation to S-RNA" and that, finally, GTP mediated the "transfer" to the microsomes "by a mechanism as yet unknown."[42] A new "intermediate reaction" in protein biosynthesis had been constructed in the test tube.

A Context for Competition

In the course of this year, 1956, evidence for in vitro nucleotide incorporation into RNA was being gathered at Wisconsin as well.[43] The group at MGH kept close contact with Van Potter of the University of Wisconsin, while still waiting for Liza Hecht to join them. Also during that late summer, Hoagland was asked to referee, and thus became aware of, a paper by Robert Holley.[44] Holley, from Cornell University, then on a Guggenheim Fellowship at the Division of Biology of the California Institute of Technology, had found a ribonuclease-sensitive step presumably following the amino acid activation process, which, however, seemed only to work for the amino acid alanine.[45] No doubt, this bit of information speeded up the efforts of the Huntington group to get their results ready for publication as soon as possible. After months of relaxed puzzling at the bench, they began to feel their "competitors' hot breath on the backs of our necks," as Hoagland put it later.[46]

It is interesting to note in this context an attempt similar to Holley's that, however, failed. Paul Berg had begun to work on the activation of methionine at Washington University School of Medicine in St. Louis shortly after Hoagland had finished his pioneering essays on amino acid activation. As early as November 1955, Berg scribbled into his notebook: "Is anything formed from ATP + methionine?" In March 1956, he consequently tested the "effect of RNA'ase and DNA'ase treatment on methionine catalyzed PP-exchange." The result was negative: "No effect of the RNA'ase and DNA'ase treatment." Disappointed, Berg did not pursue the question during the following months. He took up his studies on amino acid incorporation into soluble RNA of bacterial extracts only in May 1957, a month after the publication of Hoagland, Zamecnik, and Stephenson's "Preliminary Note."[47]

Tore Hultin of the Wenner-Gren Institute in Stockholm had obtained independent evidence for an intermediate step in protein synthesis from kinetic isotope dilution studies.[48] A year earlier, he had already been ruminating on the relationship between "cytoplasmic nucleoproteins" with a high label in the RNA and the microsomes with a high label in the proteins. However, he considered it "premature at present to speculate upon possible explanations of this inverse behavior," that is, to speculate about a possible function of nonmicrosomal RNA.[49]

By October 1956, Kikuo Ogata and Hiroyoshi Nohara at the Niigata University School of Medicine in Japan had also collected hints for an RNA-connected intermediate in protein synthesis. Ogata, who had been working on the in vitro incorporation of radioactive glycine into antibodies of immunized rabbits,[50] was interested in cell-free systems for that reason, and at the beginning of 1956, together with his postdoctoral fellow Nohara, he established the fractionated rat liver system at Niigata. In following up Hoagland's amino acid activation studies, Ogata and

Nohara got radioactive alanine attached to the RNA of the soluble fraction. Ogata recalls: "Since I had known that the fraction of NaCl extracts which was precipitated at pH 5, contained ribonucleoprotein (Griffin, Nye, Noda, and Luck 1948), and considered that RNA is important in the protein biosynthesis, I assumed that RNA contained in the pH 5 fraction is the intermediate in the protein synthesis in rat liver microsomes."[51] Whereas Hultin's experiments certainly had been recognized at MGH,[52] the work from Japan came to be noted only when it was published in 1957. Hoagland, Zamecnik and Stephenson's note to *Biochimica et Biophysica Acta*, in turn, came to the attention of the Japanese workers only after their own paper had been submitted.[53] Through a remarkable coincidence, a differently motivated pursuit using the same system had led to a similar result.

Epistemological Obstacles

RNA had been part of the soluble fraction, albeit as a contamination, for roughly three years. Why had nobody at MGH noticed earlier that this fraction took up amino acids from the activated aminoacyl-AMP? Why had Zamecnik not pursued his finding with more rigor after the first hints? In trying to answer this question, I will invoke a notion of Bachelard's: An experimental system in general, and maybe even constitutively, is ambiguous with respect to its own productive potentials. It displays options and obstacles at the same time. In speaking of epistemological obstacles, I do not dwell

on a consideration of external obstacles such as the complexity or the fugacity of phenomena, or on a complaint about the weakness of the senses and of human mind. Within the act of gaining knowledge itself, in its innermost agitation, inertia and entanglement make their appearance according to a kind of functional necessity. [Empirical] reasoning is clear only *in hindsight*, when the apparatus of explanation has been set going.[54]

This quotation can be taken as an adequate description of what was going on with the protein synthesis system. Let me repeat that Hoagland had made visible amino acid activation via a model reaction that was the reversal of the metabolic process assumed to go on within the cell. He measured the attachment of radioactive pyrophosphate to AMP, not the attachment of ATP to an amino acid. The subsequent reaction he had studied via a model compound that, due to its characteristics, made its natural counterpart invisible: hydroxylamine not only accepted amino acids from aminoacyl adenylate, but also—as it turned out later—from aminoacylated S-RNA.[55]

Once again, the conditions of identifying one partial reaction—adenylation—precluded the identification of another, still unknown partial reaction—transacylation. Representing the former amounted to depresenting the latter. Perhaps most importantly, until the RNA work had started, Zamecnik had always placed a strong emphasis on circumstantially demonstrating that the amino acids were indeed caught in α-peptide bonds. Such demonstration required harsh conditions for eliminating all other possible bondings including, as it would become clear ex post, the attachment of amino acids to RNA. Zamecnik abandoned this harsh purification regime only when, for reasons of experimental symmetry, he needed identical evaluation conditions for the parallel assays of amino acid and nucleotide incorporation because the latter required gentle purification.

In addition, a style of research was at issue. By education and by temperament, Zamecnik was cautious.[56] Although aware of and effective in competition,[57] he did not like to practice research as a race for novelty and priority. He liked to have his collaborators choose their subjects according to their predilections. He thereby covered a broad field of research problems; was able to follow several issues jointly; effectively articulated different experimental skills; and, by that means, created an ambience in which unprecedented things could appear. But at the same time, this attitude created possibilities for diversion, distraction, and delay.

The displacement of soluble RNA from a contaminant in the supernatant fraction to an aminoacylated intermediate in protein synthesis is an exemplar of serendipity. Zamecnik had looked for an activity that would synthesize RNA in his fractionated rat liver system, and he had assumed it to be part of the microsomes. What he found was an already synthesized small RNA molecule, to which both ATP and amino acids became attached. In addition, the RNA was not, as expected, part of the microsomes; it was part of the enzyme fraction. A new object appeared, and it had very surprising specifications. Operationally, it was characterized as "soluble" because it emerged from the soluble fraction. Functionally, it was characterized as an intermediate in protein synthesis. The path along which epistemic things come into existence defines what they count. This experimental object appeared as the cellular counterpart of a model substance, hydroxylamine, previously used to catch activated amino acids. It emerged as a biochemical intermediate in protein metabolism. Even before this it had emerged as an RNA molecule to which ATP could be linked.

RNA began to be represented in the protein synthesis system under two different headings. Zamecnik began to consider the possibility that microsomal RNA, on the one hand, and soluble RNA, on the other, might display different functions. A new phase in the fractional representation of the system was opened. Once the workers at MGH had realized that the signal that showed up might help answer questions that had not been anticipated, the outline of the new intermediate chain of reactions was established in a few months toward the end of 1956. The last cornerstone—the experiment providing evidence for the transfer of the RNA-bound amino acids to the microsome—was laid in what Hoagland later described as "the most suspenseful and exciting few hours of my professional life."[58]

DISPLACEMENT II: From Biochemistry to Molecular Biology

In 1955, Francis Crick had jotted down "a note for the RNA tie club" entitled "On degenerate templates and the adaptor hypothesis," which, however, remained unprinted. The paper "was never published in a proper journal. [Eventually] I did publish a short remark briefly outlining the idea and tentatively suggesting that the adaptor might be a small piece of nucleic acid."[59] Crick's hypothesis basically stated that the assembly of amino acids might be guided by oligonucleotides attached to them, whereby the latter would recognize the coding units of template RNA via complementary base pairing. The manuscript circulated among the few members of George Gamow's "RNA tie club," who regarded themselves as molecular biology's avant-garde predestined to solve the riddle of the genetic code. Crick's paper had not come to the attention of the Harvard protein synthesis workers who were not members of the club.[60] Whether Lipmann, who was a member of the club, read it I was not able to determine. In any case, if he did, it had no immediate impact on his protein synthesis work.

Which Epistemic Field? Which Experimental Culture?

As I have mentioned, Hoagland, Zamecnik, and Stephenson's "Preliminary Note" presented S-RNA as "evidence for another step in the reaction sequence between amino acid activation and peptide bond condensation."[61] Not a word about molecular information transfer. By that time, however, biochemical innocence was gradually being lost. Hoagland recalls:

In late 1956, I had my first visit from a card-carrying "molecular biologist." (We at the Huntington considered ourselves biochemists.) Jim Watson had just become Professor of Biology at Harvard and was probing the structure of the ribosome [microsome at that time]. He had heard rumors of our discovery of transfer RNA [soluble RNA] and I jubilantly told him of our findings. He was restlessly attentive and when I'd finished he told me that Francis Crick had forecast the existence of transfer RNA-like molecules! Hadn't I heard of the *adaptor hypothesis*? I was astonished and admitted that I hadn't. [I] was bowled over by the ingenuity and beauty of the idea and sensed it had to be the explanation of our experimental findings.[62]

Judson quotes him as saying, somewhat more reluctantly, in an interview, "in fact, I can remember vividly leaning over a centrifuge in the particular laboratory and talking with Jim about it, and his saying, 'This is the interpretation of your results.' And, I can *sense, palpably*, my feeling of resentment, at the time, that Jim would be telling me how to interpret my results—but also the feeling that, God damn it, he's right. You know. It was just—it was just right."[63] Or again, from Hoagland's recollections accompanying a reprint of the seminal 1957 paper: "An image arose in my mind: we biochemical explorers slashing our way through a dense jungle to discover a beautiful temple while Francis Crick, floating gracefully overhead on gossamer wings of theory, waited patiently for us to see the goal that he was already gazing down upon!"[64]

So it appeared to Hoagland some thirty years later, in retrospect and from the inescapable vantage point of a solution that had already made its career. Yet, in the last months of 1956, the issue was neither a missing explanation for an unexpected result nor a psychological barrier to admitting somebody else's gossamer-winged ideas. The result that Hoagland, Zamecnik, and Stephenson had obtained *was* something else. It did not represent an adaptor. The RNA emerging from the experimental system was a link in the chain of intermediate reactions in protein synthesis—it was a biochemical thing. Changing its identity would not simply mean giving the "experimental finding" an "explanation" or another name; it would mean engaging in a different research project. This could not be done by "leaning over a centrifuge" and listening to James Watson. It would mean rearranging everything in a different space of representation. It was not just a matter of changing the frame of interpretation; it was a

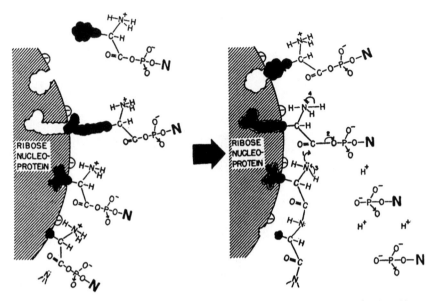

Fig. 10.3. Schematic representation of the mechanism of protein synthesis. *N* is a nucleotide residue. Reprinted from Loftfield 1957a, fig. 13, by permission of Elsevier Science Ltd.

matter of doing different experiments in a different experimental context. It would mean making S-RNA work as an adaptor of a presumed genetic code. And because this code was still a big riddle it would mean making S-RNA an instrument to unravel the code. It would require not just looking at the ribonucleoprotein particle as an ordering frame of amino acids but transforming it into a "template." And so on.

Alternatively, S-RNA could be investigated further along the lines by which it had come into experimental existence: as a molecule to which amino acids became attached before they condensed into proteins. On the one hand, the system had produced a situation where "the facts combine to indicate a new and unforeseen direction."[65] It promised the experimental implementation of the most urgent question of molecular biology—the genetic code. On the other hand, it exerted a kind of mass action, a resistance against the adaptor hypothesis. As Hoagland rightly states, "in this instance, a grand theory neither substituted for nor guided the successful analytical dissection of the machinery of protein synthesis."[66] Biochemical reasoning in terms of metabolic intermediates came to be confronted with reasoning in terms of genetic information transfer.

Consider the "schematic representation" of the mechanism of protein synthesis shown in Figure 10.3, from a review that Robert Loftfield wrote in this time of transition.

In the figure, different anhydrides between amino acids and nucleotides (designated N) align themselves on the surface of the ribonucleoprotein particle by virtue of the specific side groups of the different amino acids.[67] The activated amino acid comes into proximity with the already assembled peptide, and peptide bond formation follows. In this picture, the nucleotide (N) has no other function than activating the amino acid for subsequent peptide bonding. How did this fit with the "intriguing suggestion [advanced] by Crick, Griffith and Orgel (unpublished)"[68] that the RNA template might consist of triplets to which the "adaptor" trinucleotides would "mate [by] interaction of the nucelotide with the ribonucleoprotein," as discussed in the accompanying text?[69]

Worse still, by and large the Hoagland-Zamecnik-Stephenson S-RNA did not match the characters of the Crick-Griffith-Orgel trinucleotide-adaptor. The molecule seemed to be too large for that function. A preliminary calculation by Zamecnik indicated that "if all these [incorporated AMP-moieties] are 'terminal,' then the chain is 60 monomers long."[70] Loftfield's "thoughts on mechanism" were a juxtaposition of mutually nonfitting pieces of a puzzle.[71] A trade between two communities had been initiated, but there was as yet no common currency. The biochemists' molecule, which bridged the gap of the metabolic reaction chain between free amino acids and protein, was not quite the thing molecular biologists were desperately after to bridge the gap between genetic code and protein code. The biochemists regarded information-talk as supplementary. The molecular biologists had difficulties understanding the experimental logic of a biochemical in vitro system, the requirements of its stabilization, and its operational terminology. Different laboratory practices were embodied in different languages. Everything depended on how these differential readings would become translated into effective "differantial" action.

At MGH, it took the first part of 1957 to accomplish a more detailed biochemical account of soluble RNA, and its publication was delayed.[72] Time enough to accommodate Crick's adaptor hypothesis? Hoagland and Crick had met around the beginning of the year, and Crick had invited Hoagland to spend a term in Cambridge.[73] Meanwhile, they exchanged letters. In a letter from January 1957, Crick opposed Watson's view that S-RNA would go "intact into the microsomal particle." Instead, Crick

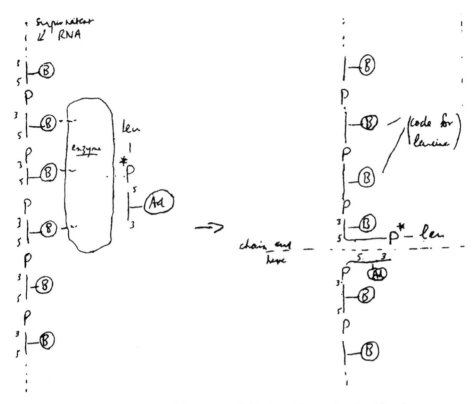

Fig. 10.4. Crick's scheme of chopping soluble RNA into trinucleotide-adaptors. Reprinted from letter of Crick to Hoagland, January 1957.

favored the idea that a trinucleotide with an amino acid attached to it would diffuse into the particle where the amino acids would condense. The trinucleotides would base-pair to the particle's RNA and in turn condense to a new, complementary RNA.

Crick's letter continued: "How can we fit your results into this scheme? The idea is that your 'activating enzymes' cut up this new RNA into trinucleotides at the same time attaching an amino acid to each one. Once this idea is stated it becomes almost obvious." And then he went on figuring how this could happen (see Figure 10.4).[74] At the Huntington Laboratories, however, things proceeded step by step. Zamecnik remained hesitant; Hoagland did too, but he seemed to be more attracted by Crick's way of reasoning.

The soluble RNA offered a plethora of experimental options: purification according to amino acid specificity, analysis of physical and chemical parameters, sequencing, secondary structure elucidation, crystallization, X-ray diffraction, functional interaction with microsomes, interaction with activating enzymes, and so forth. But the importance of these approaches, the prospective significance of their effects, was hard to anticipate. There was no predetermined strategy to be followed in dealing with soluble RNA. The different research trajectories were not clearly delineated. They had themselves to be created in the process of doing the experiments and intuitively adjusted according to whatever happened. At a stage at which the research options had bifurcated well beyond the capacities of any single laboratory, there was no guarantee that following a particular trajectory might not lead a group into a dead end and that others, who had pursued another path, would not eventually take over. The balance between settling extant issues and creating unprecedented ones, which determines how long an experimental system stays at the forefront of research, had to be re-established with every new turning point.

The overarching problem was which epistemic field to opt for, and thus which experimental culture to adopt: that of "metabolism" or that of "instruction"; protein synthesis as an enzymatic chain of biochemical reactions or as a translational event in an as yet obscure process of genetic information transfer. This would ultimately determine the spaces of representation, the design of the assay systems, and even the identity of their constituents.

A Supplement After All

James Watson and Francis Crick were among the first to become aware of the potential impact of S-RNA, as an adaptor of a presumed code, in bridging the gap between the genes and their expression as biologically functional proteins. The biochemists at MGH gradually realized that they had to trade with another scientific community. Translating their results into the framework of molecular biology became a necessity if they wanted to occupy the space of protein sequence specificity left open by classical biochemistry as well as by the actual genetic approaches to the problem by molecular biologists. Such translation did not happen without friction. With S-RNA, the shift from biochemistry to molecular biology had been made possible from within a particular biochemical system, which in turn had arisen from cancer research, and which itself had been established at MGH over the course of ten years. At the same time, a shift was suggested from without, by the parallel coding discussion. This was

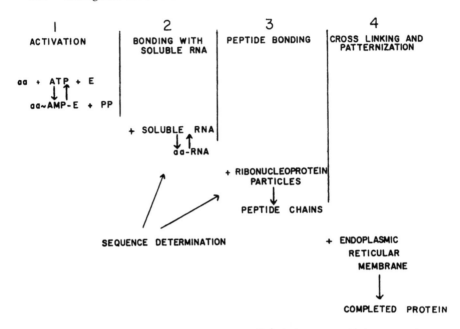

Fig. 10.5. Postulated steps in incorporation of labeled amino acids into proteins. Reprinted from Zamecnik, Stephenson, and Hecht 1958, fig. 1.

the specific form of that conjuncture. The surprising things that had emerged as biochemical intermediates in protein metabolism began to be translated into a different context, that of genetic information transfer. At the beginning, this translation took the form of a *supplement*. Its character as a supplement becomes evident when we look at one of the first reviews, whose title, two years after S-RNA had come into experimental being, still reads, "Intermediate Reactions in Amino Acid Incorporation."[75] The problem of protein synthesis as a process guided by genetic instruction surfaced only in the last sentence preceding the review's summarizing paragraph. Here we read: "It is *also* possible that determination of the amino acid sequence may be a two-step RNA-participating process."[76] The *also* indicates an alternative to viewing the whole process as an enzyme-directed metabolic reaction chain. Consider the diagram shown in Figure 10.5. The diagram neither represents a metabolic pathway governed by energy dissipation and enzymatic mediation, nor a flow of genetic information. It is a diagram representing steps into which experimentally discernible components and the reactions associated with them are entered. We may therefore term it "modular." The supplement takes the form of a double arrow (lower left of Figure 10.5) originating from the expression *sequence determination* and pointing to the components thought to be involved. It is a supplement in precisely the sense Derrida confers on this notion.[77] It is a simple addendum, but it has the potential to reorient the system, to direct the movement of its *différance*. Derrida develops the "economy of the supplement" through an analysis of Rousseau's theory of writing as a supplement of language. The comparison with the process of experimentation, which I stipulate here, is a structural one. The subversion of the protein synthesis system by the supplement of coding shows both aspects characteristic for the process of supplementation: it tends to change the identity of its components, and it fails to do so by its very presence as a supplement.

The Coding Problem

One thing was clear. Soluble RNA was an obligatory component of protein synthesis. But how did it function? Was it a "primary template" that interacted with a "secondary template,"[78] the RNA of the ribonucleoprotein particle, via complementary base-pairing? During the fall of 1956, Jesse Scott, a Vanderbilt M.D. with strong biophysical inclinations, joined the MGH team in its effort to characterize S-RNA. Reasoning along the lines of Crick's suggestion,[79] Scott had come up with the following scheme early in 1957:

Assume the RNA of the pH 5 Enz. to be a double helix, intact in the inter synthetic period and attached to protein.

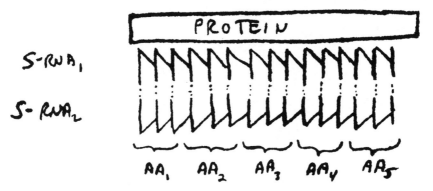

Fig. 10.6A. Possible generation mechanism of adaptor molecules. February 1957. Reprinted from laboratory drawing of Jesse Scott.

S-RNA$_2$ is the sense code specifying the amino acids 1, 2, 3. [S-RNA$_1$] is the complement to the sense code. ATP and AA react to form the AMP~AA. AMP~AA in turn reacts at the site on RNA$_2$ specified by the code to yield—

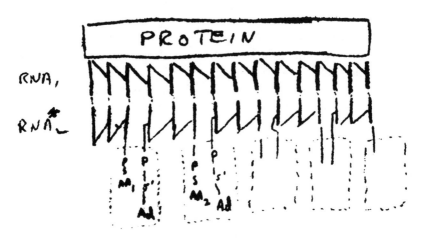

Fig. 10.6B. Possible generation mechanism of adaptor molecules. February 1957. Reprinted from laboratory drawing of Jesse Scott.

[How] can the RNA~AA go to microsome protein? One hypothesis can be that there is on the microsome for (a) specific protein(s) an RNA identical to S-RNA$_1$—M-RNA$_1$. The trinucleotides making up S-RNA$_2$* are transferred to M-RNA$_1$, without entering the acid soluble pool of the cell. Conditions on the microsome are such that in the presence of GTP the AA's of the S-RNA$_2$* are polymerized to polypeptide, AMP is released, and S-RNA$_2$ is reformed as the high polymer.[80]

From such a molecular biologist's point of view, the interaction between aminoacylated RNA and the RNP-particle became crucial, as can be seen in Figures 10.6A and B. During the summer of 1957, Hoagland had left Boston to spend a year with Crick in Cambridge.[81] The hope that had brought him to England was "that tRNA [S-RNA] might help reveal the genetic code."[82] His idea was "to isolate individual tRNA species and identify the nucleotide sequence responsible for binding a particular amino acid."[83] The collaboration did not materialize into any significant achievement. Crick's vision of trinucleotide adaptors did not lend itself to a workable experimental implementation.[84]

S-RNA: Shaping the Identity of a Molecule

There was a lot of space left between "a *chemical* association of RNA and amino acids," on the one hand, and a possible "*complementarity*" between S-RNA and microsomal RNA on the other.[85] Between the extremes of these two ways of viewing things, soluble RNA gained identity. Above all, a major technical achievement must be mentioned here. Following the protocol of K. S. Kirby, as well as that of Alfred Gierer and Georg Schramm, S-RNA was now separated from the proteins of the supernatant by phenol extraction.[86] The procedure was to replace the precarious solubilization of the precipitated RNA by a hot salt solution. The introduction of this technical innovation can be followed through the publications. For a while, the laboratory crew adopted phenol extraction for large-scale preparations, whereas it continued to use the sodium chloride treatment in the assays. For reasons of consistency, simultaneous application was vital. For instance, the first ^{14}C-ATP-labeled S-RNA isolated by way of phenol extraction contained by far "too many counts," as compared to the usual charging level.[87] The difference was as large as an order of magnitude. In tedious attempts to solve the discrepancy, it turned out that this was not an improvement, as might have been expected, but simply the effect of free radioactive ATP being carried along through the preparation procedure.

The first experiments attacking the problem of how amino acids might become attached to S-RNA date back to July 1956.[88] They resulted in the following reversible reaction scheme:

$$ATP + leucine\text{-}^{14}C + E <-> E(AMP{\sim}leucine\text{-}^{14}C) + PP$$
$$E(AMP{\sim}leucine\text{-}^{14}C) + RNA <-> RNA{\sim}leucine\text{-}^{14}C + E + (AMP)$$

As far as the nature of the chemical bond between the amino acid and the RNA is concerned, Zamecnik envisaged several possibilities late in 1956.[89] "There seemed to be three possibilities: the ring, the internucleoside phosphate, or the ribose. But we didn't have any record, except that Dan Brown who had been working with Todd had said, oh no, I don't think that the inter-nucleoside phosphate would be stable enough. I mean the amino acid would not stay there, it would be too unstable for what you find. So then, it was either in the ring, or the ribose."[90] The first encompassing paper on S-RNA, ready for the press on September 27, 1957, states: "Thus far we cannot assign a specific structure to the amino acid-RNA linkage."[91]

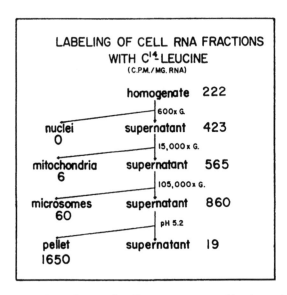

Fig. 10.7. Fractionation scheme of rat liver, as represented by the end of 1957. Reprinted from Hoagland, Stephenson, Scott, Hecht, and Zamecnik 1958, fig. 1.

The fractional representation of the RNA-labeling capacity led to the picture shown in Figure 10.7. The figure shows that after each centrifugation step the activity resided in one fraction. The representation of S-RNA and the fractional representation of the system thus mapped onto each other. At present, however, there was no way to push the fractional representation of the system one step further. The reason was that although active pH 5 RNA could be separated from the enzymes via phenol extraction, the reverse—separation of active enzymes from RNA—did not hold.

The leucine-label had also been monitored kinetically. Whereas the proteins continually acquired new label, the RNA-label quickly reached a plateau on the particles. S-RNA appeared to turn over, thereby delivering its amino acids to the protein. With these experiments, the collaborative effort of the MGH researchers to manipulate their system approached virtuosity. They complemented in vivo observations with in vitro data, and according to the purposes of the assay, they interchanged the two cell-types at hand—rat liver and mouse ascites cells—in a mutual fitting procedure. In the course of these experiments, the unknown point of intervention of GTP also became focused. Its "locus of action" was "narrowed down to the area of interaction between pH 5 RNA-amino acid and microsomes," indicating that "a new transfer enzyme" was required for this interaction.[92] With that indication, the pH 5 enzyme-fraction itself began to be dissected into different components.

Bifurcations

When Zamecnik had put RNA on his agenda in the autumn of 1955, his expectation had been that he would find an RNA synthesis activity in the protein synthesis system. What he and his colleagues actually found was an RNA molecule that incorporated amino acids. Now, a whole set of new enzymes emerged, and all of them appeared to be present in the pH 5 precipitate. This fraction exploded. The differential reproduction of the system now gave rise to so many questions that it quickly became the working place of a whole protein synthesis industry. The point had been reached where the further differentiation of the cell-free system became coupled to the differentiation of research groups. By the end of 1957, different aspects of amino acid-oligonucleotide compounds were being investigated by at least six other groups: Kikuo Ogata's at Niigata University in Japan; Victor Koningsberger's, who had returned from Lipmann's lab to the Netherlands, at the Van't Hoff Laboratory in Utrecht; Paul Berg's at Washington University in St. Louis; Richard Schweet's at the Biology Division of the California Institute of Technology in Pasadena; Fritz Lipmann's, who had moved from MGH to the Rockefeller Institute in New York; and Robert Holley's at Cornell University.[93] All of them had entered the field from slightly different starting points and joined the race to add items to the list of what these RNA molecules and their related enzymes did, how they behaved, and what they failed to do. Their reports soon filled the pages of the prestigious *Proceedings of the National Academy of Sciences*. This was a major point of bifurcation and truly one of those moments "where the dream of novelty suddenly takes on consistency without being fully assured of becoming reality."[94]

New Skills: Nucleotide Attachment Pursued

In the beginning of 1957, Liza Hecht came from Wisconsin. One of the big questions that remained to be teased apart was the chemical nature of the attachment of ATP to the RNA and the possible relation of this reaction to the amino acid attachment. Right from the beginning, Zamecnik had been aware that approaching this puzzle required more sophisticated RNA biochemistry. As with amino acid activation,[95] he had been looking for new skills from outside. He recalls: "It seemed to me that this was a new area, and that we had to put someone full-time on it."[96]

Contrary to Zamecnik's original assumption, the incorporation of ATP was obviously not part of a de novo RNA synthesis. It appeared to be a modification reaction. Was it specific? In February 1957, Mary Stephenson compared the uptake of radioactive ATP to that of UTP. She noticed a difference of a factor of 100.[97] Once again, the experimental exploration came to be organized by exhaustive symmetry considerations. Liza Hecht took up the task and set out to systematically test all the other nucleotides. She shared her lab with Jesse Scott at the beginning and soon developed a close collaboration with Stephenson.[98] Hecht quickly found that CTP, too, served as a precursor for RNA modification.[99] Moreover, CTP stimulated the incorporation of ATP, and the incorporation of radio-labeled leucine depended on the prior incorporation of these nucleotides.

The problem had become rather complex. What was the sequence of these different reactions; what was their extent? Fortuitously, Hecht had realized that with respect to ATP-incorporation "old preparations of pH 5 fractions" depended upon CTP.[100] Accordingly, she precipitated a 105,000 × g supernatant of ascites cells, "aged" it for some time at 37°C, and reprecipitated it prior to the incorporation reaction.[101] As a result, "what her experiments showed was that if you used that purified preparation and just ATP, you didn't get any amino acid in that. But if you added C and A, then you did. And if you added the C's and no A's, you didn't."[102] Thus, a sequence of reactions was established: first CTP, then ATP, then the amino acid. The conclusion was at hand that an "end unit containing cytosine and adenine nucleotides [provides] a functional grouping which is required for the binding of activated amino acids to RNA."[103]

A small event thus had become crucial and turned into an instrument of research: the incidental use of an old preparation. A slightly different preincubation and reprecipitation of the pH 5 fraction in a slightly different experimental context—its differential reproduction—was sufficient for a major achievement. With its help, the incorporation reactions could be sequentialized. This is another example of an approach that may look deceptively simple. However, looking for such "simple approaches," according to Hecht, was Zamecnik's strength.[104]

Hecht used ascites cells in these experiments. Initially, they were introduced, and then abandoned, as a model for cancer growth. Subsequently, they served as a source for preparing membrane-free ribonucleoprotein particles. Now, their supernatant proved useful as a source for S-RNA because it was deficient in ATP-degrading enzymes. These cells thus acquired the status of a multivalent research tool.

Meanwhile, just as with the amino acid incorporation reaction, the RNA-nucleotide interaction had become the focus of competitive efforts by a number of investigators, most of them from the Wisconsin connection. Among them, besides Potter, were Alan Paterson and Gerald LePage from the McArdle Laboratory in Wisconsin, Mary Edmonds, who had moved from Wisconsin to Pittsburgh, Edward Herbert, who had moved to MIT, and Evangelo Canellakis at Yale.[105]

By the end of 1957, Liza Hecht and her colleagues supposed that three nucleotide incorporation reactions would occur on the soluble RNA:[106]

1. CTP + RNA <-> CMP-RNA + PP
2. ATP + CMP-RNA <-> AMP-CMP-RNA + PP
3. UTP + RNA <-> UMP-RNA + PP

But the situation was much less clear-cut than suggested by these formulae. What could be considered as a signal? What had to be taken as noise? Was the UTP uptake significant? It amounted to only 10 percent as compared to ATP, but it could be measured without ambiguity. And there was a stimulatory effect of GTP on the ATP incorporation that remained completely obscure. "An explanation for this observation cannot be given at present."[107] Hecht complemented the assays by a mild alkaline hydrolysis of the products. The degradation pattern suggested that the end group of the RNA was RNA-U in the presence of UTP, RNA-CC in the presence of CTP, and RNA-CCA in the presence of both CTP and ATP. Finally, an amino acid could be attached to the end group -CCA, yielding an RNA-CCA-(amino acid). Whether both enzymatic processes—nucleotide attachment, amino acid attachment—had anything to do with each other in terms of biological function remained a puzzle.

There were now good reasons to assume that soluble RNA was different from microsomal RNA. Interestingly, soluble RNA derived from mouse and rat could be substituted for each other, and so, to a certain extent, could RNA from baker's yeast! Therefore, functional specificity rather than species specificity seemed to be involved. The fractional representation of these RNAs and especially their modifying enzymes, however, required more sophisticated procedures. It was obvious that a lot of analytical protein chemistry had to be brought into play. Until then, the fractional representation of the system had happily evolved from and lived with centrifugation and acid precipitation. Here, it came to a stop. Other technologies had to be tried and developed.

Alternative Representations

Everything revolved around S-RNA. It mediated the transfer of amino acids to the ribonucleoprotein. It was supposed to mediate the transfer of genetic information. But it could also be involved in something else: regulation. There was RNA-U, RNA-C(C), and RNA-CCA, and the reactions could be reversed. These features would "suggest that the soluble RNA may serve as a storehouse for releasable nucleotide coenzymes, which play a directive role in regulation of the metabolism."[108] Such a function would have been in line with the "ultimate goal" of MGH's research venture invoked again at a congress of the International Union Against Cancer, namely, to "throw light on the regulation of protein synthesis in the normal growth process and on its possible aberrations in the neoplastic state."[109] Given the later developments, which will briefly be reviewed in Chapters 12 and 13, invoking regulation was more than a lip service paid on the occasion of a cancer congress. It amounted to a third option.

The resolution of the fractionated system with respect to the enzymes that were involved in the four-step process depicted in Figure 10.8 was at issue, too. These enzymes could now be connected in the form of a cascade comprising at least four different species.

The enzyme chain was constructed from an articulation of evidences rather than being the direct product of the fractional representation itself. With respect to these enzymes, the fractions were all "dirty." None of them corresponded to a single enzyme. The first enzymatic activity (E_1) was characterized by three "pieces of evidence." One of them was its insensitivity toward ribonuclease. In contrast, sensitivity to RNase was a distinguishing feature of the second activity (E_2), resulting in an amino-acylated RNA. "New evidence" for a third enzymatic activity (E_3) came from the inability of the RNP-particle to incorporate amino acids in the absence of GTP. Only "shreds of evidence" existed for a fourth activity (E_4).

The fractional representation of the system—its materially existing

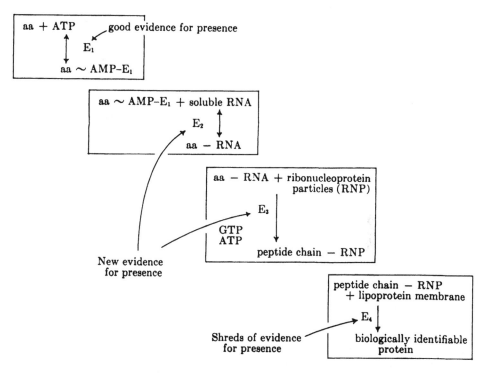

Fig. 10.8. Possible enzymes involved in protein synthesis. Reprinted from Za-mecnik, Stephenson, and Hecht 1958, fig. 3.

representation—did not discriminate against the different postulated enzymes. The fractions contained the enzymes in a mixture of overlapping combinations as can be seen in Figure 10.9.

Enzyme 1 was present both in the pH 5 precipitate of the 105,000 × g supernatant and in the microsomes. Enzyme 2 occurred in both of these fractions and in the supernatant of the acid precipitation. Enzyme 3 was predominantly found in the pH 5 supernatant and the corresponding precipitate. Finally, enzyme 4 could be supposed to be in the microsomes. The fractional representation did not match the partition of the catalytic processes. For the time being, the enzymes remained without a defined existence in the experimental space.

In a review in 1958, Hoagland depicted four stages of dissection of the rat liver protein synthesis system (see Figure 10.10). The chart in this figure superposes the fractionation pattern and the intermediate steps of the biochemical reaction chain. The transition of stage II to stage III had

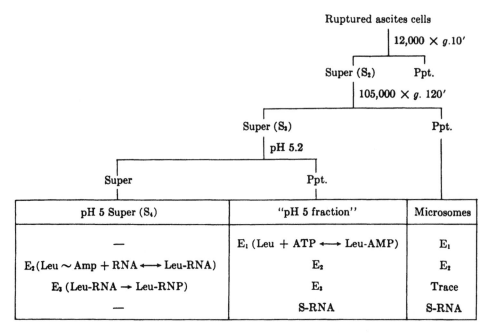

Fig. 10.9. Location of enzymatic components of the amino acid incorporation system. Ppt., precipitate; E_n, enzymes. Reprinted from Zamecnik, Stephenson, and Hecht 1958, fig. 4.

been made possible through the introduction of the ATP-phosphate exchange procedure. The prominent feature of stage III, the conversion of microsomes into functionally active RNP-particles, had been made possible by switching to another model organism, the Ehrlich ascites tumor cells, and by introducing high-salt washing procedures for purification.[110]

A Question-Generating Machine

Soluble RNA (stage IV in Figure 10.10) had been an outstanding event in the stepwise resolution of the process. But despite the efforts of Novelli and Berg,[111] the existence of aminoacyl-adenylates in a biological system still lacked direct evidence as manipulable compounds. The RNP-particles obtained so far were shaky. The system at MGH still did not respond to a full complement of amino acids. Was there a different enzyme for every amino acid? Was the coprecipitation of S-RNA and enzymes at pH 5 a fortuitous event? How large was S-RNA? How were the amino acids connected to the RNA? Was there more than one amino acid bound to a single S-RNA molecule? Was there a special transfer enzyme involved in the transport of the aminoacyl-RNA to the RNP-particle? Was GTP connected to such an enzyme? How did S-RNAs and RNP-particles interact?

The questions abounded. The protein synthesis system was en route to assuming the quality of a huge research attractor, of a high-speed future-generating machine. Besides some crude outlines, everything was open to investigation. At the same time, the fragmentation of the system promised a rapid clarification of all sorts of open questions. Moreover, it permitted the *production* of questions. In Hoagland's words, "the fact that we have raised more questions than we are able to answer encourages us to think that our studies are leading us to a clearer understanding of protein synthesis."[112] In a note added in proof to his review Hoagland listed no less than seven "new developments" that occurred after he submitted the manuscript in January 1958 and before he received the galley proofs.[113] The paper appeared in the July issue of *Recueil des Travaux Chimiques des Pays-Bas et de la Belgique*.

At the Huntington Laboratories, the coding problem remained one option among others. The situation was comparable to a battle on multiple fronts. The relation between the attachment of nucleotides and amino

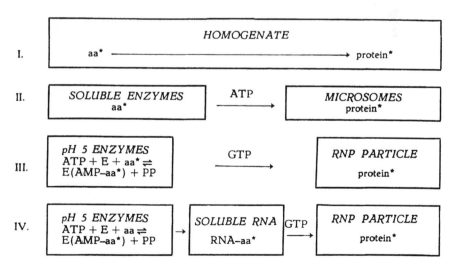

Fig. 10.10. Stages in the dissection of the rat liver cell-free system for the incorporation of amino acids into protein. Reprinted from Hoagland 1958, fig. 1.

acids to S-RNA was unsolved. The enzymes catalyzing the different reactions had to be sorted out. An S-RNA specific to a particular amino acid had not yet been isolated. Experimental progress, Zamecnik judged, would depend on the pursuit of all these options.

So far, the transfer of radioactive amino acids from the soluble RNA to the microsomes was defined in purely operational terms: it was measured as the amount of radioactivity cosedimenting with the microsomes.[114] No other differential signal could be produced from the interaction of microsomes with S-RNA. There was a transfer, and transfer meant cosedimentation.

Concerning the S-RNA molecule itself, some interesting details emerged. Loftfield had begun to investigate the accuracy of the amino acid-RNA attachment.[115] Clearly, the attachment of amino acids to S-RNA was a selective step in the pathway from amino acids to protein. Concerning the nature of the linkage between S-RNA and amino acid, a list of "what it does" and "how it behaves" had been assembled over three years. All the observations were compatible with, but not demonstrative of, a link between the amino acid and the 2′ or 3′ hydroxyl group of the terminal ribose of the RNA. With respect to the chemistry of that link, Lipmann's laboratory won the race—at least in publishing the results.[116] In January 1958, Zamecnik had had a conversation with Crick and Lipmann during a meeting at the Waldorf Astoria, New York, in honor of Basil O'Connor's sixty-fifth birthday. During breakfast, "Lipmann became instantly curious, and asked where I thought the covalent attachment might be. I replied that there were three choices on the 3′ terminal adenylyl residue: (1) the 2′ or 3′ ribosyl position, (2) somewhere on the adenine ring, and (3) the first internucleoside phosphate position. I said we were going to try to distinguish among these possibilities. Unknown to me, Lipmann's laboratory then immediately undertook the same quest. We both reached the same conclusion—the 2′ or 3′ ribosyl location."[117] Hans Zachau and Lipmann had used ribonuclease to split off the terminal aminoacyl-adenosine, Zamecnik and his colleagues had used periodate to distinguish protected from unprotected 2′ or 3′ hydroxyl groups of the terminal ribose.[118] At the same time, in the course of an attempt to purify a methionine-activating enzyme, Paul Berg and James Ofengand had come to the conclusion that amino acid activation and the attachment of the amino acid to its S-RNA was catalyzed by one and the same enzyme.[119]

Code, Template—So What?

Let me summarize the state of the art with respect to S-RNA at the end of 1958, after three years of a more and more competitive and explosive search. All the different RNAs appeared to have the same terminal nucleotide sequence: -CCA. Obviously, this sequence was a general, either structural or functional feature, but not a specifying one—anything but a code. On the other hand, every one of the twenty or so amino acids appeared to have its own S-RNA carrier. Thus, Zamecnik conjectured that "each RNA molecule *coded* in some fashion for a specific amino acid, and in turn perhaps *coded* for a specific, complementary site on the ribonucleic acid portion of the ribonucleoprotein particle which serves as a protein synthesizing *template*."[120] "Coding" became the term for two different processes. Zamecnik began to see S-RNA coding for amino acids and ribonucleoprotein-RNA in turn coding for S-RNAs. The soluble RNA was considered to carry the recognition signal for a specific amino acid; the ribonucleic acid of the microsome was considered to act as the template that specified the sequence of a particular protein. The language of molecular information transfer began to inscribe itself into the metabolic representation of protein synthesis. But it neither did nor could replace the biochemical framework immediately. It delineated the contours of a different space of representation, but this space did not yet have its proper experimental embodiment. Why not? Mainly because the template was identical with the RNA of the ribonucleoprotein particle in terms of experimental existence. Manipulating the template meant manipulating the microsomal particle. This particle was the component that was most difficult to handle. I shall return to the detachment of "template" from "microsome" as a prerequisite for experimentally tackling the code in Chapter 13.

CHAPTER 11

Historiality, Narration, and Reflection

❖❖❖   [One] would no longer be able to call it "origin" or
"ground."
             —Jacques Derrida, *Of Grammatology*

My remarks in this chapter revolve around an epistemology of time to
which I have already alluded on several occasions in this book. The title of
the chapter avoids the notion of history or historicity. Instead, I speak of
"historiality." Why this apparent game with words? Moreover, how to
speak about an issue that, provisionally and some 30 years ago, Derrida
approached in his *Of Grammatology*, a text that since then has become the
locus classicus of deconstruction? The issue and the question is: how can
we speak of history without invoking "origins" and "grounds"? "An un-
emphatic and difficult thought that, through much unperceived mediation, must carry the entire burden of our question, a question that I shall
provisionally call *historial* [historiale]."[1] Historians of science, like historians of other cultural systems, are constantly confronted with an unsur-
mountable obstacle. What haunts them is the specter of reflexivity: What
are they dealing with? Are they looking at a past that is the transformation
of another, foregoing past? Or are they looking at a past that is the product
of a past deferred, that is, of a future present?

A Glimpse at an Epistemology of Time

Early in 1991, I happened to be with a group of historians of biology
assembling at the Natural History Museum in Berlin (East Berlin less than
two years before). A paleontologist who had spent his whole professional
life ordering and reconstructing fossil material pointed to the famous
Solnhofen Archaeopterix of the Berlin collection and summarized the

experience of 40 years of work with the following words: "At the point of the emergence of the new, the new is not the new. It becomes a novelty only by a transformation which makes it a trace of something to which it has given rise."[2] What an unwitting, but inescapable, irony lies hidden behind this coincidental complicity of political history and paleontological practice! As a rule, new developments are at best an irritation at the point where they first appear: they can be approached only in the mode of a future perfect. Of course, we may try to unearth the conditions of their emergence. But these conditions, and so the new, seem accessible only by way of a recurrence that requires the existence of a product as a prerequisite for assessing the conditions of its production. Who would have predicted the collapse of the Berlin Wall even in the fall of 1989, when I left West Berlin—where I had lived for more than twenty years—for a sabbatical at Stanford? Such recursive assessments hold for the monuments of natural history, as well as for the documents of the history of science. The same may be said of "all the new varieties of scientific thought, which, after all, come to project a recurrent light on the obscurities of knowledge unaccomplished."[3]

According to Georges Canguilhem it is exactly at this point that the roads of the historian, in the positivistic sense, and of the reflexive epistemologist, in the sense of Bachelard, divide:

The historian proceeds from the origins toward the present in such a way that today's science is always to a certain degree announced within the past. The epistemologist proceeds from the actual toward its beginnings in such a way that only part of what took itself as science yesterday finds itself founded within the present. So, in founding—never of course for ever but over and over again—the science of today also destroys—forever.[4]

The movement of the historian is sometimes criticized as whiggish in its spontaneous form, or else it is denounced as an idle search for forerunners, according to whether the historian looks for "grounds" or for "origins." The movement of the epistemologist calls for and implies an inevitable bending of thought, which is often misunderstood as a form of teleological retrojection. Attempts at its linearization constitute the classical illusion of the task of a historical narrative: to tell—whatever methodological refinements are added to the core of the argument—the story of what, then, really happened. Such an attitude presupposes the existence of an undistorted past "behind there" that, from the present viewpoint of a detached spectator, can be grasped in principle by means of an analysis whose results are supposed not to be pollinated by what is going to be retold.

Within the frame of art history as a history of formal sequences of things, George Kubler insisted on pointing to what André Malraux called the "Eliot effect"—which is here referred to as recurrence: "Every major work of art forces upon us a reassessment of all previous works."[5] German scholars might add that Goethe has addressed this relationship on several occasions with respect to the sciences when speaking of a "provisional rearrangement" that may become necessary "from time to time," and even of a "rewriting" of the course of science in the aftermath of path-breaking events.[6] To dismiss this rearranging force is to perpetuate the illusion that the task of the historian is to tell "real history" as opposed to telling stories.[7]

Historial thinking, in contrast, not only has to accept such recursive forces as being inherent in any hindsight—and hence, in the interpretation, or iterative action on the part of the historian. From a historial point of view, we also have to assume that recurrence, in terms of rearrangement and reorientation, is at work as part of the time structure of the innermost differential activity of the systems of investigation themselves. What we call their history is deferred in a constitutive sense: the recent is made into the result of something that did not so happen. And the past is made into a trace of something that had not (yet) occurred. Such is the general temporal structure of the production of a trace. "The trace is not only the disappearance of origin—within the discourse that we sustain and according to the path that we follow it means that the origin did not even disappear, that it was never constituted except reciprocally by a nonorigin, the trace, which thus becomes the origin of the origin."[8]

History of science has a long tradition of viewing at least modern science as a continuous, unified, and cumulative undertaking. In our century, this view has seriously and lastingly been challenged by models that see the course of science as a sequence of punctuated equilibriums, marked by a series of more or less radical breaks. However, both the revolutionary and the gradual, the discontinuous and the continuous, conception of scientific change generally share the assumption of a global epistemic structure, called "The Science," which as a whole either asymptotically grows—toward truth—or is periodically reconstructed according to a new paradigm. Although there is a pretty heavy load of relativism in the second view, especially Thomas Kuhn's, a paradigm at a given time is still assumed to have enough power to coordinate and make coherent the activity of a whole—and potentially *the* whole—scientific community. Even in the denial of a continuum of rationality and of commensurability as far as the time axis is concerned, there remains an element of totalization. Science remains a process whose norms and values govern the ensemble of its actors and their practices in a common secular endeavor. And associated with this endeavor remains the general view of an overarching chronological coherence stamped by a handful of epistemological breaks.

The closer we look at the microdynamics of scientific activity, the more problematic these views seem. In recent years, Kuhn too has come to stress not only the diachronic incommensurability of paradigms but also the synchronic incommensurability of bits and pieces of the ever more compartmentalized scientific workshop, thereby characterizing the whole scientific enterprise as a process of proliferation.[9] We note that this is yet another way of speaking about history without grounds and for that reason without telos. From the perspective of assessing the scientific research process at the level of the basic, functional units of scientific activity—experimental systems—it is clear that the monolithic macroscopic appearance of "The Science" is decisively and sustainably subverted.

At issue is the fragmentation of the sciences into disunified areas,[10] into nonoverlapping clusters of experimental systems with their corresponding time characteristics. I am not advocating the introduction of a distinction here that has pervaded late-nineteenth- and early-twentieth-century physics, that is, the distinction between the microscopic indeterminacy of the elements of a system and its macroscopically determinate phenomenological parameters. This would be too easygoing an analogy for what I have in mind. Nevertheless, I do not want to exclude any resonance with the notions of uncertainty and indeterminacy that have come to undermine the scientific self-perception of a whole epoch. Moreover, I do like to invoke a resonance with recent developments in the field of the thermodynamics of irreversible processes. Among others, Ilya Prigogine has shown that new possibilities arise from modeling dissipative structures for conceiving of what might be called localized and situated time. Prigogine has suggested defining time, not simply as a parameter (the little t of Newtonian-to-Einsteinian physics), but introducing an "operational" time into the modeling of irreversible processes—that is, determining time as an operator (capital T).[11] Viewed formally, an operator is a prescription for manipulating, that is, for reproducing a function such that the function itself survives the operation but at the same time is changed by some factor or factors. The proper context of thermodynamics is not relevant to the present discussion. Of relevance is the idea of "operator"-time, or internal time on a deliberately metaphorical level. Let us assume that with respect to the movement of material systems, systems of things, or systems of actions, time can be viewed as an operator and not simply as a chronological axis of extension in a system of coordinates. In this sense, time is a structural, local, and intrinsic characteristic of any system maintaining itself in a stationary state far from thermodynamic equilibrium and reaching from time to time, as a result of turbulences, a point of bifurcation.

Thus, every system of material entities, and therefore every system of actions concerning such entities that can be said to possess reproductive qualities, may also be said to possess its own intrinsic time. Internal time is not simply a parameter of the system's existence in space and time. It characterizes a sequence of states of a system insofar as it undergoes continuing cycles of nonidentical reproduction. Research systems, with which I am here concerned, are characterized by a kind of differential reproduction by which the generation of previously unknown things through unprecedented events becomes the reproductive driving force of the whole machinery. As long as this movement goes on, we may say that the system remains "young." Being young, then, is not a result of being located near zero on the time scale; it is a function—if you will—of the very functioning of the system. The age of the system is measured by its capacity to produce differences that count as unprecedented events and keep the machinery going. In a quite similar vein, Kubler has described artistic activity "as a linked progression of experiments," whose "characteristic spans and periods" cannot be grasped by mere "calendaric time."[12]

If there is any pertinence to such a tentative and somewhat formalistic description, we can look at the events in a particular research field as an assembly, or an ecological patchwork, of experimental systems, each with its own time requisites. Some of them are close enough so that their reproductive cycles can become operationally coupled by exchange of subroutines, of epistemic entities, and of tacit knowledge built into their performance. Others are far enough from one another to have their operational transformations performed independently, which in itself is a matter of the actual transformations going on within the different systems. Thus, we end up with a field of systems that has a rather complex time structure, or shape of time. The systems or reproductive series retain their own "ages" as long as they differentially replicate, and the epistemic field can no longer be seen as dominated by an overarching chronotopic paradigm. This makes an ensemble of experimental systems similar to a field of discursive practices in the sense Foucault has attributed to this expression, with the strength of discursive couplings between experimental systems constantly changing through space and time. Treating these systems as "regular series and distinct events" should permit us "the introduction, into the very roots of thought, of notions of *chance, discontinuity* and *materiality*."[13] Basically, there is no all-encompassing theoretical framework, no overarching political program, no homogenizing social context effective enough to pervade and coordinate this universe of drifting, merging, and bifurcating systems. Where the systems do get linked, the links do not form stable connections; rather, transient interfaces are generated by the differential reproduction of the systems and the constellation of their ages. There is no common ground, source, or principle of development from which they would all spring, no hierarchy in which they would all be encapsulated. The constitution and constellation of differently aged experimental systems as a whole is u-topic and a-chronic. It is a de-centered reticulum, a rhizomic structure in which connecting capillaries and anastomoses constantly are formed and dissolved, and where attractors permanently shift.

The multiplicity of experimental systems endowed with their own times and, of course, their rationales, shifting and drifting in an open horizon, constitutes a historial ensemble. Such ensembles escape the strong notions of social history such as linear causation, retroaction, influence, dominance, and subordination. They also escape the notion of a purely contingent or stochastic process. The term *history* has been connected with both clusters of notions, either by law or by singularity. Instead, I propose to follow the traces that will be transformed as time goes on and that will create, through their action, the origin of their nonorigin. This constitutive belatedness, or deferred action, that "knotting of time" is inscribed into the character of a trace. For it has to de-double in order to become what it has been. According to this temporality, "the 'after' becomes constitutive of the 'before.'"[14] It is to the macroscopic effects of this microscopic structure that we can attach the notion of the unheard-of, or scandal. All history of science consists in the vain effort to prevent the scandal in bringing it about. The conquest of the origin, in a gesture that is as hallucinatory as it is inescapable, remains bound to the trace that it will have left. Therefore, there can be no once-forever canonical history, as there can also be no global foresight. François Jacob has mercilessly expressed this situation in the chapter entitled

"Time and the Invention of the Future" at the end of his essay on *The Possible and the Actual*:

What we can guess today will not be realized. Change is bound to occur anyway, but the future will be different from what we believe. This is especially true in science. The search for knowledge is an endless process and one can never tell how it is going to turn out. Unpredictability is in the nature of the scientific enterprise. If what is to be found is really new, then it is by definition unknown in advance. There is no way of telling where a particular line of research will lead.[15]

With "differential temporality," we are further than ever from the romantic illusion of history as an all-pervading "totality" dominated by mimetic, metamorphotic, or "expressive" relations of the parts within an ensemble.[16] The figure of differential reproduction of serial lines nestling and wrestling in a landscape of research activities creates another, perhaps no less pervading, but much more fragile and productive coherence. This coherence is no longer based on the simultaneity or the successive deployment of all possible metamorphoses of an "ur-form," or paradigm. The structure of this coherence is based not on expression, reflection, or mirroring but on the tensions of an ecological reticulum, on resonance and dissonance in a patchwork of precocious and deferred actions. Its extinctions and reinforcements, interferences and intercalations, resist a unifying time of history. If we take this statement seriously, then we have to cope with and envisage the in-principle impossibility of an algorithm or a logic of scientific development that in its historical course is causally grounded.

Under such conditions, there is little place left for telling the history of the sciences as a process being influenced, directed, hampered, or promoted by whatever instances, be they internal or external, simply because there is no preexisting "outside" here to a preexisting "inside." Outside and inside are everywhere. Topologically speaking, we are dealing with a Moebius strip; geometrically, with a fractal constitution. We could characterize the inside/outside view of the history of science as a Lamarckian one, and it is astonishing how long it has survived its biological counterpart, that is, an account of evolution along the lines of what we are accustomed to labeling "Darwinian." Thinking along the latter lines, the epistemologist, in his or her realm of empirical investigation, has to account for contingent events, which result in a scattered field of variants that establish their own filtering regime on the basis of their finite possibilities of extension.

Once Again: Differing/Deferring

Can we, if only metonymically, come back to Derrida's notion of *différance* in approaching the dynamics of research fields? Let us take *différance* as an "economic concept designating the production of differing/deferring."[17] In a paper elaborating on this notion, Derrida explains:

In the delineation of *différance* everything is strategic and adventurous. Strategic because no transcendent truth present outside the field of writing can govern theologically the totality of the field. Adventurous because this strategy is not a simple strategy in the sense that strategy orients tactics according to a final goal, a *telos* or theme of domination, a mastery and ultimate reappropriation of the development of the field. Finally, it is a strategy without finality, what might be called blind tactics, or empirical wandering if the value of empiricism did not itself acquire its entire meaning in its opposition to philosophical responsibility. If there is a certain wandering in the tracing of *différance*, it no more follows the line of philosophical-logical discourse than that of its symmetrical and integral inverse, empirical-logical discourse. The concept of *play* keeps itself beyond this opposition, announcing, on the eve of philosophy and beyond it, the unity of chance and necessity in calculations without end.[18]

A strategy without finality, at first glance, is a *contradictio in adjecto*. We might well need another expression to convey the sense of a movement that is not goal directed, but nevertheless anything but chaotic for that reason. The rationale of such "wandering" is well expressed in an aphorism of Goethe: "One never goes further than if one does not know where one is going."[19] This movement is intimately linked to the nature of the means by which an experimental texture gets scribbled and written over and over again. It is in the nature of these means, materials, graphematic entities, that they are in the capacity of an excess. They contain more and other possibilities than those to which they actually are held to be bound. The excess embodies the historial movement of the trace: the trace transgresses the boundaries within which the game appears to be confined. As an excess, it escapes the definition of the system. On the other hand, it brings its boundary into prominence by cutting breaches into it. It defines what it escapes. The movement of the trace is recurrent.

In describing a field of experimental systems as pervaded by *différance*, this point is crucial. It stresses that the systems in question undergo a play of differences and displacements governed by their operator-times and at the same time that they displace what at any given moment appear to be their boundaries. These *décalages* and displacements imply that other experimental systems impinge on each other and take on meaning in terms of these shifts. What goes on, then, could be characterized by the concept of "grafting" as invoked by Derrida—once more—as a model of working on and with textual structures. "One ought to explore systematically not only what appears to be a simple etymological coincidence uniting the graft and the graph (both from the Greek *graphion*: writing instrument, stylus), but also the analogy between the forms of textual grafting and so-called vegetal grafting, or even, more and more commonly today, animal grafting."[20] Interestingly, and probably not by chance, the notion is derived from a biological background. Grafting keeps alive, as a support, the system upon which one grafts. At the same time, it induces the supporting system to produce, not only its own seeds, but also those of the supported graft. On the one hand, the relation of graft and support is that of a tight insertion; on the other hand, it is the continuation of a manifest separation. The graft is an inverted excess: an intrusion. It brings the boundary into existence by transgressing it in the reverse direction. But its very functioning as a graft also shows the feasibility of the support to be intruded. Thus, there is a fundamental complicity. What is inside and what is outside? This ceases to be a question that can be answered in any meaningful way for the process at issue. The adventure of grafting as a special form of iteration lies not in that it leads to progress—it leads to "pro-grafts."

Research processes are, much like the concepts by which they are accompanied, processes and concepts "in flux."[21] At every step, what is about to take shape creates unforeseen and alternative directions for the next step to be taken. Experimental systems concatenate into a constantly changing signifying context. Experimental systems oscillate around epistemic things that escape fixation by transplanting and grafting new methods, instruments, and skills into the setup; or, by altering the location of their embodiment, the systems themselves, constantly shift their boundaries. There is, as a rule, no unique perspective that could account for the research movement with all its possible turns, no definite direction to its "blind tactics," its "empirical wandering."

Recurrent Narration

Recombination and reshuffling, bifurcation and hybridization within and between experimental systems, are prerequisites for producing unprecedented events. Such events could not happen if the lines of descent were bred too "pure." The historial movement of the *différance* is always impure; it is a hybrid creation, it works by transplantation. So is the movement of narration. If *we* historians want to know what a particular epistemic thing represented at a given time, the material signifiers of the experimental game will already have turned into something that, at the time, they could not (yet) have been. Canguilhem rightly warns the historian: "The past of a science of today should not be confounded with that science in its history."[22] The transitory significance organizing such recurrence belongs to successively different spaces of representation. Being "*dans le vrai* (within the true)" of a science at a given time and thus "obey[ing] the rules of some discursive 'policy' " means something radically different from "speak[ing] the truth in a void," as Foucault has put it in referring to Canguilhem.[23] "Within the true" of a particular ongoing piece of research there exist always only the minimal constraints for the coherence of a significant chain. Those who work on it endow it with the dignity of a scientific object. Thus, what André Lwoff of the Pasteur Institute in Paris said about viruses in 1957 is not to be read as a tautological joke, but points precisely to the argument I am trying to make: "The conclusion of this lecture will be prosy, coarse, and vulgar: viruses should be considered as viruses because viruses are viruses."[24] Here, the signified has been crossed out, and the referent itself has become a signifier. Narration has always lived from such replacements. Due disenchantment with an elusive referent in science may help us to get acquainted with the consoling message that science, too, does not escape the space of narrative.

With that, I come back to my remarks concerning recurrence in the history of science. Scientists tend to account for their discoveries in the frame of what one might call the "spontaneous history of the scientist."[25] "Woe betide us," I exclaim with William Clark, "if historians of science plot the very same fictions as the scientists do."[26] In the spontaneous history of the scientist, the most recent developments tend to take on the meaning of something always already present, albeit hidden, as *the* research goal from the beginning: a vanishing point, a teleological focus. But as we have seen, the new is not the new at the beginning of its emergence. That does not mean, however, that we should look at the recollections of scientists as a mere idealization, or even a malevolent distortion. The retrospective view of the scientist as a spontaneous historian is not only concealing but in many respects also revealing. It reminds us that an experimental system is full of stories, of which the experimenter at any given moment is trying to tell only one. Experimental systems not only contain submerged narratives, the story of the repressions and displacements of their epistemic concerns; nor, as long as they remain research systems, have they played out their potential excess. Experimental systems contain remnants of older narratives as well as shreds and traces of narratives that have not yet been related. Grasping at the unknown is a process of tinkering; it never proceeds by completely doing away with old elements or introducing new ones ex nihilo, but rather by removing and reorienting given elements by an unprecedented concatenation of the possible(s). It differs/defers. If the scientist, as a spontaneous historian, presents the latest story as the one that has always already been told, or at least has had an effort made to be told, this is not a deliberate dissimulation; it reflects a process of marginalization of earlier concerns that is borne into the ongoing research movement itself. But, on the other hand, it reflects the rebuilding, the replacement, the patching, the brushing aside—in short, the deconstruction of the research meandering, as a carefully crafted construction; and it thus remains within the demiurgic illusion that is inherent in philosophical constructivism as well. In the spontaneous history of the scientist, the present appears as the straightforward result of a past pregnant with what is going to be. Strangely enough, in a kind of double reversion, spontaneous histories inevitably also present the new as the result of a prehistory that did not so happen. The historical narrative, without realizing it, both obeys and discloses the figure and the signature of the historial.

In the remaining chapters I struggle with and against this inevitability. "Transfer RNA," "ribosomes," and "messenger RNA" were entities that did not shape the discursive framework of protein biosynthesis from its beginning. Once these entities were established, however, they changed the experimental and intellectual practice of doing protein synthesis in the test tube to such an extent that, a few years later, a newcomer to the field had difficulties understanding what the generation before her or him had been talking about. Those who had participated in the molecular biology turn came to memorize what they had done before along the lines of later achievements. Concomitantly, "transfer RNA," "ribosomes," and "messenger RNA" turned "soluble RNA," "microsomes," and "templates" into their precursors, making them links in a chain of transformations that endowed the new entities with the dignity of being the reward of a long-standing desire.

CHAPTER 12

Names Matter: Transfer RNA and Ribosomes, 1958–61

❖ A formidable intellectual machine, Francis Crick played a
major role in the development of molecular biology. He
had no taste for experimentation, for manipulation.
—François Jacob, *The Statue Within*

With the emergence of a small, soluble RNA, the experimental "machine for making the future" of the group at MGH and the "intellectual
machine" of the Cambridge connection, especially Francis Crick, had
come to interact. However, this interaction was by no means a matter of
perfect matching. Many of Crick's suggestions failed experimentally, and
there was, as we have seen, a considerable hesitation on the part of the
Boston researchers to endorse a molecular biologist's view of soluble
RNA. Their "finding" was embedded in a research program saturated
with considerations of biochemical energy flows and the search for intermediates of metabolic reaction chains. Reasoning in terms of molecular
information transfer, therefore, did not and could not restyle the system as
a whole and at once. On the contrary, it invaded the experimental space
locally and partially: as a possibility of rearranging the research agenda
after the event, as a theory in the operational and practical sense of the
experimenter's use of the term, that is, a hypothesis bridging gaps between "facts." But even this did not work from the beginning. As long
as the informational perspective remained without experimental packaging, it acted as a supplement to the established biochemical space of
representation. More important perhaps, many questions remained to be
answered within the established framework. Local supplementation of
biochemical reasoning by informational vocabulary was therefore characteristic of the state of affairs at MGH when S-RNA began to assume the
quality of a research attractor. It is in the nature of a supplement that it
joins a system, but, as a joint, it remains foreign to it: an addendum.

Experiment and Orientation

In May 1959, Zamecnik had been invited to deliver a Harvey Lecture. He chose to speak about "Historical and Current Aspects of Protein Synthesis." He traced history back to "careful, patient studies" extending, as he said, "over half a century."[1] He began with Franz Hofmeister and Emil Fischer, who recognized the peptide bond structure of proteins, went on to Henry Borsook, who realized the endergonic nature of peptide bond formation, and then to Fritz Lipmann, who postulated the participation of a high-energy phosphate intermediate in protein synthesis. He traced these studies further back to Max Bergmann, who determined the specificity of proteolytic enzymes; to Rudolf Schoenheimer and David Rittenberg, who pioneered the use of radioactive tracer techniques in following metabolic pathways; to Jean Brachet and Torbjörn Caspersson, who became aware of the possible role of RNA in protein synthesis; to Frederick Sanger, who unraveled the first primary structure of a protein to show the specificity and uniqueness of the amino acid composition of insulin; and finally to George Palade, who provided visual evidence of the particulate structures in the cytoplasm that acted as the cellular sites of protein synthesis.[2] This is an impressive list of pioneers, who all, according to Zamecnik, "blazed the trail to the present scene," which in retrospect and quite inadvertently had assumed the character of a royal path to present knowledge. It was not until the very end of the lecture that Zamecnik came back to what really matters: "From a historical vantage point, too simple a mechanistic view of it [protein synthesis] has been taken in the past. [The] details of the mechanisms at present unfolding were *largely unanticipated*." How, then, was the unanticipated brought into being? The answer can be read as an homage to the unknown soldier: "By the direct experimental approach of the foot soldiers at work in the field."[3] But does that not render questionable the field marshal's report of how his generals won the battle? According to the report made to headquarters, they proceeded step by step and at every move made sure they received support from the surrounding allies. The history of quirks and breaks I have been trying to tell here tends to disappear between the lines. What constitutes the open-ended dynamics of the experimental process has shrunk to a final remark: the thwarted plans, the chaotic moves at the front line; the intrusions, displacements, and recurrences; the reinforcements; the relief attacks, running fights, and sudden advances; in short, everything that makes up the work on the battlefield was rendered invisible. The scientist tells his or her story from the point of view of those selected insights that have been crowned with success.[4]

Names Matter: Transfer RNA and Ribosome

Two new terms were making their way into Zamecnik's narrative. Both of them had been coined by others and began to leak into the discourse of protein synthesis. The RNA to which the activated amino acids became attached had originally been dubbed pH 5 RNA ("soluble RNA"), according to the fraction of the protein synthesizing system in which it had shown up: in the soluble supernatant fraction and its pH 5 precipitate, respectively. We are inclined to look at these terms, with Hoagland, as being "operational" rather than "functional."[5] Quite surprisingly, Zamecnik referred to them as reflecting a "clear-cut biological property."[6] *His* "biology" of the process was its division into fractions that represented the components of the system. Given this apparent identity of fractional representation and biological meaning, there was no need for another term. Nevertheless, when in 1959 Richard Schweet proposed the notion of "transfer RNA,"[7] which reflected the amino acid carrier function of the molecule as well as its role in the transfer of genetic information, it took over and gradually came into general use. Although Zamecnik found no "overwhelming evidence" for this "interpretation," he accommodated himself to the "apt designation," as he called it, and began to use it himself.[8] Why these philological details? They exemplify Zamecnik's reserve toward a vocabulary that was in the course of intervening in the representational framework of protein synthesis. They also make clear that labels are not usually neutral designations. They feed back on the renamed entity and finally on the whole network in which the entity is embedded. Instead of an "interaction" between "microsomal RNA" and "soluble RNA-amino acid," we have now a "transfer RNA" that brings the amino acid to an "acceptor site" on the ribonucleoprotein particle RNA, thereby transferring its amino acid to the growing peptide chain.[9] Although the process was visualized with exactly the same drawing of a year before, we can now view it in a different light (see Figure 12.1).[10]

The second term making its way into Zamecnik's narrative concerned the protein synthesizing particle. It already had a remarkable career behind it, starting as a sedimentable entity invisible in the light microscope—

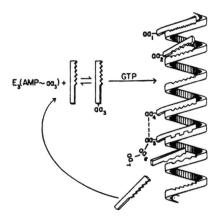

Fig. 12.1. Scheme for the interaction of microsomal RNA and soluble RNA–amino acid. Reprinted from Hoagland, Zamecnik, and Stephenson 1959, fig. 2; Zamecnik 1960, fig. 5.

the "microsome"; continuing as a granular cytoplasmic constituent that was "deoxycholate-insoluble" or "salt-insoluble"; and mutating into a "ribonucleoprotein particle" consisting of half protein and half RNA and visible under the electron microscope as a dense granule.[11] Again, these labels translated the technical constraints of widely different representations of this subcellular constituent. Gradually, its RNA moiety had attracted more attention. By the middle of the 1950s, researchers generally assumed that microsomal RNA provided the template upon which the amino acids were assembled into protein threads. Around 1958, Howard Dintzis and Richard Roberts cast the term *ribosome*, which quickly made its way into the laboratories and into the literature. It conveniently distinguished between purified ribonucleoprotein particles devoid of reticular membrane fragments and crude microsome preparations that, of course, also continued to be investigated.[12] Although the biological reasons for changing the name of the RNP particle were somewhat obscure, the new designation clearly no longer reflected a mere technical representation but a biological function linked to RNA. Like "transfer RNA," the "ribosome" began to subvert the biochemically characterized protein synthesis system as part of what Crick had called the central dogma of molecular biology, which subsumed protein synthesis as the final step in the overarching process of gene expression.[13] There are no noncommittal terms in science.

The Adaptor Impasse

With respect to the coding problem and the "adaptor hypothesis,"[14] Zamecnik had the following to say in that same 1959 Harvey Lecture:

> We have most lately been concerned with the possibility that at least a portion of the soluble RNA molecule to which the amino acid is attached is transferred along with the amino acid to the ribonucleoprotein particle, aligning itself in some base-pairing arrangement with the microsomal RNA, prior to formation of a peptide chain. This concept agrees with the proposal of Crick that the soluble RNA molecule may serve as an adaptor in a base-pairing arrangement which determines [the] amino acid sequence.[15]

What had been found in the Huntington Laboratories, in fact, did not look very much like Crick's adaptor. Crick had envisaged it as a trinucleotide representing the sort of code he had proposed to explain how a piece of RNA might convey sequence specificity to a protein. The Bostonian catching device for amino acids was much larger, perhaps even 45 to 60 nucleotides long, and the overall base composition of microsomal and soluble RNA was distinctly different, in that S-RNA had a high content of nonstandard bases.[16] One feasible way to "adapt" Crick's adaptor to the properties of S-RNA had been to have the latter break down into (a) small amino acid-activated fragment(s) before it assembled on the ribosomal template. On the other hand, Liza Hecht's invariable CCA-end was far from being a distinctive code![17] Thus, if there was a coding portion of the molecule, it had to occupy a "situation more central than the three (or alternatively two) terminal residues common to the whole family of transfer RNA molecules."[18] Why was S-RNA so bulky? There was another problem Zamecnik called attention to: the molecule had to be recognized by an enzyme attaching the correct amino acid, and for this recognition step, Zamecnik guessed, "a fairly large number of mononucleotides" might be necessary.[19] In any case, the bulkiness of S-RNA made it an unlikely candidate for a Crick-type adaptor fitting a code of adjacent triplets.

Hoagland and Zamecnik assumed, as Figure 12.1 suggests, that the process of adaptation took place along an oligonucleotide stretch of an RNA helix, the amino acids condensing along the vertical axis of a helical template. When Hoagland showed this drawing for the first time at a conference in 1957, he had chosen the flattened helix to accommodate a bulky adaptor. At that time, it was the recognition process between the enzyme and S-RNA that he conceived of as a triplet function: "Thus the sequence *AGU*, for example, reacts only with activating enzyme 1; *GAC* with enzyme 2; and so on up to 20."[20] When Zamecnik presented the same figure in his Harvey Lecture in 1959, he considered the "choice of the helix in depiction of the acceptor site" to be "of no particular significance." Just the other way round, he now assumed that the coding units were smaller and the enzyme recognition units larger. The problem of bulkiness remained, however: "According to Crick's hypothesis, and to our views, the soluble RNA molecule appears to be too long and complex to serve in its entirety as a suitable transfer agent."[21]

From the viewpoint of Crick's adaptor hypothesis, small active amino acid-carrying fragments of RNA would have been the ideal solution. During the Molecular Biology Symposium at the University of Chicago in March 1957, Hoagland had already announced the transfer of an active fragment: "Preliminary experiments [do] suggest such a limited transfer."[22] Vexed by Crick's adaptor hypothesis, Hoagland and Zamecnik tried for almost two years to find a small S-RNA fragment—without success. They got stuck in an impasse. A conjuncture is not something that just happens.

Template, First Code, and Second Code

Let us stay with the template for a moment. As we have seen (see Chapter 8), the early attempts to account for the specific assembly of amino acids into proteins had all been based on the assumption of a direct physicochemical interaction between the amino acids and some template— not necessarily RNA or RNA solely.[23] According to the status of the adaptor hypothesis in 1959, this solution was no longer plausible. Now the interaction was split in two parts: first, a *chemical reaction* between an amino acid and a small oligonucleotide and, second, a *base-pairing interaction* between the adaptor oligonucleotide and the template RNA.[24] But, finally, what was a template? For some ten years, the template had been figured out as a rigid matrix onto which the amino acids could settle. The activated amino acids would align themselves along the template RNA as shown in Figure 12.2.

According to Figure 12.2, the broken signs representing the code are restored along the full length of the template thread. The experiments, however, gave "the *appearance* of a steady state phenomenon," that is, of a

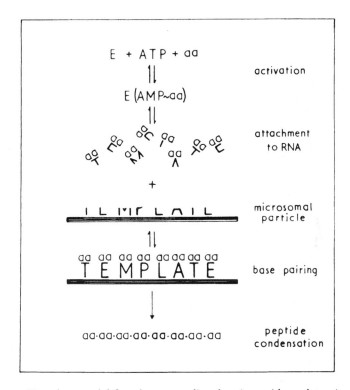

Fig. 12.2. Template model for adaptor-mediated amino acid condensation. Reprinted from Hoagland 1959a, fig. 1.

sequential process.[25] A small, but constant amount of aminoacylated transfer RNAs remained associated with the microsomal RNA, whereas the amino acid incorporation into protein went on in a linear fashion. This finding suggested a much more dynamic picture of the whole transfer reaction.[26] Once more, the experimental representation was at odds with its pictorial counterpart, let alone its conceptualization. The acknowledged template hypothesis obscured rather than elucidated the experiments. The experiments suggested a bulky adaptor that turned over quickly. The hypothesis dealt with small adaptors that assembled simultaneously. The adaptor remained a supplement. It did not fit.

Quite obviously, rendering the language of molecular information transfer operational in terms of experimentation did not proceed straightforwardly. Hence Zamecnik's cautious wordings. But Hoagland, addressing the larger public in an article in *Scientific American*, no longer doubted

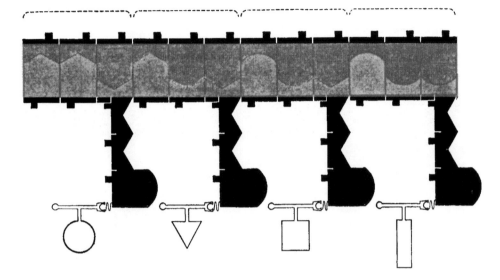

Fig. 12.3. Final step in the assembly of protein molecules through the attachment of transfer RNA to the template-RNA chain. Reprinted from Hoagland 1959b. Copyright © 1959 by Scientific American, Inc. All rights reserved.

that proteins were the result of "translating" a "plan" from DNA via RNA into protein.[27] Neither he nor Zamecnik had previously used such linguistic and textual icons. Some steps in the coding process were "still hypothetical." The remaining gaps, Hoagland contended, had to be bridged by "ideas," and even if some of them turned out to be wrong, "they will have been invaluable as a guide to further research." The transfer RNAs represented the words of the code (see Figure 12.3). Transfer RNA promised to become "the Rosetta Stone that will decipher the language of the gene."[28]

The strategy seemed obvious: "pure types" of transfer RNA had to be isolated and then taken apart "nucleotide by nucleotide." The charging enzymes could be assumed to "recognize the correct sequence of bases on transfer RNA," for they were able to put a specific amino acid on one particular kind of transfer RNA. Following this line of reasoning, however, the "*a priori* argument for the adaptor hypothesis," as Hoagland rightly stated, lost "some of its impact."[29] If the charging enzymes did the *encoding*, the subsequent *decoding* on the ribosomal template was a secondary event. Thus, it became "of fundamental importance to determine the minimal structural requirements of transfer RNA underlying the specificity with which it reacts with amino acids."[30] This "second code" attracted Zamecnik.[31] Unfortunately, this code turned out to be a highly specific interaction between each pair of enzyme/transfer RNA, and its clarification kept RNA workers busy for the next two decades.

Why not, as an alternative to working on transfer RNA, introduce pure types of templates into the assay system? Zamecnik and Hoagland knew about the results of Lazarus Astrachan and Elliot Volkin concerning a short-lived RNA intermediate in the synthesis of phage as well as Gale's studies on the synthesis of RNA concomitant with the induction of enzymes in *Staphylococcus*.[32] These observations, however, were at odds with what went on in the liver: differentiated liver cells continuously made the same set of proteins.[33] There were thus good reasons to conceive of templates in the liver homogenate as integral parts of a stable ribosomal particle. "One would expect," Hoagland reasoned, "that a template might have to be particulate, thus permitting a reasonably rigid spatial arrangement of the RNA." Therefore it seemed "highly probable that the RNA of the ribosomes is the cytoplasmic template for protein synthesis."[34] As a consequence, to work with model templates would have amounted to constructing model ribosomes. There were attempts, of course, to have reticulocyte ribosomes producing hemoglobin.[35] But compared to the challenge of modeling ribosomes, the purification of transfer RNAs appeared to be an easy undertaking.

Shaping the Ribosomes

With respect to their physical parameters, the protein synthesizing particles began to change their appearance. As early as 1956, Fu-Chuan Chao and Howard Schachmann from Wendell Stanley's Virus Laboratory in Berkeley had found yeast microsomes sedimenting with a velocity constant (S) of 80 and dissociating into two unequal portions of 60S and 40S.[36] Similarly, Mary Petermann and her coworkers took 78S liver ribonucleoprotein particles apart into 62S and 46S particles.[37] Alfred Tissières and James Watson at Harvard had started to work with *Escherichia coli* and had their ribosomes sediment with 70S. They dissociated their bacterial particles reversibly into a 50S and 30S component.[38] Gradually, in the course of years of painstaking isolation attempts, the confusion around the size of the RNP-particles began to clear up:[39] the secret of the stabilization and destabilization of the RNP-complexes resided in the concentration of divalent Mg^{2+}-ions. Work on a variety of particles from other sources made two distinguishing features prominent: bacterial ribosomes (roughly 70S) were consistently smaller than their eukaryotic counterparts (roughly 80S), but both could be separated into a small and a large subunit. Correspondingly, isolated RNA from pea seedlings and rabbit reticulocytes sedimented as two major 28S and 18S peaks.[40] That these peaks represented two single, large RNA molecules was established shortly thereafter for bacterial ribosomes as well. Here they came as 23S and 16S entities, respectively.[41] Whether the protein moiety of the particle was made up of a single species or of many different proteins, and whether all ribosomes had the same protein composition, however, remained an open question.

Puzzles

Concerning the in vitro amino acid incorporation systems, things were moving too. After years of vain efforts, Schweet had been able to recover a hemoglobin-like protein from reticulocyte ribosomes in the test tube. "Amino acid incorporation" and "protein synthesis" were becoming synonyms. In addition, the hemoglobin experiments were a strong argument in favor of tissue-specific ribosomes with an inbuilt template.[42] George Webster's pea seedling particles also effected a net synthesis of protein.[43] The controversy over whether the amino acid uptake in cell-free protein synthesis was reversible or irreversible had come to a surprising solution as well. Gale's "reversible" uptake of amino acids into the protein fraction of disrupted *Staphylococcus* cells could now be attributed to the attachment of these amino acids to transfer RNA. At the Fourth International Congress of Biochemistry in Vienna, in September 1958, Gale reported that "a preparation of the 'soluble prn [polyribonucleotide]' from liver, provided by Dr. Hoagland, is as effective as staphylococcal nucleic acid in promoting glycine incorporation in the disrupted staphylococcal cell." Shortly thereafter, Gale abandoned his system altogether because he realized, as he put it later, "that the system we had chosen to study had shown itself to be complex and difficult to analyze."[44] Thus, an intriguing signal that had led Gale already earlier to abandon the tracer technique because he judged it unreliable finally caused him to leave the field. In Zamecnik and Hoagland's system, a similar signal had set off an avalanche of new differentiations.

The soluble enzyme/transfer RNA-complex meanwhile had acquired the status of an experimental system on its own. It became detached from the incorporation system. After this bifurcation, the former "soluble RNA" differentiated into a number of distinct molecules, including material that was referred to as "contaminating RNA." As Waldo Cohn of the Oak Ridge National Laboratory once sarcastically remarked, "I think it was Gulland who said that 'nucleic acids are not compounds, they are methods of preparation.' "[45] Actually, none of the experts in the field thought of assigning a function to this new contamination. It was considered to be junk, just as a few years earlier the later transfer RNA had been considered a contamination of the enzyme supernatant.[46] Such assignments are characteristic of emergent entities. These entities acquire meaning through deconstructive groping rather than through straightforward constructive efforts or purposeful and deliberate interventions. In a differentially reproducing, sufficiently complex experimental system, a permanent game of presentation/absentation is taking place. Representing one component inevitably means de-presenting another one. It is much like a game of wedges. If you drive in one, you throw out another. In an ongoing research process, the actors usually do not know which cleft to widen and which one to narrow. Epistemic things oscillate between different representations. Today, the historian is in a position to judge the junk of the past as an early sign of messenger RNA. This does not mean, however, that it could have played this role except in retrospective fantasy.

Toward the end of the 1950s, nobody knew any more which laboratory would come up with the next surprise. New groups joined the field, established groups were picking up new threads, competition became dominant, and convergences and redundancies resulted.[47] In the first two volumes of the newly founded *Journal of Molecular Biology* (1959 and 1960), almost one-third of the regular papers were concerned with protein synthesis and ribosome research. Three major experimental prospects could be delineated in the midst of this bewildering reticulum of research activities around transfer RNA. These prospects were, first, refinement of the existing eukaryotic and development of alternative, bacterial protein synthesizing systems; second, fractionation, purification, and sequence determination of amino acid-specific transfer RNAs; and, third, investigation of the amino acid transfer process to the ribosome.

From Rat Liver to *Escherichia coli*

During the 1950s, the bacterium *Escherichia coli* (*E. coli*) had become the predominant model organism for genetic analyses.[48] But although

Tissières and Watson had started to investigate *E. coli* ribosomes, and Berg and Ofengand were concentrating their efforts on the aminoacylation of its transfer RNAs, no suitable fractionated protein synthesizing system existed for *E. coli*.[49] In 1958, Marvin Lamborg, a postdoctoral fellow with a Ph.D. in biology from the Johns Hopkins University, had come to work with Zamecnik. Zamecnik had tried to obtain an extract from broken *E. coli* cells as early as 1951 but had failed to get the homogenate sufficiently free from intact bacteria.[50] Lamborg made a new effort to prepare a bacterial system comparable to the well-established mammalian system. It took him two years to make it work.[51] Getting rid of whole cells and membrane fragments from animal tissue had been relatively easy. With bacteria, this proved to be a major problem. The bacterial cell walls were much more difficult to disrupt, and remaining cells easily reengaged in multiplication when contaminated extracts were incubated. Most bacterial systems that had been described thus far were dependent on protoplasts or on membrane fractions for activity.[52] To eliminate the "risks of whole cell artifact," Lamborg and Zamecnik found it necessary to have preparations with no more than 1 times 10^5 viable cells per milliliter of incubation extract.[53] The notion of "whole cell artifact" deserves some attention here. Whether an entity is considered natural or artificial depends on what one is doing with that entity: If one works with an in vitro system, every whole cell therein behaves as an artifact. If one fractionates nature, unfractionated nature has to be excluded from the space of representation.

Lamborg grew cells, harvested them in the early growth phase, washed them, and ground them with alumina. In three centrifugation steps, he removed the alumina and cell debris and separated the cytoplasmic contents into an enzyme supernatant and a ribosomal pellet. The incorporation of radioactive leucine into the $30,000 \times g$ supernatant depended on GTP, ATP, an energy regenerating system, and amino acids. Neither the isolated $100,000 \times g$ supernatant nor the ribosomal pellet alone showed activity. Incorporation activity, however, could be restored upon recombination of these two high-speed fractions.

To the satisfaction of Zamecnik, the incorporation of leucine finally was stimulated by a mixture of the other amino acids. He and his colleagues had never observed such a dependence in the rat liver system, and this failure had been a source of irritation for almost a decade. Now, the missing effect could be judged to be due to a peculiar experimental difficulty. It was not an intrinsic peculiarity that would characterize in vitro protein synthesis as generally weak and unreliable. Both model systems reinforced each other: they were sufficiently similar to be comparable, but they were also sufficiently different to remain interesting per se. They complemented each other's deficiencies and thus constituted significance by triangulation in a common metric space.

Most importantly, the E. coli system provided the material basis for a major expansion of protein synthesis research. Within a short time, it became a kit for molecular biologists. It did not require the whole infrastructure of a medical laboratory, where animals had to be kept and tumors transferred. It could easily be set up anywhere. And it connected functional protein biosynthesis research with the frontiers of structural ribosome research. Watson was at Harvard at this time. He and Alfred Tissières frequently came over to Zamecnik's lab. "Oh, if you can only get a good bacterial system, things will really boom," Watson is reported to have exclaimed.[54] Tissières obtained a preprint of Lamborg's paper,[55] and together with David Schlessinger and François Gros, he went ahead with the optimization of the system. A careful control of the magnesium concentration turned out to be crucial.[56] As a consequence, in a very short time the E. coli system came into general use within the growing scientific community of molecular biologists. The rat liver system declined. Model systems and model organisms have their historically assigned times and places. They can grow old, fall out of use, and be replaced.[57]

Bacteria, especially E. coli, are often referred to as the model organisms of molecular biology. But their enormous metabolic turnover capacities, which made them ideal for in vivo manipulation rather early, also made it much more difficult to detach the basic features of protein synthesis from the rest of the metabolic network in vitro. Once, however, there was a certain scaffold from which to begin, bacterial extracts took over the function of the rat liver system. Their protein synthesis machinery was more robust, they were easier to handle routinely, and they could be distributed without precaution. Thus, they found their way into more laboratories without a bacteriological background or a history of keeping laboratory animals.

The Isolation of Individual Transfer RNAs

A second line of inquiry aimed at the characterization of specific, individual transfer RNAs (tRNAs). Sorting them out and getting access at their molecular structure promised to elucidate the translation signatures of these molecules and, finally, the code. First, however, several tedious prerequisites had to be fulfilled, such as isolating crude RNA on a large scale and designing methods for separating specific amino acid-accepting molecules.

In 1958, Robert Monier, a Rockefeller Foundation Fellow for 1958–59, came to work with Zamecnik. He set out to find the conditions for a large-scale preparation of transfer RNA from baker's yeast. Yeast was a "readily available source."[58] While the conventional techniques of centrifugation and precipitation had the advantage of prefractionating the total cellular RNA, they were time consuming and usually did not yield large quantities of RNA. Once again, chance lent a hand. "One lazy day we added 50 percent aqueous phenol to intact yeast rather than to broken yeast. This treatment appeared to permeabilize the yeast cell wall, and cell contents containing tRNA leaked out, while ribosomes and ribosomal RNA were retained within the yeast cell."[59]

After Monier had shown that low molecular weight RNA was comparable to conventionally prepared S-RNA, and that it could be obtained easily and in large quantities, soluble RNA from yeast quickly became a kit for molecular biologists too. Its use was not restricted to a yeast system. It could be charged with amino acids by enzymes from other organisms, and it was able to transfer amino acids to rat liver microsomes.[60] This opened up the possibility of combining conveniently available components from different sources, and, at the same time, it opened up the prospect of using *heterologous* systems for the differential characterization of homologous compounds.

Zamecnik's idea was to develop a general method for the isolation of any given transfer RNA. His strategy was to charge the desired RNA with the appropriate amino acid, to modify the noncharged RNAs at their free 3'-end with a dye, and to remove the modified material by physical separation methods. The effectiveness of the method depended on the completeness of the reaction at each step involved in the procedure. The first fractionation attempts resulted in an enrichment of charged RNA over the original mixture by a factor of ten.[61] Although this was promising, the resulting material was far from being pure.

In view of the expected twenty or so different transfer RNAs, Zamecnik's was, in the long run, certainly an economical way to go. But it was not the most economical procedure to begin with. It involved a whole set of experimental techniques, of chemical modification, chromatography, and separation procedures. The development of this new expertise took time, and it tended to develop its own dynamics. To give just one example: In the beginning, Zamecnik and Stephenson had used periodate oxidation to evidence an ester linkage between the amino acid and the terminal ribose of RNA. Subsequently, they had adapted it to remove uncharged RNA from charged RNA. Now, in still another combination of operations, they envisaged the reaction as a means of the sequential degradation, that is, the sequence determination of transfer RNA.

The problem was to obtain both a high degree of purification and high yields. A variety of laboratories, including Lipmann's, Holley's, and Schweet's, had engaged in developing separation methods that were based on the slightly different physical properties of the different transfer RNAs.[62] Holley started with Monier's yeast extraction procedure, and using countercurrent distribution of individual tRNAs in combination with partial ribonuclease digestion, he finally took the lead and won the race for sequencing the first transfer RNA molecule in 1965.[63] Zamecnik, as he remarked later, had put his bets on the wrong horse.[64] Three years of work had been invested, if not wasted. Although with the advent of commercial Sephadex material column chromatography had undergone something like a revolution, there seemed to be no general and easy way to fractionate the different tRNAs. The original impact of the purification attempts had been to get access to the "translation signature" of transfer RNA, yielding the key to the genetic code. As time went on, the work grew into a more and more tantalizing and disappointing organic chemistry. Zamecnik still hoped to make progress, and so, in 1961, he went on to see the nucleic acid expert Alexander Todd in Cambridge and learn more RNA chemistry.

The Pursuit of Ribosomal Function

The third experimental prospect consisted in following the functional interaction between transfer RNA and ribosomes. This was Hoagland's option after his return from Cambridge. This field, too, was rapidly expanding.[65] There had been indications that S-RNA might cycle the microsome on and off steadily, while the amino acid incorporation proceeded at a constant rate.[66] Hoagland and Lucy Comly approached the problem with an elegant double-label experiment.[67] They labeled the RNA with ^{32}P in vivo and added a ^{14}C-amino acids after isolation. The assay system was built from three sources: the tRNA came from yeast, the microsomes were from rat liver, and the enzyme fraction was derived from Ehrlich mouse ascites tumor cells. When Hoagland and Comly used doubly labeled tRNA for the incorporation reaction, the ^{32}P-counts of the RNA moiety reached a plateau, whereas the ^{14}C-counts of the amino acid continued to accumulate on the microsomes. The kinetic data pointed to a dynamic turnover of the S-RNA moiety.

In the course of these experiments, Hoagland observed a "background" phenomenon that proved "difficult to reduce."[68] The "binding" of transfer RNA to the ribosomes occurred at "zero time," and it took place before the amino acid incorporation reaction. He considered this to be an artifact due to the "prolonged centrifugations at 4°."[69] A "control" experiment revealed that even uncharged and therefore nonfunctional tRNA found its way into the particles. It was altogether unclear how to evaluate these observations. At worst, the parallelism between the attachment of tRNA and the incorporation of amino acids might turn out to be a mere coincidence.

In many respects, the situation was comparable to the uncertainties about the "incorporation" reaction in the early days of the fractionated in vitro system. There was an activity, but its nature remained doubtful. Was it connected with protein synthesis at all? Was it an unspecific signal raised above background by the resolving power of isotope tracing? The double-labeling technique resolved more details than could be fitted into the existing representation. It produced more differences than could actually be handled. It produced an experimental excess. There was no possibility of immediately deciding whether the differences were protein synthesis signals or whether they were mere visualized noise. The power of the graphematical device was ahead of the interpretation of its traces. Within a short period, it would lead to a new category of experiment: the transfer RNA binding assay. One thing, however, had become clear: S-RNA was not broken down into small adaptor fragments prior to its attachment to the microsomes. Quite the contrary to constituting a productive excess generated by the system, the pursuit of a small adaptor remained a supplement. All the available evidence pointed to a stable but transient binding of the intact, bulky aminoacylated tRNA molecule.

These assays were performed throughout 1960. They marked an important transition. The in vitro protein synthesis system was on its way to becoming a model system of binding tRNA to ribosomes. Again, this transition exhibited the classical features of a subversion: The uncoupling of the incorporation reaction from the binding of transfer RNA had its origin in a corollary experiment. The corollary experiment itself responded to something that was perceived as an irregularity: a high back-ground that proved to be difficult to reduce. The answer obtained by the control was left in suspense. It might be something that had nothing to do with protein synthesis or it might assume significance in a space of representation not yet unfolded. Again, a differential signal had shown up that, if it could be stabilized, would orient the experimental system in an unforeseen direction.

CHAPTER 13

Play-off: Messenger RNA and the Genetic Code

❖❖   In the last analysis, however, what mattered was that X,
the unstable intermediary, was materializing.
—François Jacob, *The Statue Within*

In this chapter, I will briefly describe two experimental events that set the stage for deciphering the genetic code and for protein synthesis research between 1960 and 1965.[1] With these events, the historical function of the rat liver protein synthesis system for the emerging field of molecular biology came to closure. This brief survey of how messenger RNA and the genetic code came into experimental existence will also serve as a closure for my case study.

A Cytoplasmic Messenger

Besides transfer RNA, another new epistemic entity began to enter the experimental discourse of molecular biology toward the end of the 1950s. Although *messenger* RNA did not emerge from the rat liver system, it had its roots in studies on protein synthesis as well. One of these experimental systems was centered on the genetic regulation of enzyme induction in bacteria.[2] In the autumn of 1957, Jacques Monod and François Jacob of the Pasteur Institute in Paris, together with Arthur Pardee of Wendell Stanley's Virus Laboratory at Berkeley, had started a series of experiments on the induction of β-galactosidase in *E. coli* cells, which became known as the PaJaMo-experiment.[3] These experiments led the Paris group to the assumption that the genetic "i-factor" of the galactosidase system, already known from Monod's previous mutant analyses, was responsible for the production of a cytoplasmic substance, which in turn affected the structural gene for β-galactosidase. It is for this special, regulatory substance that Pardee, Jacob, and Monod used the term *cyto-plasmic messenger* for the first time in a published paper in 1959.[4] It fits perfectly into this history of experimental displacements that "messenger" arose as a concept for a narrowly circumscribed phenomenon of regulation. It took another year before it became generally attached to an RNA-intermediate in the flow of information from DNA to protein, as will be briefly summarized on the following pages.

During the continuation of these experiments on the regulation of inducible enzymes in Paris and in Berkeley, an additional intriguing observation arose: after induction, β-galactosidase made its appearance without a measurable time delay, and the inactivation of the gene stopped the synthesis of the enzyme immediately. Monod, Pardee, and their colleagues tentatively conjectured that there existed an additional, "functionally unstable intermediate" that in turn would be responsible for the expression of the structural gene. "The experiment does not exclude the possibility," they stated, "that an information-bearing RNA closely associated with the DNA of the gene is transferred as a part of the genetic unit."[5] In the absence of any chemical or molecular characterization, which could not in any event be obtained from the in vivo system at Pasteur, Jacob termed the component *X*. When he mentioned it for the first time in public, at a colloquium in Copenhagen in September 1959, the reaction of the participating molecular biologists, he recalled, was anything but enthusiastic. Nobody seemed interested or willing to engage in discussion. "No one reacted. No one batted an eyelash. No one asked a question. Jim [Watson] continued to read his newspaper."[6]

During Easter 1960, Jacob stopped over at King's College, Cambridge, for an informal meeting with Francis Crick, Sidney Brenner, Leslie Orgel, Alan Garen, and Ole Maaløe. The episode has been recounted on several occasions, so I may be brief here.[7] During the meeting, Crick and Brenner bridged a gap. They compared the molecule of the Pasteur group in Paris and of Pardee and Monica Riley in Berkeley with the quickly metabolizing RNA that Elliot Volkin and Lazarus Astrachan of Oak Ridge National Laboratory had observed upon infection of their bacteria with T2-phages.[8]

For several years, experimental hints had circulated that pointed to a quickly metabolizing RNA fraction distinct from the bulk of cellular RNA. But these findings remained more or less isolated. At least, and quite obviously, they had not been taken seriously by the Pasteur-based network and by Crick and his friends in Cambridge. This is astonishing in that all these observations were related to research on phage and induced enzyme synthesis.[9] As early as 1955, it was clear to the Cambridge-based microbiologist Ernest Gale, a neighbor of Crick's, that "in inducible systems at any rate, protein synthesis is accompanied by, if not dependent upon, RNA synthesis." Gale had even discussed the issue with Monod during an international symposium on enzymes held at Henry Ford Hospital, Detroit, in November 1955. At that time, Monod appears to have been anything but enthusiastic about the suggestion "that inducible enzymes are formed by 'unstable' RNA templates."[10] Sol Spiegelman, recalling the experiments of Gale and Joan Folkes on the production of β-galactosidase in disrupted *Staphylococcus* cells, had concluded at a CIBA symposium in March 1956 that "the RNA templates of induced enzymes are unstable."[11] Studies of yeast "petite colonies" had led Hubert Chantrenne to speculate that under conditions of induction "some change in pre-existing RNA" would occur or even "specific RNAs" would be assembled that were "involved in the synthesis of the individual proteins."[12] And as I reported in Chapter 10, similar suggestions had been made in 1956 by Walter Vincent. Indeed, considerations of a similar kind had led Zamecnik to look for an RNA synthesis activity in the rat liver protein synthesis system.

The concept of microsomes that had gained common currency toward the end of the 1950s, however, had emerged from eukaryotic systems with reduced metabolic activity, and it was clearly at odds with these observations on the metabolism of lower organisms. For all those working on cells from higher organisms, microsomes represented "a stable factory already containing an RNA transcript of DNA," and obviously, these particles were continually "making the same proteins."[13] In Crick's words of 1958: "'Template RNA' is located inside the microsomal particles."[14] Indeed, this concept was so firmly established that Hoagland, who had visited the Pasteur Institute in January 1958, did not even think of correlating the PaJaMo regulation phenomena with the eukaryotic machinery of protein synthesis.[15] Moreover, bacterial in vitro assays had a bad reputation in the leading circles of protein synthesis: they were considered dirty systems in which everything was possible.[16] Thus, it is no exaggeration to say that the conjuncture of phage genetics, enzyme induction, and protein synthesis, which led to messenger RNA, took shape in a field of idiosyncrasies, experimental redundancy, and conceptual underdetermination.

The new entity that tentatively came to be addressed as an "information-bearing RNA" arose from studies on bacterial genetics and differential enzyme regulation that were much more phenomenological and less molecular and mechanistic than the experimental space of mainstream in vitro protein synthesis research.[17] The template concept of the latter had all the characteristics of an epistemological obstacle. There was no place for the new RNA entity, unless the microsomes lost their intrinsic template specificity, unless they were degraded to "simple machines for assembling amino acids to form proteins of any kind."[18] The tacit assumption of "one microsome-one enzyme" would have to be dropped. A separate, unstable RNA would have to assume the role of the template, a magnetic tape carrying the instruction for whatever protein music was going to be played.[19]

Jacob and Brenner had both been invited (by Matthew Meselson and Max Delbrück, respectively) to visit the California Institute of Technology during the summer of 1960. They decided to do the "crucial" experiment together: grow bacteria on heavy isotopes in order to tag the ribosomes; infect the E. coli cells with a virulent phage in the presence of radioactive isotopes; and see whether a posteriori produced radioactive phage RNA became associated with a priori existing heavy ribosomes. Brenner and Jacob judged that for the identification of both heavy and radioactive ribosomes Meselson's density gradient ultracentrifugation might be appropriate. Accordingly, they performed the experiments in his laboratory. At the end of a four-week experimental effort, Jacob and Brenner got the signal, crude as it was, they were looking for: new phage RNA became associated with old ribosomes.[20] X had become "structural messenger,"[21] and, leaving the context of specific regulatory models of protein synthesis, it began to acquire the status of a general molecular information transmitter.

That same year, 1960, in Sol Spiegelman's laboratory at Urbana Masayasu Nomura and Benjamin Hall characterized, besides a "ribosomal" RNA, a "soluble" form of RNA synthesized in E. coli after bacteriophage T2 infection. It became associated with ribosomes in the presence of high magnesium concentrations.[22] The Urbana group, however, drew no conclusions with respect to the function of the new form of soluble RNA. As Nomura recalled, he was "unaware of the new developments, both experimental and conceptual, that were taking place in Cambridge, England, as well as in Paris."[23] Shortly thereafter, in James Watson's laboratory at Harvard, François Gros from the Pasteur Institute, Walter Gilbert, and Chuck Kurland showed that unstable "messenger RNA templates" also belonged to the metabolic makeup of uninfected E. coli cells.[24]

Protein Synthesis and the Genetic Code

The differentiation of bacterial in vitro assays happened to occur parallel with, but independently of, the experimental context I just described. In a rapid dissemination, the Lamborg–Zamecnik type of system made its way into other laboratories and assumed the role of a leading model system. It sprouted everywhere. Among the first using bacterial extracts in protein synthesis were David Novelli at the Oak Ridge National Laboratory, Daniel Nathans and Fritz Lipmann at the Rockefeller Institute in New York, Kenichi Matsubara and Itaru Watanabe at the University of Tokyo and Kyoto University, and James Ofengand, then on a fellowship at the Medical Research Council Unit for Molecular Biology in Cambridge.[25] In Watson's group, which at that time included Alfred Tissières, David Schlessinger, Chuck Kurland, François Gros, and Walter Gilbert, the structure and function of the ribosome and of unstable RNA had moved to the center of attention. But the *E. coli* system was also being introduced at the National Institutes of Health (NIH) in Bethesda. The days of the rat liver system as a pacemaker for unprecedented events were over. Its role was displaced from innovative representation to demonstration: it became marginal. The *E. coli* system shifted in the opposite direction: originally it had been set up as a demonstration of the general applicability of the notions developed in the rat liver system. Now it acquired its own dynamics. It brought the genetic code into the realm of experimental manipulation in a surprising shift that left behind all those who had tried to tackle the code by means of mutation procedures. The shift came about with the replacement of endogenous, heterologous messenger RNA by exogenous, homologous *model substances*: synthetic polyribonucleotides.

Marshall Nirenberg came to the NIH in 1957, shortly after receiving his Ph.D. in biochemistry from the University of Michigan. After two years of postdoctoral training, he joined the department of Gordon Tompkins at NIH. He was in the course of establishing a cell-free bacterial system when Heinrich Matthaei came to work with him in the fall of 1960. As Nirenberg saw it in retrospect, he had set himself the task of investigating "the steps that connect DNA, RNA, and proteins" and of synthesizing a specific protein in the test tube.[26] He decided to try it with penicillinase, a small protein that in contrast to most other proteins lacked the amino acid cysteine. By leaving out cysteine from the reaction mix-

ture, he hoped to be able to create conditions under which only penicillinase was synthesized. Despite all the efforts mentioned in previous
chapters, the synthesis of a defined and complete protein in vitro had
continued to be a challenge for all those concerned with protein synthesis
ever since the end of the 1940s. Matthaei had similar ideas in mind when
he came to the United States on a NATO fellowship in 1960. Trained as a
plant physiologist in Bonn, Germany, he had thought of expressing a
carrot protein. However, this plan did not match the interests of Frederick
Steward of Cornell, with whom he had planned to stay during his fellowship. Matthaei saw himself as being forced to look for another place.
Finally he came to Nirenberg and immediately, at the beginning of
November 1960, plunged himself into the optimization of an in vitro
system based on E. coli extracts.[27]

The search for *template specificity* appears to have been the crucial clue
to Nirenberg and Matthaei's moves and advances, as far as can be judged
from Matthaei's laboratory notebooks.[28] In a first set of preliminary experiments, in which ^{14}C-valine served as a radioactive tracer, Matthaei set
out to explore the general assay conditions for defining the specific action
of template RNA. Among the conditions he tested were the inhibition of
the extract's activity by the antibiotic chloramphenicol, a specific inhibitor of bacterial protein synthesis, and by the RNA-degrading enzyme
RNase. In addition, he tested the quenching effect of DNA-degrading
DNase. The effect of DNase, however, was far less clear than that of
RNase, and it varied from experiment to experiment.[29] The amino acid
incorporation signal was extremely faint and barely above background in
these first exploratory assays. It amounted to a few dozen counts per
minute. It was only after Matthaei had fine-tuned several parameters of
the system and tried different preparation methods that the incorporation
signal became high enough for reliable radioactive counting. Based on my
analysis of Matthaei's notebooks, there are good reasons to assume that
during this initial phase of the work Nirenberg and Matthaei shared the
prevailing picture of the ribosome, whose RNA—or part of it—they
believed played the role of a template in the hovering sense this notion
had assumed in 1960.

Matthaei exploited two sources of RNA for his first set of stimulation
experiments: the supernatant RNA (called "S-RNA"), and the RNA
extracted from ribosomes (called "mRNA").[30] However, despite relent-
less efforts, he was not able to record a stimulatory effect of a ribosomal

RNA preparation on the incorporation of radioactive amino acids until February 1961.[31] Three minor procedural tricks, taken together, had set the stage for this accomplishment. First, Matthaei had learned to store the enzymatic components of his system in a deep freezer. This allowed him to work on batches of material with known characteristics over a whole series of experiments. Every new batch could be tested and calibrated against a previous one. Such minutiae are far from being trivial for the kind of bioassays at issue here because the quality of the preparations used to differ considerably from day to day.[32] Second, he found a speedy method for efficiently precipitating and filtrating radioactive proteins. This allowed him to increase the number of assays he could perform simultaneously and thus extend the network of his control experiments.[33] The third procedural change consisted of a preincubation of the bacterial cell extract:[34] Matthaei and Nirenberg put the system to work until its endogenous protein synthesis activity came to a halt. Only then did they add the exogenous RNA. The resulting effect due to ribosomal RNA was small, but it was specific. Without delay, they sent off a preliminary report to *Biochemical and Biophysical Research Communications* on March 22, 1961.[35]

It is characteristic of the sequence of events documented in Matthaei's notebooks that he had already introduced the first synthetic RNA polymer (polyadenylic acid, or poly [A]) in an experiment in December 1960. However, it served a completely different purpose at the beginning: Matthaei did not use it as a template but as a *polyanion* thought to affect the action of DNase.[36] In the preliminary report of March 1961, Matthaei and Nirenberg described poly(A) as a negative control, that is, as a polyanion *unable* to mimic the stimulatory effect of ribosomal RNA (which is also a polyanion) on protein synthesis. On the second of March, we find polyadenylic acid again in a stimulation test, this time in a series with salmon sperm DNA and polyglucose.[37] Without effect.

Starting in late February 1961, Matthaei introduced additional RNAs into the system while keeping all the other conditions constant. S-RNA continued to serve him as a negative control: it did not stimulate. Among other RNAs, he tested yeast RNA, ascites RNA, "David's RNA," and "Crestfield-RNA."[38] Finally, at the beginning of May, he added an RNA sample of tobacco mosaic virus to his list. It affected the activity of the system to different degrees, but no clear-cut pattern arose.

In the context of these tests of different RNAs, the labeling strategy was also varied. Matthaei's standard system was based on the incorporation of radioactive valine. Late in March, Matthaei tested the "mRNA-effect" with several different radioactive amino acids, among them phenylalanine.[39] In the presence of mRNA, he had them all incorporated into protein. As an alternative, Matthaei used a mixture of radioactive amino acids derived from the hydrolysis of algal proteins.

At this point, the decisive series of experiments begins to take shape. On May 15, Matthaei devised an experiment to "test synthetic polynucleotides." Poly(A), poly(U), poly(2A/U), and Poly(4A/U) were among them. As a source of radioactive amino acids, he decided to use the unspecific "algal hydrolysate." On May 22, Matthaei noted an 11.8-fold stimulation of his activity in the poly(U)-containing sample.[40] In the protocols, no explicit rationale for the introduction of these polymers is given, but the outline of the ensuing assays is systematic and contains a number of cautiously chosen controls. "Compare always," we read in the notebook, "complete [system], no RNA, poly(U), UMP, Crestfield RNA."[41] On May 25, 1961, Matthaei used a mixture of eleven different radioactive amino acids instead of his algal hydrolysate and found a huge stimulation of a factor of 20 in the presence of poly(U). Now it was a question of two days until he had identified the amino acid responding to poly(U). On May 26, Matthaei had narrowed the activity down to a mixture of tyrosine and phenylalanine. On May 27, 1961, at three o'clock in the morning, he performed the assay with poly(U) and phenylalanine alone. The incorporation was stimulated by a factor of 26, as Figure 13.1 shows.

No doubt, it was a happy coincidence that Leon Heppel, who was the head of the laboratory in which Nirenberg and Matthaei were working, was an expert in the synthesis of artificial RNAs. The polymers poly(A), poly(U), poly(A/U), poly(C), and poly(I) were ready on his shelf to be used. Once one of them had shown a signal, it was only a matter of days for Matthaei, by systematically varying the radioactive amino acids at hand, to decipher the first code word: the homopolymer polyuridylic acid appeared to be translated into the artificial protein polyphenylalanine, although this conjecture remained to be tested. Assuming a triplet code, UUU coded for phenylalanine (Phe).[42]

Remember that the original purpose of both Matthaei and Nirenberg had been to optimize an in vitro system for the expression of a specific protein. On the way to tackling this task, the deciphering of the first code

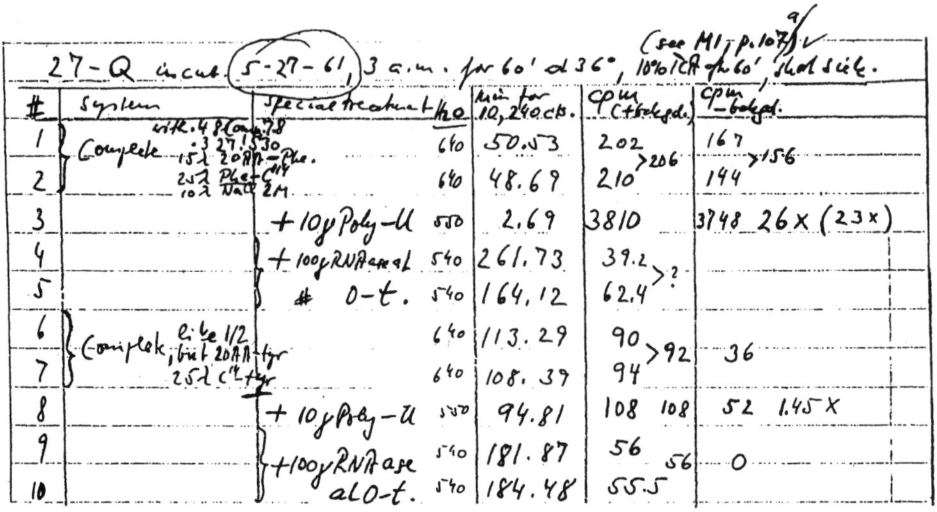

27-Q incub. 5-27-61, 3 a.m. for 60' at 36°, 10% TCA for 60', (see MI p. 107)

#	System	Special treatment	H_2O	Min for 10,240 cts.	cpm (+bckgd.)	cpm −bckgd.
1	Complete		640	50.53	202 >206	167 >156
2			640	48.69	210	144
3		+10γ Poly-U	550	2.69	3810	3748 26X (23X)
4		+100γ RNAase	540	261.73	39.2 >?	
5		0-t.	540	164.12	62.4	
6	Complete, but 20AA-tyr		640	113.29	90 >92	36
7			640	108.39	94	
8		+10γ Poly-U	550	94.81	108 108	52 1.45X
9		+100γ RNAase	540	181.87	56 56	0
10		at 0-t.	540	184.48	55.5	

Fig. 13.1. Experiment that led to the identification of the first code word. Reprinted from laboratory notebooks of Heinrich Matthaei, experiment 27Q, May 27, 1961.

word certainly was not a chance event. It was not a mere instance of a "control" that became the "real experiment." But it was not a planned event either, in any case not in the common sense of an anticipation of what would happen. We might say that what increasingly tended to appear as an *intention* to the participants after the event accompanied the research process as an *intuition* before the event.[43] Matthaei and Nirenberg were exploring the experimental space of cell-free protein synthesis according to the cutting edge standards of the biochemical state of the art. On the priority list of things to check on their way to synthesizing a protein was the "specificity of the template." Matthaei and Nirenberg's concept of messenger took shape from probing different RNAs as "templates." But apart from the observation that turned out to be the decisive displacing event, other things happened as well that caught the attention of the two researchers: for instance, a record of the incorporation of leucine into proteins "in the absence of ribosomes." This sounds slightly odd to the ears of a molecular biologist of today. In 1961, Nirenberg and Matthaei esteemed their finding important enough to present it before a major international audience.[44] In which direction the system would move was a matter of probing the scope of its space of representation. It was not an a priori decision that had only to be executed.

Nirenberg and Matthaei both claim to have been unaware of the messenger-experiments being carried out at the Pasteur Institute and at Harvard when they were optimizing their assay system at NIH late in 1960.[45] Yet in their first report dated March 1961 we find, in addition to the concept of template, the notion of "messenger."[46] The concept of messenger was obviously "in the air" in these days. It was bandied about in conferences during the fall of 1960.[47] It emerged simultaneously from different experimental contexts: among them was a delicate, genetically triggered in vivo system of enzyme regulation and a comparatively modest, fractionated in vitro system of protein synthesis. In any case, we cannot simply follow the legend of the club of molecular biologists, according to which the messenger from the Pasteur-Harvard-Cambridge connection delivered the golden key for the code. The decisive components by which the flow of genetic information from nucleic acids to proteins revealed itself to be mediated—transfer RNA, messenger RNA, and the code—were above all the products of biochemically oriented in vitro systems. The distance between "microsomal template" and "messenger RNA" bridged at NIH was as large as the gap between the "unstable intermediate" of induced enzyme synthesis and the "information-bearing RNA" filled at Paris.

After the Fifth International Congress of Biochemistry in August 1961 in Moscow, where Nirenberg reported the news from his laboratory, Sidney Brenner, Heinz Fraenkel-Conrat, and Heinz-Günter Wittmann could drop their sophisticated attempts at deciphering the code by genetic and chemical analyses of phage and virus mutants.[48] The subsequent race for the remaining code words was a matter of refining the experimental conditions of the E. coli system, and it was headed by the laboratories of Nirenberg and of Severo Ochoa in New York.[49] Neither Zamecnik nor Hoagland participated in the race, although they had made substantial contributions to the experimental infrastructure that had rendered this decoding strategy possible. At the time of the breakthrough, Zamecnik was staying with Alexander Todd in Cambridge, and Hoagland was building up a new laboratory in the Department of Bacteriology and Immunology at Harvard Medical School. Zamecnik felt that "there was no reason to add another laboratory to the chase"[50] and continued to work on the characterization of individual transfer RNAs. Hoagland stayed with the regulatory aspects of regenerating rat liver. A brief look at their activities during the following years will indicate how they reacted to the "recent excitement in the coding problem."[51]

The Rat Liver System, Regulation, and Messenger RNA

The test-tube representation of rat liver protein synthesis was due for an adaptation. At a Cold Spring Harbor symposium on cell regulatory mechanisms in 1961, Hoagland had already taken up the messenger RNA challenge. Until 1960, there had been no fractional entity in the rat liver system that corresponded to a messenger. At least in its Paris variant, the notion of mRNA was tied to a context of protein synthesis *regulation*. Accordingly, Hoagland embedded his search for a messengerlike component in the mammalian cell-free system in a context of control. Experimentally, he approached the problem of regulation by comparing normal and regenerating rat liver. Regenerating liver had long been known to be considerably more active in protein synthesis than normal liver, both in vivo and in vitro.[52] Hoagland undertook a series of experiments in which fractions of normal and regenerating rat liver were cross-reacted, including a fraction centrifuged at high speed for up to half a day. His "tentative conclusion" was that "both normal and regenerating liver contain a sedimentable fraction, distinct from ribosomes, which is precipitable at pH 5, contains RNA, and markedly stimulates the in vitro incorporation reaction."[53] The only problem was that X, as Hoagland called his new fraction, following the example of Jacob, was much less sensitive to ribonuclease than expected if the active component was an RNA.

Because X from regenerating liver was more active, and microsomes from regenerating liver were more responsive to X, it occurred to Hoagland that he might be looking into a mechanism of "control by repression" "at the ribosome level," in contrast to what Jacob and Monod had found at the level of the gene. But something did not fit. "We meet with some difficulty when we try to apply the messenger concept to the behavior of the microsomes in this system."[54] If the activity was strictly correlated to the amount of messenger produced, the ribosomes from both normal and regenerating liver should behave identically with respect to the message. And this was not the case.

Part of the difficulty was due to the fact that Hoagland was pursuing two goals at the same time. On the one hand, he wanted to give X a fractional representation by refining the technique of differential centrifugation. In doing so, he obtained a "new" fraction, but the evidence that it was mRNA was inconclusive. On the other hand, Hoagland's interest—a long-standing interest of the Harvard group—was not only in protein synthesis per se, but in its regulation. For him, the concept of "messenger" retained its meaning as a regulatory entity rather than as an integral component of protein synthesis. He saw "some promise that certain regulatory processes may be studied in cell-free preparations," which differed from the model of the Pasteur group,[55] and he felt that investigating them might be more rewarding than merely confirming the existence of mRNA in eukaryotic cells.

But was mRNA a bona fide component of eukaryotic protein synthesis after all and had Hoagland simply failed to represent it properly? The actual situation, in 1961, was much less clear than molecular biological mythology would like to have it. No one had yet isolated a cytoplasmic messenger and put it into the fabric of proteins. Nobody could exclude the possibility that there were two classes of ribosomes: messenger-requiring regulatory particles, and constitutive, specialized particles with an inbuilt template. The "provocative hypothesis" of Jacob and Monod would have to await the in vitro identification of the controlling elements, before "fact and theory" could be "reconciled."[56] Messenger was not yet a regular thing, a powder that could be poured into a reaction vessel, from which proteins could be extracted after stirring and incubation.

At Harvard Medical School, Hoagland continued to work on regenerating rat liver as a model system for the analysis of protein synthesis control in vitro. The X-fraction gradually acquired more messenger-like qualities. But as late as 1963, it still came under the heading of a control element of protein synthesis.[57] The RNA properties of the fraction were still at issue. Attempts to further purify the active component revealed nothing with the unambiguous characteristics of a defined RNA. Taken together, all effects pointed to a ribonucleoprotein, whose RNA component, in combination with some proteins, was responsible for the regulatory "biological action."

It is obvious that the way in which messenger RNA took circumstantial shape in the rat liver in vitro system was quite different from the way in which the system had given rise to transfer RNA. In the case of transfer RNA, an entity had emerged that those who worked with the system had not thought of. The system had begot an unprecedented event. In the case of mRNA, an entity gradually gathered contours whose presumed characteristics were derived from other systems and could serve as a guideline for the shaping process. Assume Hoagland had known nothing about an RNA acting as messenger. How might he have been able to identify an RNA component that, even in the enriched X-fraction, amounted only to 15 percent of the total RNA present in this fraction and that, in its native state, was insensitive to ribonuclease? The rat liver in vitro protein synthesis system did not generate the messenger. There is a difference between a machine for making the future and an identification procedure, between a research system and a screening program. This is not, however, a difference *between* systems. It is a difference that lies in the use made of a system. Experimental systems oscillate between these extremes: phases of generative representation and phases of confirming demonstration.

Hoagland's messenger experiments remained inconclusive. A number of observations led him to suspect that the activity of ribosomes associated with membranes (microsomes) might be controlled by additional factors. An inhibitor seemed to be involved that was antagonized by GTP.[58] A detailed mode of action of GTP in protein synthesis was still missing. Was it a regulatory rather than a constitutive element? At the Cold Spring Harbor symposium in 1961, Robin Monro, from Lipmann's laboratory, had described a partially purified rabbit reticulocyte system, in which GTP was required for the transfer of ^{14}C-leucine from S-RNA to protein.[59] The transfer depended on a partial fraction of the pH 5 enzyme supernatant.[60] But attempts to clarify the role of GTP as a cofactor for the presumed "transfer factor" had failed so far.[61] Hoagland's approach was different. He thought that excessive purification might lead him to overlook some crucial catalytic factors. In 1964, he could see "no correlation between the degree of 'purity' of a system and its GTP requirement."[62] Lipmann and his coworkers did see such a correlation. In that same year, 1964, Monro and Jorge Allende identified an enzyme fraction in *E. coli* whose transfer activity overlapped with a GTPase activity.[63] It became quickly known as "G factor (G)."[64]

One of Hoagland's concerns had been to isolate microsomes that still responded to the modulatory action of regulation. Gently isolated larger ribosomal particles had become one of the great attractions of the early 1960s in the burgeoning field of ribosomology. They were variously termed "ribosomal clusters," "active complexes," "ergosomes," and "aggregated ribosomes," until the notion of *polysomes* came into general use.[65] Zamecnik had visualized such particles as a thread and beadlike structure in the electron microscope as early as in 1960–61. But because he could make no sense of the observation, he left his pictures unpublished.[66] Polysomes appeared to consist of strings of ribosomes aligned on a particular mRNA. Special isolation procedures were needed to prevent them from breaking down to monosomes during fractionation. Hoagland, too, made his contribution to the field. Essentially all cytoplasmic ribosomes of rat liver appeared to be assembled into such polysomes.[67]

Over the years, the fractional representation of the rat liver cell sap had grown into a sophisticated partition diagram. It can be summarized as shown in Figure 13.2.

Minimal Requirements versus Integrated Requirements

Two strategies could be pursued in the further elaboration of in vitro protein synthesis systems. The first was to try to identify the *minimal requirements* for any given activity of interest. This was the option followed by most of the groups that had adopted the *E. coli* system.[68] They focused on the constituents that were necessary and sufficient for different partial reactions. They isolated homogeneous molecular components and recombined and reconstituted them in the test tube. Hoagland chose a different strategy. He wanted to proceed along "more physiological" lines. He and his colleagues fractionated the liver cell sap into as many constituents as possible and thereby used techniques as smooth and gentle as possible. In principle, none of the constituents were omitted from the overall reaction scheme. Partial reactions were visualized by using appropriate tracers, and not by the isolation of pure molecular components. This was a more physiological, but also a more phenomenological, approach, and thus a strategy of *integrated requirements*.[69] Hence the hesitation to use purified ribosomes instead of microsomes that had the ribonucleoprotein particles still attached to fragments of the endoplasmic reticulum. Hence the inclination to fractionate by centrifugation, where the sedimented material could, at any time, be complemented with the supernatant. Nothing was lost here during irreversible discarding steps. Thus, the system remained complete but at the cost of remaining also "dirty." This had been an appropriate strategy for arriving at signals that would otherwise have gone unnoticed. It had been optimal at the beginning. The problem was that the very results of this strategy now made it historically obsolete.

Bacterial in vitro systems aiming at a representation of partial reactions

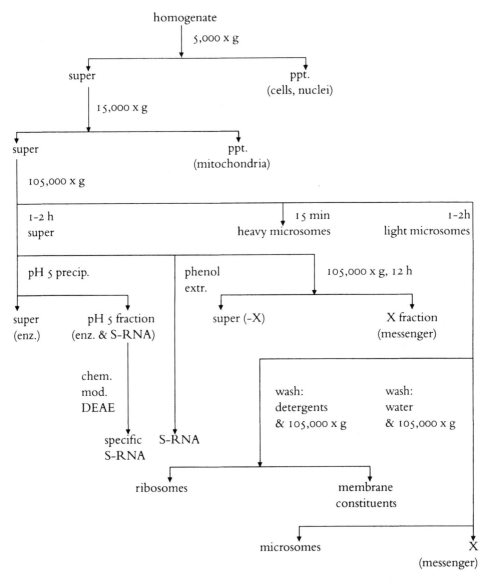

Fig. 13.2. Fractional representation of the rat liver cell sap, ca. 1963. ppt., precipitate; super, supernatant; enz., enzymes; chem. mod., chemical modification; DEAE, diethylaminoethyl-cellulose chromatography. Author.

of protein synthesis had become the new reference point. During the 1960s, most "new findings" came from *E. coli* systems. After the initial difficulties had been overcome, they were less complex, therefore easier to dissect and simpler to maintain. Perhaps most importantly, on a genetic level, *E. coli* cells were easier to define in terms of metabolic deficiencies and above all in terms of genetic markers. In fact, they had already become a model organism for bacterial genetics in the late 1940s.[70] In the end, this granted them their future as a molecular biological in vitro system that dominated the 1960s and 1970s. They became the settings in which, gradually, the structure-function relationships between ribosomes, messenger RNA, and transfer RNAs could be attacked.[71]

The language of the phenomenological approach remained operational. Hoagland and Zamecnik generally did not isolate ribosomes, add transfer RNA and messenger RNA, and synthesize proteins. They took a "heavy microsome fraction," added "pH 5 supernatant" containing "soluble RNA," incubated the mixture with an "X-fraction," and got a "stimulation" of the "amino acid incorporation" into "hot trichloroacetic acid insoluble material." Their language remained largely within the technicalities of the experimental space of representation, where the required entities appeared as peaks, fractions, and precipitable materials and not as entities of a molecular function in terms of a conceptual construction.

Unsettled Questions

Zamecnik continued to pursue "unsettled questions" in the field of protein synthesis.[72] Together with David Allen, he was among the first to investigate the effect of the antibiotic puromycin on protein synthesis.[73] After the introduction of poly(U) by Matthaei and Nirenberg, Zamecnik looked at its characteristics as a template.[74] But his main interest continued to be the purification of individual S-RNAs, the characterization of their amino acid-accepting properties, and the recognition properties of synthetases.[75] For a while, he concentrated on physical measurements of the conformations of transfer RNA, especially of differences between charged and uncharged forms, and then went on with specific questions about their charging enzymes.[76]

Zamecnik had dedicated much of his effort to S-RNA, in the hope of being able to sequence the molecule and elucidate its "translation signature." With respect to sequencing, he had unfortunately put his bets on chemical degradation, and Robert Holley's combination of countercurrent distribution and nucleolytic fragmentation "plus a clever team," which had been joined by Betty Keller from Zamecnik's laboratory, made the race.[77] With respect to coding, Nirenberg as well as Ochoa and collaborators had taken the lead in using all sorts of synthetic polynucleotides as messenger. Philip Leder at NIH introduced the triplet assay for transfer RNA binding, which became a model for many subsequent binding studies.[78] What remained to be clarified was the signature by which the amino acid–activating enzymes recognized their transfer RNA molecules. To the disappointment of those, like Zamecnik, who took up the challenge, that signature turned out to be highly specific and different for every couple of transfer RNA and enzyme, respectively. The hope for quick and general solutions had to be buried. Biochemistry lay ahead for the next twenty years.

Concluding Remark

Between 1955 and 1961, two molecular objects emerged from different experimental systems and were rapidly transformed into central objects of molecular biology in the 1960s: soluble RNA into transfer RNA, and microsomal template RNA into messenger RNA. It was the conjuncture of these two epistemic things, their transposition, grafting, dissemination, hybridization, and bifurcation that brought the genetic code into experimental existence. Transfer RNA had arisen from test-tube experiments on rat liver protein synthesis. Messenger RNA had one of its embeddings in bacterial regulation in vivo, another in bacterial protein synthesis in vitro. The *genetic code* could be attacked from an in vitro combination of both epistemic things, turning them into tools of research. The deciphering of the code was based on an in vitro system of protein synthesis, which, over the years, had been stabilized to such a degree that it could serve as a technical package, or kit, for a whole protein synthesis industry. In order to attack the code, a strategy of "minimal representation" was required that was opposed to the strategy of "integrated representation," which had led Zamecnik and his colleagues to the establishment of the system. The differentiation of the experimental system itself, its differential reproduction, had made this transition possible, albeit not in an automatic fashion, and not within the confines of a single laboratory. The *E. coli* system had been a prerequisite for this conjuncture, but the exploration of its space and scope was not a mere continuation of work under the conditions of the ancien régime. The concepts accompanying this turn in the experimental history of molecular biology were concepts of information and communication: transfer, message, transcription, translation, and coding.[79] They subreptitiously had crept into the experimental discourse of biochemistry rather than governing it from the beginning.

Epilogue: Science and Writing

❖❖❖ For a writing to be a writing it must continue to "act" [even] when what is called the author of the writing no longer answers.

—Jacques Derrida, "Signature Event Context"

This book is not an encompassing history of protein biosynthesis, let alone of molecular biology. Nor is it a collective biography of Paul Zamecnik's protein biosynthesis research group at the Massachusetts General Hospital.

My purpose has been other. I have followed, as closely as the available sources allowed me to, the history of an experimental system. I have tried to recapture its establishment and its reproductive dynamics as well as the conjunctures, filiations, and substitutions it underwent during the course of fifteen years, from 1947 to roughly 1962. I have been led to admit that such systems can generate conjunctures in which they subsequently play out their inherent capacities, in time and in space. They constitute historial trajectories, much as objects and forms of art constitute such trajectories. But I have also been led to understand that experimental trajectories, at one and the same time, insert themselves into and lead to the transformation of particular technical, instrumental, representational, institutional, and broader cultural environments or contexts.

Let us take the notion of context literally for a moment. To do so means that an experimental system in turn legitimately can be seen as having the characteristics of a text. Text and context thus become conflated. It is the grounding feature of a text, of an *écriture* broadly conceived, that it becomes and remains a text only through being continually reread, rewritten, detached from its pretext and author, and recontextualized:

[The] structural possibility of being weaned from the referent or from the signified (hence from communication and from its context) seems to me to make every mark, including those which are oral, a grapheme in general; which is to say, as we have seen, the nonpresent *remainder* [*restance*] of a differential mark cut off from its putative 'production' or origin. And I shall even extend this law to all 'experience' in general if it is conceded that there is no experience consisting of *pure* presence but only of chains of differential marks.[1]

This is exactly what holds for a productive experimental system with its multiply textured spaces of representation. The enterprise dubbed "modern science" derives its power from its peculiar spaces of representation. The forces and the reasonings they enact, as well as the rules by which they are structured, are less those of Cartesian egos than of mentalities and textures. Accordingly, our methodologies for assessing what happens where the generation of unprecedented events has taken over must change. And so must our views of what it means for scientists, as "authors," to be engaged in the production of scientific novelty. During the past 150 years, our view of the world has been profoundly altered. We have witnessed economies that are no longer centered around an ego: a Darwinian economy of nature, a Marxian economy of production, a Nietzschean economy of the moral, a Freudian economy of the unconscious, an Einsteinian economy of relativity, a Saussurean economy of the sign, a Foucauldian economy of discourse. What we need, in history of science, is an economy of scientific change, an economy of epistemic things.

This book is no more than a prolegomenon to such an endeavor. I have been concerned with experimental reasoning in a historically and locally circumscribed setting at the intersection of biochemistry and molecular biology. Briefly, I have argued along the following lines. Experimental systems are the basic units of the scientific tracing-game. Within a framework of technical things taken for granted, they provide the conditions for the generation of epistemic things. Such systems must be capable of being differentially reproduced in order to serve and behave as machines for generating the future. This does not simply mean that they must allow differences to occur; they must be organized in such a way that the production of differences becomes the reproductive driving force of the whole machinery; the system, then, may be said to be governed by *différance*. I have further argued that experimental systems display their dynamics in a space of representation in which material graphemes are articulated and disconnected—placed, displaced, and replaced. And finally, I have tried to show that it is through events of conjuncture, bifurcation, and hybridization that particular experimental systems rhizomize and get lumped and linked into larger experimental cultures.

Experimental scientists, in their daily bench work, deal with material units, with traces to which they convey the significance of being the "reals" of their particular practice. They move these marks around in spaces of representation, produce new entities by suppressing others, concatenate them into ever changing chains. They sort them out, let them act on each other, and from their responses, engage them in re-action. These significant units, or epistemic things, do not stand for immutable referents. In this respect, they are not signs for given objects, representatives of natural entities. They mean what they mean as far as they can be concatenated in spaces of representation. Concatenation counts. Experimental systems, like other systems of signification, are diacritically constituted as differential entities. Theoreticians, I claim, do nothing else when they move around graphic objects or numbers in a mathematically definable space. Brian Rotman argues that the "insistence on the priority of certain 'things' to mathematical signs is a misconception, a referentialist misreading, of the nature of written signs," and he shows that constellations of mathematical entities basically signify just their "capacity to further signify."[2] An experimental system only exists as a research activity as long as such a process of material signification, of decontextualization and recontextualization, continues. Experimental scientists do not read the book of nature, they do not depict reality. But they do not construct reality either. They are not engaged in platonistic exercises, in copy theory-guided asymptotic approximations to reality, or in bluntly social constructivist endeavors.

In configuring and reconfiguring epistemic things, scientists meet with resistance, resilience, recalcitrance. Not anything goes. If there is construction, it is constrained. The lines along which researchers are trying to partition their worlds only take on meaning from resonance between the partitions. Without a minimum of resonance there would be no durable technical objects.[3] But if there is a scientific real, it is multiple. Without a minimum of plurality, there would be no changeable epistemic objects either. What saves us from flat realism is that the resonances that are built up in a particular historical conjuncture are not the only ones possible and conceivable. Resonances are transient, although sometimes long lived. Their transitivity institutes the endless game of realization of the possibles. As Francis Crick once pointed out, in a typical research situation, "*anything* [should] be pursued as far as it can be, since one can never be sure in advance what may turn up."[4] Scientific objects, not objects or reality per se, but objects insofar as they are epistemic targets, are unstable concatenations of marks or representations. At best, epistemic things become metastable for some historically bounded period. It is not that there was no materiality there before they came into being or that they would dwindle altogether and shrink to mere illusions one day. But they can become, within an altered scientific context, for instance, altogether marginal because nobody expects them any more to be generators of unprecedented events. They can also become silenced as objects of research and live their lives as unquestioned technicalities. To understand the strange and fragile reality of scientific objects in the long run, it is crucial to take into account this double movement of becoming central and fading into marginality within the realm of a particular experimental culture.

Scientists are engaged and involuted in graphematic activities, not in purely technical endeavors. They configure and reconfigure recalcitrant entities along partitions that are neither laid bare nor simply imposed by the one material world, nor baldly created through our actions. Scientists are engaged in the paradox of cognition: com*passionate* dis*engagement*. The resulting "sutures" constitute a graphematic texture, and the graphematic texture constitutes the sutures. I see no escape from this circle. The use of the notion of suture here is inspired by the use Jacques Lacan makes of it when he says that "the subject [remains] the correlate of science, but an antinomial correlate since science turns out to be defined by the deadlocked endeavor to suture the subject." Hence, "there is something in the status of science's object which seems to me to have remained unelucidated since the birth of science."[5] Scientific objects have captured my attention throughout this book, and I hope to have been able to cast some light on their coming into being, their status, and their dynamics. Sutures are the conditions of possibility of things becoming objects of science, and they are the visible signs of a mutilation. Sutures are the lines along which dissections were tried, and at the same time they are the never healing traces of their failure. They make up the experience of *versäumen* in the double sense of this old-fashioned German word: to patch and miss.

It is in a similar vein that I have claimed, "the writing writes."[6] It is the differential iteration of chains of signification that renders an experimental texture prolific. Should we speak, therefore, of "significtion"? The scientist, as an authoritative speaker, is not the ultimate master of the game. But as a humble subject, he or she finds him- or herself captured in an inextricable relation of internal exclusion with his or her objects. He or she makes them, but only insofar as they make him or her make *them*. This movement is continually deferred and subverted by its own products. I ask and answer with Marjorie Grene: "*Why* can't we check our beliefs against reality? Not, as sceptics believe, because we can't reach out to reality, but because we're part of it."[7] For want of better imagery, we might speak of an ecohistorical nexus in an environment of potential traces. Experimental systems are comparable to ecological niches. A niche is neither defined by its inhabitants nor by the physical parameters of the habitat. It is an ensemble of changing interactions. It is a nonreferential entity.

Prolific experimental systems operate most productively at the fuzzy boundary *between* the trivial and the complex. Seen from within a particular experimental system, complexity functions as an epistemic horizon. What guarantees that the experimental breakdown of complexity proceeds into acceptable directions? My claim is that there is no general guarantee or rule for this. It is the network of surrounding experimental systems that makes each of its elements take on its epistemic value. If ontical complexity has to be reduced in order to make experimental research possible, this very complexity is *epistemically* retained in the rich contexture of an experimental landscape in which new connections and disconnections can happen at any time and where the eruption of one "volcanic system" can change the whole landscape, through passage and propagation.

With this, we come very near to an idea that Stuart Kauffman develops in his recent book *At Home in the Universe* and which he calls the patch procedure.

The basic idea of the patch procedure is simple: take a hard, conflict-laden task in which many parts interact, and divide it into a quilt of nonoverlapping patches. Try to optimize within each patch. As this occurs, the couplings between parts in two patches across patch boundaries will mean that finding a "good" solution in one patch will change the problem to be solved by the part in the adjacent patches. Since changes in each patch will alter the problems confronted by the neighboring patches, and the adaptive moves by those patches in turn will alter the problem faced by yet other patches, the system is just like our model coevolving ecosystems. [We] are about to see that if the entire conflict-laden task is broken into the properly chosen patches, the coevolving system lies at a phase transition between order and chaos and rapidly finds very good solutions. Patches, in short, may be a fundamental process we have evolved in our social systems, and perhaps elsewhere, to solve very hard problems.[8]

Let us call this the patchwork view of research. We have to look for a logic that lies somewhere other than in the rationality of actors and the ambitions of a community. The patches, that is, the experimental systems, are the subcritical elements of a network that, as a whole, takes on the features of a supracritical process we call science in the making. For the time being, of course, this does not amount to much more than a seductive metaphor. I hope, however, that it will be a good heuristic guide to what will result, one day, in a comprehensive history of molecular biology that has not yet been written. This book is itself a patch, a shred of historical work that may hint at the directions to be taken. Epistemology is the quest for tools that help us to understand a little better the historical evolution of the sciences. We only have started to look at the non-Cartesian, emergent properties of this inextricable web that unfolds in utterly unforeseeable ways and nevertheless shows pattern.

At the end of his book, Kauffman ponders, "I wonder if we really understand very much of what we are creating." And he continues: "All we can do is be locally wise, even though our own best efforts will ultimately create the conditions that lead to our transformations to utterly unforeseeable ways of being."[9] Local wisdom is what characterizes the practice of the sciences. Similarly, instead of hunting after universal theories, the order of the day for epistemology is to learn to understand how local wisdoms, entrenched in research attractors such as experimental systems, get connected to knowledge patchworks. Fragmentation, far from being deleterious, appears as one of the basic conditions of unprecedented development. Fragmentation, aiming at simplicity, finally recreates complexity. Understanding the dynamics of these interactions and transformations is what a philosophy of the epistemological detail will have to address. Epistemology, too, is called to become experimental if it wants to be an endeavor congenial with the practices it tries to analyze. There is still a long way to go in this direction.

Our modern world, with all its postmodern amoeboid protrusions, is shaped and dominated by a plexus of rhizomic technical systems. We rightly speak therefore of a technological civilization. It is not the sciences that have been the founding forces of modern technology. Just the opposite: it is a technological form of life that gave that particular epistemic activity we call "science" its historical impact and its quasi-irresistible drive. In the last resort, scientific systems derive their importance, their dignity, and their valuation from this superspace. But let me pronounce clearly and distinctly—if only once and at the very end of a book dedicated to what could be called an epistemology of the vague in science—that this does not mean that the sciences' potential for signification, and their epistemic power as well, derives from this technical superspace. A permanent threat of technological deconstruction goes along with experimental systems. Where should change come from otherwise? The significance of the experimental game derives from its reproductive decoherence: from being articulated horizontally and vertically, synchronically and diachronically, in space and in time, as a generator of surprises. In contrast to students of technoscience and "technowledge," respectively, I therefore think that we have reasons not to abandon altogether the distinction between the sciences and technology. We should better try to understand their mutual generation and degeneration.

Throughout this book, I have referred to the different epistemic stages of and events in a particular research system: an in vitro system of protein biosynthesis. I have localized this system within the floating history of cancer research, biochemistry, and molecular biology. Naturally, the reorientation at work in such a localization will not come to a halt before my own narrative. In fact, if this narrative gives rise to recurrent assessments at all, it will have performed its service. Thus, I return to Freud. The conceptual structure that has served to frame my historical interpretation remains open to refinement, replacement, and discharge.

Let me anticipate the objection that the institutional setting, the societal interests, the political power game involved in making science in post–World War II America, and above all the actors have not been given enough voice. To this I answer that my purpose has been different: I have tried to convey a sense of how experimental systems become articulated, boom, and come of age; how they eventually may shape a whole laboratory culture; and how they become, remain, and finally cease to be generators of epistemic novelty. My concern has been with a history of traces and things rather than with theories and interests. For those who have followed to this point, this will be an unnecessary statement. For those who read books by looking for conclusions, it stands as a necessary caveat.

REFERENCE MATTER

Glossary

A glossary inevitably codifies the meanings of scientific concepts—past and present—through definition and by some accepted standards. I have hesitated to add this largely anachronistic index to the book. I do it in the hope that it may help historians and philosophers of science unfamiliar with the terminology of biochemistry, especially protein synthesis, to find their way through the technical parts of the text. For scientists and historians unfamiliar with poststructuralist terminology, shorthand definitions of some major epistemological concepts are included.

acceptor site. Site on the ribosome to which incoming transfer RNAs become attached, thereby delivering their amino acid to the growing peptide chain.

acetate. Negatively charged moiety of acetic acid, CH_3COO^-.

acetyl. Chemical group derived from acetic acid, CH_3CO-. Important in metabolism.

acetyl-CoA (coenzyme A). Molecule containing a high-energy sulfur bond, $H_3C-CO \sim S\text{-}CoA$. Used in the enzymatic transfer of acyl groups in the cell.

activating enzyme. Any one of at least twenty different enzymes that catalyze (1) the reaction of a specific amino acid with ATP to form aminoacyl-AMP (activated amino acids) and pyrophosphate and (2) the transfer of the activated amino acid to transfer RNA resulting in aminoacyl-tRNA and free AMP.

activation. Process in which amino acids are prepared to become attached to their respective transfer RNAs.

adaptor molecules. Small RNA molecules that guide amino acids to their proper positions on an mRNA template during protein synthesis. Each adaptor is specific to both an amino acid and a template codon. See also *S-RNA, transfer RNA*.

adenine (A). A nitrogenous base, one member of the base pair A=T (adenine-thymine).

adenosine monophosphate (AMP). One of the four nucleotides in an RNA molecule. Two phosphates are added to AMP to form ATP.

adenosine triphosphate (ATP). Building block of RNA composed of adenine, ribose, and three phosphate groups. Serves also as the principal carrier of chemical energy in cells. The terminal phosphate groups are highly reactive in the sense that their hydrolysis, or transfer to another molecule, goes along with the release of a large amount of free energy.

adenyl-amino acids. In protein synthesis, an activated compound that is an intermediate in the formation of a covalent bond between an amino acid and its transfer RNA adaptor. In short, aa~AMP.

adenylation. Process by which amino acids become connected to adenosine monophosphate.

aerobic. Describes a process that requires, or occurs in the presence of, gaseous oxygen (O_2).

alumina. Powder of Al_2O_3 used for adsorption of biomolecules.

amino acids. The building blocks of proteins. Proteins are composed of twenty common amino acids. All amino acids have the same basic structure (H_2N-CHR-COOH), containing both an amino group and a carboxyl group, but they differ in their side groups (R). The twenty amino acids said to occur "naturally" are alanine, arginine, asparagine, aspartic acid, cysteine, glutamine, glutamic acid, glycine, histidine, isoleucine, leucine, lysine, methionine, phenylalanine, proline, serine, threonine, tryptophane, tyrosine, and valine.

amino group. Chemical moiety ($-NH_2$) that is an invariant part of amino acids. Characteristically basic because of the addition of a proton to form $-NH_3^+$.

ammonium sulfate precipitation. Technique for fractionating proteins according to their solubility in varying concentrations of $(NH_4)_2SO_4$ (ammonium sulfate).

analogue. Chemical substance that can replace a similar substance in a particular reaction.

angstrom (Å). Unit of length used to describe atomic dimensions. Equal to 10^{-10} m or 0.1 nm.

anhydride. Chemical substance generated by the fusion of two molecules of acid, whereby one molecule of water is eliminated.

anion. Negatively charged atom or molecule.

antibiotic. Substance usually produced by microorganisms (bacteria, fungi) in order to fight competitors in an environment of limited nutrient supply.

antibody. Y-shaped protein molecule that binds to and neutralizes a foreign molecule (antigen).

apoenzyme. Core protein of an enzyme complex.

archaeology. A type of historical analysis interested in discourse-objects and the strata-specific relations they entertain between each other (Foucault).

ascites cells. Cells derived from the fluid of the peritoneal cavity under certain pathological conditions.

autocatalysis. Reaction that is catalyzed by one of its products, thus creating a positive feedback (self-amplifying) effect on the reaction rate.

autoradiography. A technique that uses X-ray film to visualize radioactively labeled molecules or fragments of molecules.

autotroph. Organism that uses exclusively anorganic material for growth.

bacterium. Common name for any member of the diverse group of prokaryotic organisms. Most bacteria are single cells and lack a nucleus and cellular organelles.

bacteriophages. Viruses that multiply in bacteria.

base. Molecule (usually containing nitrogen) that accepts a proton in solution. Often used to refer to the purines and pyrimidines in DNA and RNA. Guanine and adenine belong to the purines; cytosine and thymine (uracil in ribonucleic acids) belong to the pyrimidines.

base-pairing. Specific (complementary) hydrogen bonding between pairs of nucleotides. A pairs with T (or U in RNA); C pairs with G.

bifurcation. Irreversible branching point in a historical trajectory.

buffer. Solution whose pH, upon addition of hydrogen or hydroxyl ions, changes only insignificantly. Used for keeping biomolecules in native state in vitro.

Butter Yellow. A cancer-inducing chemical, para-dimethylaminoazobenzene.

^{14}C. A radioactive carbon isotope emitting weak beta particles (electrons). Its half-life is 5,700 years.

carboxyl group. Chemical moiety (-COOH) that is part of many organic compounds. It is an invariant part of amino acids. Characteristically acidic as a result of the dissociation of the hydrogen (H) to form $-COO^-$.

carcinogen. Agent, such as a chemical or a form of radiation, that causes cancer.

carcinoma. Cancer of epithelial cells; the most common form of human cancer.

catalyst. Substance that accelerates a chemical reaction without itself undergoing a change. Enzymes are usually protein catalysts.

cell-free extract. A fluid containing most of the soluble molecules of a cell, made by breaking open cells and getting rid of remaining whole cells.

charging enzymes. Enzymes that load activated amino acids on their respective transfer RNAs.

chloramphenicol. Antibiotic that interferes with protein synthesis in bacteria.

chloroplast. Specialized organelle in green algae and plants that contains chlorophyll and performs photosynthesis. It is a specialized form of a plastid.

cholesterol. Lipid with a steroid component, found in many membranes.

chromatography. Biochemical technique in which a mixture of substances is separated on the basis of charge, size, or some other property by allowing the mixture to partition between a moving phase and a stationary phase.

code. Triplet pattern of nucleic acids that serves to specify the order of amino acids in proteins. For example, the triplet UUU codes for the amino acid phenylalanine. Also, as "second code," the molecular properties of a transfer RNA specifying the type of amino acid attached to it.

coenzyme. Small molecule tightly associated with an enzyme. Participates in the reaction that the enzyme catalyzes, often by forming a transient covalent bond to the substrate.

coenzyme A. See *acetyl-CoA*.

cofactor. Inorganic ion or coenzyme required for an enzyme's activity.

complementarity. Stereochemical fit between nucleotide bases $C \equiv G$ and $A = T$ ($A = U$), based on specific hydrogen bonding. Permits faithful replication of DNA, transcription of RNA, and translation of RNA into proteins.

condensation. Process of polymerization of macromolecules in which water moieties are liberated.

conformation. Three-dimensional shape of a macromolecule. As a rule, macromolecules can adopt different, active and inactive, conformations.

conjuncture. The irreversible turns and/or fusions in a historical trajectory.

countercurrent distribution. Preparative procedure for separating macromolecules on the basis of small differences in solubility.

covalent bonds. Strong chemical bonds formed by the sharing of one or more pairs of electrons between atoms.

cytidine triphosphate (CTP). Building block of RNA composed of cytosine, ribose, and three phosphate groups.

cytoplasm. Contents of a cell that are contained within its plasma membrane but, in the case of eukaryotic cells, outside the nucleus.

cytosine (C). A nitrogenous base, one member of the base pair $G \equiv C$ (guanine and cytosine).

cytosol. Contents of the main compartment of the cytoplasm, excluding membrane-bounded organelles such as endoplasmic reticulum and mitochondria. Originally defined operationally as the cell fraction remaining after membranes, cytoskeletal components, and organelles have been removed by low-speed centrifugation.

deoxycholate. Detergent used for the solubilization of lipid membrane components.

deoxyribonucleotide. A compound that consists of a purine or pyrimidine base bonded to the sugar, 2-deoxyribose, which in turn is bound to a phosphate group.

detergent. Type of small molecule that tends to coalesce in water, with its hydrophobic tails buried and its hydrophilic heads exposed; widely used to solubilize membrane proteins.

dialysis. Procedure whereby small components of a solution can diffuse through a membrane and equilibrate with the surrounding medium.

différance. The movement and force of differentiation of any process of signification (Derrida).

dinitrophenol (DNP). Aromatic compound interfering with oxidative phosphorylation.

discourse. An ensemble of practices including linguistic ones that convey a horizon both of extended meaning and of constraints to scientific and other cultural activities (Foucault).

DNA (deoxyribonucleic acid). A generally double-stranded, helical polymer of deoxyribonucleotides. The genetic material of all cells.

DNase. Enzyme that degrades DNA.

Ehrlich ascites. See *ascites cells*.

electron microscopy. A visualization technique that uses beams of electrons instead of light rays and that permits magnifications beyond the realm of optical microscopes. Resolutions of about 10 angstroms are attainable with biological materials.

electrophoresis. A method for separating large molecules (such as nucleic acids and proteins) from a mixture of similar molecules. An electric current is passed through a medium containing the mixture, and each kind of molecule travels through the medium at a different rate, depending on its electrical charge and size.

endergonic. Used to describe a chemical reaction that needs an input of (net) energy to take place.

endogenous. Used to describe a component that is a constituent part of an entity such as an organism or a cell.

endoplasmic reticulum. Labyrinthine, membrane-bounded compartment in the cytoplasm of eukaryotic cells, where lipids are synthesized and membrane-bound proteins are made.

enzyme. Protein molecule capable of catalyzing biochemical reactions.

enzyme induction. Process whereby a molecule (e.g., a sugar) induces the production of an enzyme not normally present in the cell that helps to metabolize such molecules.

epistemic thing. Scientific object, that is, an entity whose unknown characteristics are the target of an experimental inquiry.

ergastoplasm. The cytoplasmic ground substance of a cell.

Escherichia coli (*E. coli*). Rodlike bacterium normally found in the colon of humans and other mammals. Widely used in biomedical research because of its genetic properties, normal lack of pathogenicity, and ease of growth in the laboratory.

eukaryote. Organism composed of cells that have a membrane-bounded nucleus, membrane-bounded organelles, and 80S-type ribosomes.

experimental system. A basic unit of experimental activity combining local, technical, instrumental, institutional, social, and epistemic aspects.

expression. Production of an observable trait using the information contained in a gene—usually the synthesis of a protein.

fatty acids. Compounds that have a carboxylic acid attached to a long hydrocarbon chain.

ferritin. An iron storage protein primarily found in the liver and spleen.

g. Gravitation constant. The strength of centrifugal fields is measured as a multiple of g.

galactosidase (β-galactosidase). An enzyme catalyzing the decomposition of the sugar lactose into glucose and galactose; in *E. coli*, the classic example of an inducible enzyme.

G factor. Protein that brings about the movement of transfer RNA on the ribosome.

gram-positive, gram-negative bacteria. Typology of microorganisms based on staining characteristics of the cell wall.

grapheme. Any kind of material signifier. Scientific objects are, first and foremost, articulations of graphemes (Derrida). See also *inscription* and *trace*.

guanine (G). A nitrogenous base, one member of the base pair G≡C (guanine and cytosine).

guanosine triphosphate (GTP). Building block of RNA composed of guanine, ribose, and three phosphate groups. Nucleoside triphosphate used in the synthesis of RNA and in some energy-transfer reactions.

GTPase. Enzyme capable of hydrolyzing GTP.

hemoglobin. Protein carrier of oxygen found in red blood cells.

Henriot-Huguenard centrifuge. Small, air-driven ultracentrifuge that can attain very high speeds and centrifugal fields.

hepatoma. A specific form of liver cancer.

heterologous. Used to describe components stemming from different sources. See *homologous*.

high-energy bond. A chemical bond that contains a relatively high amount of free energy. The energy can be liberated in hydrolysis or transfer reactions.

histochemistry. Localization of chemically defined components and reactions in different tissues.

historiality. The temporal characteristics of any signifying activity as a process without definitely assignable origin or ground (Derrida).

homogenization. Procedure whereby cells of various sources are broken and their contents separated from the cellular envelopes.

homologous. Used to describe components stemming from the same source. See *heterologous*.

homopolymer. Macromolecule composed of an array of identical building blocks.

hydrolysis. The breaking of a molecule into two or more smaller molecules by the addition of a water molecule.

hydroxamate. Compound composed of an amino acid and a hydroxylamine molecule.

hydroxyl. Chemical group (-OH) consisting of a hydrogen atom linked to an oxygen.

hydroxylamine. Chemical compound (NH_2OH) able to react with an activated amino acid.

incorporation of amino acids. Operational definition of protein synthesis based on the observation that radioactive amino acids become incorporated into proteins.

inorganic phosphate. Single phosphate molecule, PO_4^{3-}.

inscription. The process of producing experimental traces with the help of instruments and other devices; also the products of this process (Latour).

in situ. Pertaining to experiments done on intact tissue and cells usually fixed and/or stained.

intermediate. Chemical compound between the educt and the product of a metabolic reaction chain.

internucleoside phosphate. Phosphate molecule connecting two nucleotides within a nucleic acid chain.

in vitro. Pertaining to experiments done in a cell-free system.

in vivo. Pertaining to experiments done in a system such that the organism remains intact, either at the level of the cell (in the case of bacteria) or at the level of the whole organism (in the case of animals).

isotope. One of several forms of an atom that have the same number of protons and electrons but differ in the number of neutrons. May be either stable or radioactive.

jaundice disease. Liver disease, here of silkworms, caused by a virus.

ketoglutarate (α-ketoglutarate). Negatively charged moiety of ketoglutaric acid, $^-OOC\text{-}CO\text{-}CH_2\text{-}CH_2\text{-}COO^-$. Intermediate in the citric acid cycle.

kinetics. Experimental procedure whereby samples are successively taken and measured over a defined period of time in order to analyze the time characteristics of a biochemical process.

label (radioactive label). A radioactive atom, introduced into a molecule to facilitate, for example, the observation of its metabolic transformation.

lipids. A large, varied class of water-insoluble organic molecules. Includes steroids, fatty acids, waxes, and the like.

macromolecule. Molecule with a molecular weight ranging from a few thousand to millions dalton. Includes nucleic acids and proteins.

messenger RNA (mRNA). RNA that becomes associated with ribosomes and serves as a template for protein synthesis.

metabolic pathway. A set of consecutive intracellular enzymatic reactions converting one molecule to another.

metabolism. The sum total of the various biochemical reactions occurring in a living cell, required for growth and the maintenance of life.

micron (micrometer). Unit of measurement often applied to cells and organelles. Equal to 10^{-6} m.

microsomes. First observed as particulate structures of eukaryotic cells sedimenting at high speed in a centrifuge. Later identified as consisting of chunks of endoplasmic reticulum and ribonucleoprotein particles. The latter became known as ribosomes.

microtome. Apparatus able to cut very thin sections of embedded tissue and other biological material.

mitochondrion. Membrane-bounded organelle found in the cytoplasm of all aerobic eukaryotic cells. Carries out oxidative phosphorylation and is the center of ATP-generation.

mitosis. Process whereby chromosomes duplicate and segregate, usually accompanied by cell division.

molecular weight. The sum of the atomic weights of the constituent atoms in a molecule. The unit is dalton.

monomer. The basic building block from which, by repetition of a particular reaction, polymers are made. For example, amino acids (monomers) condense to yield polypeptides or proteins (polymers).

monosome. Single ribosome devoid of messenger RNA.

nanometer (nm). Unit of length commonly used to measure molecules and cell organelles; 1 nm = 10^{-9} m.

neoplasm. Tumor caused by progressive uncontrolled cell division.

ninhydrin reaction. Color reaction involving the free amino group of an amino acid or protein. Serves to identify free amino acids.

nucleic acid. A large molecule composed of nucleotide subunits. See also *DNA* and *RNA*.

nucleolus. Round, granular structure found in the nucleus of eukaryotic cells, usually associated with specific chromosomal sites. Involved in ribosomal RNA synthesis and ribosome formation.

nucleolysis. General term designating the degradation of nucleic acids.

nucleotide. Building block of nucleic acids composed of a base (purines A and G or pyrimidines C, T, or U); a sugar moiety (ribose or deoxyribose); and one, two, or three phosphate groups (NMP, NDP, or NTP, respectively; N standing for a base and M, D, and T for a mono-, di-, or triphosphate, respectively).

nucleus. Membrane-bounded organelle in eukaryotic cells that contains the chromosomes.

oligonucleotide. Short piece of DNA or RNA.

oncogenesis. The process of the development of cancer.

organelle. Membrane-bounded structure found in eukaryotic cells containing enzymes for specialized functions. Includes mitochondria and chloroplasts.

oxidation. Chemical reaction that involves electron transfer from a reductant to an oxidant. The reductant is said to be oxidized and the oxidant reduced as a result of the transfer.

oxidative phosphorylation. Process in bacteria and mitochondria in which ATP formation is driven by the transfer of electrons from food molecules to molecular oxygen, mediated by the generation of a proton gradient across the membrane.

^{32}P. A radioactive isotope of phosphorus that emits strong beta particles and has a half-life of 14.3 days.

pancreas. Secretory organ of vertebrates. Produces insulin.

pantothenic acid. Vitamin belonging to the vitamin B_2 complex.

penicillinase. Enzyme capable of degrading the antibiotic penicilline. Found, for example, in *staphylococci*.

pentose nucleic acid. Formerly used to designate RNA.

peptidases. Enzymes that are able to cut peptide bonds.

peptide bond (α-peptide bond). A covalent bond between two amino acids in which the α-amino group of one amino acid is bonded to the α-carboxyl group of the other to yield a special form of amide linkage.

peptides. Stretches (usually short) of amino acids linked together via peptide bonds.

periodate. Chemical differing from iodate by the presence of an extra oxygen atom. Useful for oxidizing and cutting open the terminal sugar of an RNA.

petite colonies (yeast). Abnormally small colonies formed by certain mutant yeast strains.

pH. Common measure of the acidity of a solution.

pH 5 enzymes. Enzymes that can be precipitated from a high-speed supernatant by adjusting its pH to approximately 5.

pH 5 precipitate. Any material precipitating from a high-speed supernatant at pH 5.

phage. See *bacteriophages*.

phenol extraction. Procedure whereby proteins can be separated from nucleic acids. After shaking and subsequent phase separation, the proteins remain in the organic phenol phase, whereas the nucleic acids remain in the water phase.

phosphate-ATP-exchange. Reaction in which phosphates are incorporated into ATP, usually monitored by incorporation of radioactive ^{32}P.

phosphatase. Enzyme that removes phosphate groups from substrates such as proteins and nucleic acids.

phospholipids. Lipids that contain charged, hydrophilic phosphate head groups. Phospholipids are a primary component of cell membranes.

phosphorylation. Reaction in which a phosphate group becomes covalently coupled to another molecule.

photosynthesis. Process by which plants and some bacteria use the energy of sunlight to drive the synthesis of organic molecules from carbon dioxide and water.

plasmagenes. Nucleoprotein of the cytoplasm assumed to have genetic properties and to be capable of self-replication. Notion in use for cytoplasmic nucleoprotein mainly during the 1940s.

polyadenylic acid (poly[A]). Nucleic acid consisting solely of phosphate-linked adenosine residues.

polyanion. Molecule carrying multiple negative charges, for example, a nucleic acid.

polyglucose. Macromolecule composed of glucose units.

polymer. A regular, covalently bonded arrangement of small subunits (monomers) that is produced by repetitive occurrence of one or a few chemical
reactions.

polymerization. Chemical process of forming polymers from monomers.

polynucleotide. A linear sequence of nucleotides in which the 3' position of the
sugar of one nucleotide is linked through a phosphate group to the 5' position on the sugar of the adjacent nucleotide.

polynucleotide phosphorylase. A bacterial enzyme that catalyzes the polymerization of ribonucleoside diphosphates to yield free phosphate and RNA.

polypeptide. A polymer of amino acids linked together by peptide bonds.

polyphenylalanine (poly[Phe]). A polypeptide composed solely of phenyl-
alanines.

polyribonucleotide. A linear sequence of ribonucleotides.

polysome. Complex of a messenger RNA molecule and ribosomes (number
depending on size of mRNA), actively engaged in polypeptide synthesis.

polyuridylic acid (poly[U]). Nucleic acid consisting solely of phosphate-linked
uridine residues.

postmicrosomal fraction. Fraction of a cell homogenate after spinning down the
microsomes.

potassium fluoride (KF). Chemical compound composed of a potassium and a
fluorine ion.

pragmatogony. The effort to understand the deployment of symbolic activities
through their entrenchment in practices (Serres).

primary structure. The sequence of monomers in a macromolecule composed
of different building blocks.

prokaryote. Generally monocellular organism lacking a nucleus and organelles.
Contains 70S-type ribosomes.

protein. A large molecule composed of one or more chains of amino acids in a
specific order. Proteins are required for the structure, function, and regulation of cells, tissues, and organs, and each protein has unique functions.

proteolysis. Degradation of a protein, usually by hydrolysis at one or more of its
peptide bonds.

protoplast. Cell devoid of cell wall but with intact cell membrane.

pulse-chase experiment. Procedure whereby cells are exposed to a short pulse of
radioactive compounds followed by a large excess of the same compound in
non-radioactive form. Serves to trace the uptake and retention of the compound, for example, an amino acid or nucleotide.

purine. One of the two categories of nitrogen-containing ring compounds found in DNA and RNA. Examples are adenine and guanine.

puromycin. Antibiotic that inhibits polypeptide synthesis.

pyrimidine. One of the two categories of nitrogen-containing ring compounds found in DNA and RNA. Examples are cytosine, thymine, and uracil.

pyrophosphate. Two molecules of inorganic phosphate linked together.

radioactive isotope. An atom with an unstable nucleus that stabilizes itself by emitting ionizing radiation. See *isotope*.

recurrence. Recursive assessment of past scientific events from the (inevitable) perspective of a future perfect, on the level of both scientific practice and historiography (Bachelard).

replication (DNA replication). The use of existing DNA as a template for the synthesis of new DNA strands. In eukaryotes, replication occurs in the cell nucleus before cell division.

respiration. General term for any process in a cell in which the uptake of molecular oxygen (O_2) molecules is coupled to the production of carbon dioxide (CO_2).

reticulocyte. Immature red blood cell active in hemoglobin synthesis.

ribonuclease (RNase). An enzyme that degrades RNA.

ribonucleoprotein. Structure composed of proteins and ribonucleic acids.

ribonucleotide. A compound that consists of a purine (A, G) or pyrimidine (C, U) base bonded to ribose, which in turn is esterified with a phosphate group.

ribose. Sugar moiety of the building blocks of RNA.

ribosome. Small cellular particle (about 200 Å in diameter) made up of ribosomal RNA and protein. Ribosomes have come to be identified as the sites of protein synthesis.

RNA (ribonucleic acid). A polymer whose building blocks are ribonucleotides. Three main kinds of RNA can be distinguished: ribosomal RNA, transfer RNA, and messenger RNA.

RNP-particle. See *ribonucleoprotein*.

rpm. Measure—in rounds per minute—for the velocity of a rotor in a centrifuge.

^{35}S. A radioactive isotope of sulfur, a beta emitter with a half-life of 87 days.

S value. The unit of sedimentation (S = svedberg). S is proportional to the rate of sedimentation of a molecule in a given centrifugal field and is thus related to the molecular weight and shape of the molecule.

sarcoma. Cancer of connective tissue.

sediment. Material settling at the bottom of a test tube at a given centrifugation speed and time.

sedimentation coefficient. See *S value*.

Sephadex. Gel material composed of cross-linked polysaccharides, used for chromatographic separation of macromolecules based on molecular weight (size).

sequence. Linear array of the (different) building blocks of nucleic acids or proteins.

sequentialization. Process that determines the linear sequence of amino acids in a protein.

soluble fraction. Fraction of a homogenate, at a given centrifugation speed, that remains in solution, that is, does not sediment.

soluble RNA (S-RNA). Class of small RNA molecules found in the soluble fraction after high-speed centrifugation of a cell homogenate. Originally identified by their capacity to bind amino acids covalently. See also *transfer RNA, adaptor molecules*.

space of representation. The coordinate system in which scientific objects become instantiated as articulations of traces. Spaces of representation are opened (not determined) by technical procedures.

Spinco. Commercial ultracentrifuge with refrigeration and a vacuum chamber.

Staphylococcus aureus. Microorganism belonging to the type of gram-positive bacteria.

stoichiometry. Determination of the equivalent and atomic weight of the elements in a molecule or the molecules in a supramolecular assembly.

sucrose. Sugar composed of a glucose and a fructose moiety.

supernatant. Part of a homogenate that remains in solution at a given centrifugation speed.

supplementation. The mutual affection of discourses at their interfaces. Results in grafting, hybridization, and displacement of discourses (Derrida).

suture. Partition line along which scientific objects are crafted by subjects remaining in internal exclusion to their objects (Lacan).

synthetase. See *activating enzyme*.

technical object. Counterpart to *epistemic thing* or object. Technical objects embody the knowledge of a given research field at a given time.

template. The macromolecular mold for the synthesis of another macromolecule.

template RNA. RNA determining the sequence specificity of a protein. Originally thought to be identical with microsomal RNA. See also *messenger RNA*.

text. Anything that conveys meaning in externalized and durable form. In addition to written sources, practices of various other kinds can be seen and treated as texts (Derrida).

thymine (T). A nitrogenous base, one member of the base pair A=T (adenine and thymine).

tobacco mosaic virus (TMV). Virus infecting tobacco plants, consisting of a ribonucleic acid core and a coat made up of many identical protein molecules.

trace. Experimentally produced signifier that becomes part of a scientific object.

tracing. Technique of following the pathway of a metabolite by means of a radioactive label.

transacylation. Here, transfer of an activated amino acid to its transfer RNA (involving a transesterification).

transamidation. Transfer of an amide linkage of a molecule from one carrier to another.

transcription. A process involving base-pairing, whereby the genetic information contained in DNA is used to order a complementary sequence of bases in an RNA chain.

transfer RNA. Any of at least twenty structurally similar species of RNA, all of which have a molecular weight of about 25,000. Each species of transfer RNA is able to combine covalently with a specific amino acid and to hydrogen bond with at least one of the 64 mRNA nucleotide triplets. See also *adaptor molecules.*

translation. The process whereby the genetic information present in a messenger RNA molecule directs the order of the amino acids in a growing protein chain.

transpeptidation. Transfer of an amino acid or peptide from one peptide to another. Sometimes also used for the transfer of a growing peptide chain to a transfer RNA-bound amino acid.

trichloroacetic acid. Acetic acid that has the three hydrogens of its -CH_3 moiety substituted by chlorine atoms.

triplet. Any combination of three nucleotides of a messenger RNA. There are 64 possible triplets.

ultracentrifuge. Analytical or preparative device that can attain high speeds (in the order of 60,000 rpm) and centrifugal fields (in the order of 500,000 × g) and thus is capable of rapidly sedimenting macromolecules.

unprecedented event. Experimental result that arises from the differential activities possible within an experimental system. Stands for the production of novel knowledge that cannot be anticipated.

uracil (U). A nitrogenous base normally found in RNA but not DNA; uracil is capable of forming a base pair with adenine.

uridine triphosphate (UTP). Building block of RNA composed of uracil, ribose, and three phosphate groups.

virus. Infectious disease-causing agent, smaller than bacteria, which always requires intact host cells for replication and contains either DNA or RNA as its genetic component.

Notes

The following abbreviations are used for archival sources throughout the Notes and References.

FCL   Francis A. Countway Library of Medicine, Boston.
GLS   Green Library, Stanford University, Special Collection.
MGHR   Massachusetts General Hospital, Research Affairs Office.
MLN   Heinrich Matthaei, Laboratory Notebooks. In possession of H.M.
ZLN   Paul Zamecnik, Laboratory Notebooks. In possession of P.Z.
ZRN   Paul Zamecnik, Research Notes. In possession of P.Z.

Prologue

1. For an outline, see Rheinberger 1992a, 1992b, 1993, 1994. See also Rheinberger and Hagner 1993, and Hagner, Rheinberger, and Wahrig-Schmidt 1994.

2. Rheinberger 1992b.   3. Hoagland 1990, p. xvii.

4. Lenoir 1993.   5. Latour 1990b, p. 66.

6. Kubler 1962, p. 10. See also Lubar and Kingery 1993.

7. Derrida 1976, p. 24 and elsewhere.

8. Hacking 1983, p. 150.

9. Stent 1968.

10. Hoagland 1990, p. 82.

11. See, e.g., Abir-Am 1980, 1985, 1992; Gaudillière 1991, 1993; Burian 1993b, 1996; Kay 1993; Chadarevian 1996; Creager 1996.

12. Jacob 1988.

13. See, among others, Zamecnik 1969 and 1979; Lipmann 1971; Tissières 1974; Siekevitz and Zamecnik 1981; Hoagland 1990; Nomura 1990; Spirin 1990. For a history of "coding" and "information" in molecular biology, see Kay, forthcoming.

14. Foucault 1972a, pp. 139–40.

Chapter 1

1. Misleadingly translated as "scientific reality" (Bachelard 1984, p. 6). French and German texts are quoted from translations where they exist and when I had access to them. Occasional alterations have been made, however, where I have judged them to be necessary. All translations from the French or German original are my own.

2. Freud 1957b, p. 117.

3. Freud 1957a, p. 77.

4. Fleck 1979, pp. 23–27.

5. Elkana 1970.

6. Löwy 1992. For the notion of "boundary object," see Star and Griesemer 1988.

7. Blumenberg 1986.

8. Serres 1989, pp. 4, 15.

9. Fleck 1979, p. 78.

10. Reichenbach 1938, pp. 6–7.

11. Holmes 1985, p. xvi.

12. See Latour 1990a, pp. 160–64, with reference to Serres 1987.

13. Kuhn 1962; Feyerabend 1975. For the post-Kuhnian engagement, see, among others, Latour and Woolgar 1979; Knorr Cetina 1981; Hacking 1983; Lynch 1985; Collins 1985; Shapin and Schaffer 1985; Franklin 1986, 1990; Galison 1987; Latour 1987; Gooding, Pinch, and Schaffer 1989; Gooding 1990; Le Grand 1990; Lynch and Woolgar 1990b; Pickering 1992, 1995; Rheinberger 1992a, Buchwald 1995; Rouse 1996.

14. Hacking 1983, p. 150.

15. Galison 1988, pp. 209, 211.

16. Lenoir 1988, pp. 11–12.

17. Lenoir 1992; see also Shapin and Schaffer 1985.

18. Pickering 1992.

19. For key texts of the strong program of sociology of science, see Barnes 1974; Bloor 1976; Barnes 1977; Collins 1985.

20. Kuhn 1992, p. 9.

21. Shapin and Schaffer 1985.

22. The notion is borrowed from William Wimsatt's account of biological models of developmental constraint. See Wimsatt 1986.

23. Latour 1987, pp. 132–44.

24. Pickering 1995. Pickering's book came to my attention only after completion of this manuscript.

25. Derrida 1978, pp. 283–84.

26. Ibid., p. 284. Note the proximity of the formulations in this passage to Freud's text on *Narcissism*; see note 3 in this chapter.

27. Derrida 1976, p. 24.

28. Latour 1993a, p. 6.

29. Latour 1990b, p. 64.

30. Latour 1993a, p. 142.

31. Hoagland 1990, p. xvi.

32. Rheinberger 1992a and 1992b.

33. Turnbull and Stokes 1990.

34. Kohler 1991b, and 1994.

35. Cf., e.g., the recent special issue on "Immunology as a Historical Object" of the *Journal of the History of Biology* 27:3 (1994); see also Rabinow 1996.

36. See Luhmann 1990, especially chapter 5, "Science as System," pp. 271–361.

37. Hoagland 1990, p. xx.

38. Bachelard 1984, p. 13; Bachelard 1951, p. 84.

39. Bachelard 1984, p. 6; I have altered the translation slightly.

40. Wittgenstein 1953, paragraph 7.

41. Ibid., paragraph 654.

42. See, e.g., Latour 1987; Wise 1992; Hentschel 1993.

43. Kant 1987, p. 264.   44. Rouse 1991, p. 161.

45. Guggenberger 1991.   46. Lacan 1989, p. 22.

47. Bachelard 1968, p. 12.

48. Polanyi 1965, fourth lecture on "The Emergence of Man," pp. 4–5. Quoted in Grene 1984, p. 219.

Chapter 2

1. Zamecnik, Keller, Littlefield, Hoagland, and Loftfield 1956. The symposium was held at the Research Conference for Biology and Medicine of the Atomic Energy Commission, Oak Ridge National Laboratory, Gatlinburg, Tennessee, April 4–6, 1955.

2. Lacan 1989, p. 10.

3. Lacan 1986, p. 122.

4. Judson 1979, p. 275. Jacob renders the dictum in the following form: "Al Hershey, one of the most brilliant American specialists on bacteriophage, said that, for a biologist, happiness consists in working up a very complex experiment and then repeating it every day, modifying only one detail" (Jacob 1988, p. 236).

5. Jacob 1988, p. 234.

6. Heidegger 1977b, p. 118. "Die Zeit des Weltbildes" might be more appropriately translated as "The Epoch of Planetary Configuration."

7. Holmes 1985, p. xvi.

8. Goethe 1988. A more accurate translation of "Der Versuch als Vermittler von Objekt und Subjekt" would be "The *Assay* as Mediator *of* Object and Subject" (emphasis added).

9. Kittler 1990.

10. Goethe 1988, p. 16.

11. Staiger 1966, letter of Goethe to Schiller, January 10, 1798.

12. Ibid., letter of Schiller to Goethe, January 12, 1798, pp. 539–42.

13. Popper 1968, p. 107.

14. Collins 1985.

15. Fleck 1979, p. 96, emphasis added. Ludwik Fleck, mainly because of his

notion of "*Denkstil*," has often been misconstrued as a forerunner of the Kuhnian way of thinking in "paradigms."

16. Fleck 1979, p. 86.

17. Kuhn 1992, p. 14.

18. Jacob 1988, p. 9.

19. Lenoir 1992. See also Lenoir 1988, where a number of related positions are discussed.

20. Bachelard 1984, p. 139.

21. Ibid., chapter 6. See also Bachelard 1968.

22. As far as the history of molecular biology is concerned, one of the most brilliant examples is Jacob 1988. In recent years, we have witnessed a rapidly accumulating body of autobiographies from molecular biologists of the first generation. See, among others, Watson 1968, Luria 1984, McCarty 1985, Crick 1988; Kornberg 1989; Hoagland 1990. For a review of some of these works, see Abir-Am 1991.

23. For a more fine-grained analysis, see Hentschel 1995.

24. Serres 1987, p. 191.

25. Latour 1987, pp. 87–88.

26. For the notion of "constraint," see Galison 1995.

27. Sanger, Nicklen, and Coulson 1977.

28. Latour 1987, p. 131 and elsewhere.

29. Fleck 1979, p. 86, emphasis in first sentence omitted.

30. For more on scientific texts in biology, see Myers 1990; see also Bazerman 1988.

31. E.g., Latour 1987, p. 174.   32. Bachelard 1984, p. 13.

33. Heidegger 1971, p. 74.   34. Ibid., p. 75.

35. Hoagland 1990, p. xvi.

36. "*Erstreckte Gegenwart*," as Helga Nowotny has called it (Nowotny 1994, p. 52).

37. See Jardine 1991.

38. Weber 1989, p. 982.

39. See Damerow and Lefèvre 1981, pp. 223–33; Rohbeck 1993, especially chapter 6.

40. Bachelard 1984, p. 8; translated here misleadingly as "revealing light." For the notion of recurrence, see also Bachelard 1951, chap. 1, "Les récurrences historiques: Epistémologie et histoire des sciences."

41. Butterfield 1957.

42. See, e.g., Mayr 1990. On the problem of "whig history," see also Clark 1995.

43. It is therefore not simply a tautological joke when Francis Crick proposed, on "doubtful grounds," as he admitted with an ironically self-deprecating gesture, that molecular biology "can be defined as anything that interests molecular biologists." See Crick 1970, p. 613.

44. See, e.g., Cairns, Stent, and Watson, 2d ed. 1992; Rich and Davidson 1968; Monod and Borek 1971; Lwoff and Ullmann 1979.

45. Burian 1996.

46. Scholarly historical work on the subject is still greatly lacking. For some information, see Portugal and Cohen 1977; Judson 1979; Bartels 1983; Rheinberger 1992b, 1993, 1995, 1996; Burian 1993a; Morange 1994, especially chapters 12 and 13.

47. Cf. Kohler 1991a.

48. For a detailed discussion, see Chapter 7 in this volume.

49. Bachelard 1984, p. 6.

50. Rotman 1987, p. 102.

51. Knorr Cetina, Amann, Hirschauer, and Schmidt 1988.

Chapter 3

1. Zamecnik and Stephenson 1978; Stephenson and Zamecnik 1978; Agrawal, Ikeuchi, Sun, Sarin, Konopka, Maizel, and Zamecnik 1989; Zamecnik and Agrawal 1991; see also Lunardini 1993.

2. Zamecnik, letter to Rheinberger, November 5, 1990.

3. Ibid.

4. Faxon 1959, p. 229; Aub, Brues, Dubos, Kety, Nathanson, Pope, and Zamecnik 1944; see also a series of six consecutive papers in the *Journal of Clinical Investigation* 24 (1945), starting with Nathanson, Nutt, Pope, Zamecnik, Aub, Brues, and Kety 1945.

5. The history of the Massachusetts General Hospital and its research facilities is well documented. Cf. Faxon 1959; Garland 1961; Castleman, Crockett, and Sutton 1983.

6. Zamecnik 1960, 1962a, 1969, 1976, 1979, 1984; Siekevitz and Zamecnik 1981; Hoagland 1990, 1996. See also Rheinberger 1993.

7. See Faxon 1959, pp. 204–7, 231–40; Castleman, Crockett, and Sutton 1983, pp. 343–50. See also Zamecnik 1974, 1983; Bucher 1987; Hoagland 1990, pp. 37–39.

8. Zamecnik, research notes (ZRN), draft "for International Cancer Research Foundation/application, 3/8/45."

9. Loftfield, correspondence on a draft of this manuscript, May 17, 1993 (abbreviated as Loftfield, correspondence).

10. Zamecnik 1950, p. 659.

11. Ibid., p. 660.

12. On the neglected history of the "multi-enzyme programme of protein synthesis," see Bartels 1983.

13. See Zamecnik and Lipmann 1947; Zamecnik, Brewster, and Lipmann 1947.

14. Faxon 1959, p. 48.

15. Lipmann 1941. See also Kalckar 1941.

16. ZRN, draft, "American Cancer Society application, March 8, 1947."

17. Castleman, Crockett, and Sutton 1983, p. 33; the quotation is from the report of the trustees.

18. See, e.g., Kay 1993.

19. Massachusetts General Hospital Research Affairs Office (MGHR), General Executive Committee, annual report, 1934, pp. 23–24.

20. MGHR, Scientific Advisory Committee, Recommendations and Comments, November 24 and 25, 1950.

21. Castleman, Crockett, and Sutton 1983, pp. 35–36.

22. Kohler 1991a.

23. Francis A. Countway Library of Medicine, Boston (FCL), Aub Files GA4, Box 3, Letter of P. W. Bridgman from Harvard University to Joseph Aub, member of the Harvard Cyclotron Committee, April 30, 1942.

24. Interview with Robert B. Loftfield, June 18, 1993.

25. The tracer he used in 1936 was naturally occurring radioactive lead. Frantz, letter to Rheinberger, July 7, 1994. See also Zamecnik 1983, p. 347. Aub had been working on the metabolism of lead since 1924 at MGH, initially under David Edsall.

26. Zamecnik 1983, p. 347.

27. ZRN, draft, "American Cancer Society application, March 8, 1947."

28. Loftfield 1947, p. 54.

29. ZRN, draft, "American Cancer Society application, March 8, 1947"; Miller 1947.

30. Loftfield, interview, 1993.

31. Frantz, letter to Rheinberger, July 7, 1994.

32. MGHR, Committee on Research Minutes, book 1, January 1947–December 1950, pp. 93–95.

33. MGHR, Committee on Research, Executive Committee Minutes, book 1, March 1948–December 1950, p. 59.

34. Frantz, letter to Rheinberger, July 7, 1994.

35. Frantz, Loftfield, and Miller 1947.

36. Jacob 1988, p. 9.

37. Loftfield, interview, 1993.

38. Those other workers included Melchior and Tarver 1947a, 1947b; Winnick, Friedberg, and Greenberg 1947. A whole-animal experiment required 100 times the radioactivity of a slice experiment. See ZRN, "Final Report to Donner Foundation, Inc., January, 1948."

39. See Zamecnik, Frantz, Loftfield, and Stephenson 1948.

40. The slices were kept in a Krebs-Ringer phosphate solution in Warburg flasks under an oxygen atmosphere.

41. Zamecnik, Frantz, Loftfield, and Stephenson 1948, p. 299.

42. Frantz, Loftfield, and Miller 1947. All parameters kept "identical," the

values ranged between 102 to 916 counts per minute incorporated into protein. The last value was considered to be erroneous but was nevertheless included "in the interest of completeness" (p. 545).

43. Zamecnik, Frantz, Loftfield, and Stephenson 1948, table 1.

44. Frantz, Zamecnik, Reese, and Stephenson 1948.

45. Loomis and Lipmann 1948.

46. Loftfield, interview, 1993. Between 1948 and 1953, MGH received an institutional grant of $100,000 per year from the American Cancer Society. Some $5,000 per year of this money went into Zamecnik's projects. MGHR, Report to the Scientific Advisory Committee Meeting, December 12 and 13, 1952.

47. Hoagland 1990, p. xix.

48. Loftfield, correspondence.

49. Melchior and Tarver 1947a, 1947b; Winnick, Friedberg, and Greenberg 1947.

50. Anfinsen, Beloff, Hastings, and Solomon 1947.

51. Melchior and Tarver 1947a, 1947b; Friedberg, Winnick, and Greenberg 1947; for an early review, see Greenberg, Friedberg, Schulman, and Winnick 1948; Borsook, Deasy, Haagen-Smit, Keighley, and Lowy 1949a; for a review of the early work, see Borsook 1950.

52. Zamecnik 1950, p. 660.

53. Frantz and Feigelman 1949, 619.

54. Zamecnik, Frantz, and Stephenson 1949.

55. Stein and Moore 1948; see Zamecnik 1984.

56. Interview with Mary L. Stephenson, March 25, 1994.

57. See, e.g., Bucher and Glinos 1948.

58. Bucher, Loftfield, and Frantz 1949.

59. Zamecnik, Frantz, Loftfield, and Stephenson 1948, p. 310.

60. Frantz, Loftfield, and Werner 1949.

61. Borsook, Deasy, Haagen-Smit, Keighley, and Lowy 1949b, p. 589.

62. Zamecnik, Loftfield, Stephenson, and Williams 1949.

63. Loftfield 1957a, p. 371.

64. The collaboration is documented by the number of jointly published articles.

65. Hoagland 1990, p. 48; interview with Mahlon B. Hoagland, March 15, 1990.

66. Stephenson, interview, 1991; interview with Nancy L. R. Bucher, July 16, 1993.

67. Loftfield, interview, 1993.

68. Zamecnik and Frantz 1949.

69. Stephenson, relying on Loftfield's expertise in ion exchange, had invested considerable time in standardizing the Moore-Stein technique for this purpose (Zamecnik, Frantz, and Stephenson 1949).

70. Zamecnik and Frantz 1949, p. 205.

71. Ibid., p. 206.

72. Ibid.

73. Ibid., p. 207; quoting from "The Secret Sits," a poem by Robert Frost (Frost 1964, p. 495).

74. Zamecnik 1950.

75. Ibid., p. 662.

76. See, e.g., Melchior and Tarver 1947b; Greenberg, Friedberg, Schulman, and Winnick 1948; Borsook, Deasy, Haagen-Smit, Keighley, and Lowy 1949a, 1949b.

77. Compared to the slice system, the activity dropped again by an order of magnitude.

78. Zamecnik 1950, p. 663.

79. Ibid., p. 659.

80. Ibid., p. 660.

Chapter 4

1. Zamecnik 1950, p. 663.

2. One of these systems monitored the incorporation of radioactive glycine into glutathione. Quotation from Zamecnik 1950, pp. 663–64.

3. Zamecnik 1950, p. 666.

4. I owe this formulation to Robert Loftfield.

5. Frantz and Loftfield 1950.

6. Linderstrøm-Lang 1952; Steinberg and Anfinsen 1952; Campbell and Work 1953; Tarver 1954.

7. Loftfield, Grover, and Stephenson 1953.

8. See Fruton 1952.

9. Loftfield, Grover, and Stephenson 1953, p. 1025.

10. Ibid.; see Sanger and Tuppy 1951.

11. Bachelard 1984, p. 139; see also Chapter 2 in this volume.

12. Stent 1968.

13. Hoagland 1990, p. 82.

14. Besides the radioactive protein, this microsome fraction contained the highest ratio of RNA (and DNA) to protein.

15. For a wider historical context, see Rheinberger 1995.

16. Rous 1911.

17. Ledingham and Gye 1935; McIntosh 1935.

18. Claude 1938, p. 402.

19. Claude 1941, p. 265.

20. Burian 1995.

21. For a review, see Brachet 1947b, p. 18.

22. Brachet 1942.

23. Ibid., p. 239.

24. Caspersson used the technique of ultraviolet absorption of nucleic acids. See Caspersson 1941.

25. Paillot and Gratia 1938.

26. Brachet and Jeener 1943–45; Chantrenne 1943–45; Jeener and Brachet 1943–45.

27. Brachet 1949, p. 863, see also Chantrenne 1991, and interview with Hubert Chantrenne, May 28, 1996.

28. Claude 1943a.

29. Claude 1943b, pp. 119–20.

30. Palade 1951, p. 144.

31. Hogeboom, Schneider, and Palade 1948.

32. Claude 1950, p. 163.

33. Chantrenne 1947, p. 445.

34. Ibid., p. 447.

35. Brachet and Shaver 1949, p. 205; Shaver and Brachet 1949.

36. Hultin 1950; Borsook, Deasy, Haagen-Smit, Keighley, and Lowy 1950b; Lee, MacRae, and Williams 1951.

37. Keller 1951.

38. For the concept of triangulation, see Star 1986; Gaudillière 1994.

39. Latour 1990b.

40. See Wahrig-Schmidt and Hildebrandt 1993.

41. Hogeboom, Schneider, and Palade 1948.

42. For a review, see Ernster and Schatz 1981; see also Rheinberger 1995.

43. Siekevitz and Zamecnik 1951.

44. α-ketoglutarate or succinate were used as energy suppliers.

45. Siekevitz 1952.

46. Possible reaction candidates were transcarboxylation, transpeptidation, and the building of S-S bridges and phosphatidic acids.

47. Melchior and Tarver 1947a; for a detailed discussion, see Tarver 1954.

48. Siekevitz, letter to Rheinberger, July 1, 1994; see also Friedberg, Winnick, and Greenberg 1947.

49. See Winnick, Peterson, and Greenberg 1949.

50. Schneider and Hogeboom 1950.

51. Ninhydrin releases CO_2 containing the radioactive ^{14}C from free, and only from free, amino acids. For the difficulties mentioned, see Tarver 1954. For a retrospective review, see Loftfield 1957a.

52. Borsook, Deasy, Haagen-Smit, Keighley, and Lowy 1950a.

53. Provided α-ketoglutarate was added as an oxidative substrate.

54. Siekevitz and Zamecnik 1951.

55. Siekevitz 1952, p. 562.

56. MGHR, Committee on Research, Executive Committee Minutes, book 1, March 1948–December 1950, and book 3, January 1953–December 1954.

57. Siekevitz and Zamecnik 1981, p. 54S.

58. See MGHR, Committee on Research Minutes, books 1–4, 1947–59. For a history of the Office of Naval Research, see Sapolsky 1990.

59. MGHR, Committee on Research, book 2, February 1951–December 1953, p. 325.

60. Loftfield, correspondence.

61. Loftfield, interview, 1993.

62. See Kay 1993.

63. Interview with Paul C. Zamecnik, March 16, 1990.

64. Siekevitz, letter to Rheinberger, July 1, 1994.

Chapter 5

1. Jacob 1988, p. 274.
2. Ibid., p. 255.
3. Kubler 1962, p. 125.
4. Heidegger 1977b, p. 124.
5. Jacob 1982.
6. Ibid., p. 11; I have altered the translation.
7. Derrida 1976, pp. 23–24 and elsewhere.
8. Ibid., p. 24.
9. Bernard 1954, p. 14.
10. Bernard 1974, p. 15.
11. Fleck 1979, p. 96.
12. Polanyi 1958, 1967, 1969, especially part 3.
13. For the quotations here, see Keller 1983, p. 198, MacColl 1989, p. 90; Elkana 1981, pp. 42–48; and Suchman 1990, p. 310, respectively.
14. Polanyi 1969, pp. 138–58.
15. Polanyi 1958, p. 49, emphasis in original omitted.
16. "It is the old story of the principle of measured sloppiness that leads to discovery" (Fischer and Lipson 1988, p. 184). Fischer quotes from a letter of Max Delbrück to his friend Salvador Luria dating from autumn 1948.
17. Here, Delbrück is quoted from a meeting in Oak Ridge in 1949 (Fischer and Lipson 1988, p. 184).
18. The quotation is taken from Dagognet's characterization of Claude Bernard (Dagognet 1984, p. 18).
19. Deleuze 1994, p. xix.
20. Bachelard 1968, p. 12.
21. Deleuze 1994, p. xix.
22. Loftfield, interview, 1993.
23. Amann and Knorr Cetina 1990, pp. 104, 111.
24. Kohler 1991b.
25. Serres 1980, p. 126.
26. Kubler 1962, p. 33 and the following.
27. Ibid., p. 85.
28. Ibid., p. 36.
29. Wise 1992, p. 34.
30. Derrida 1988, pp. 18–19.
31. Derrida 1976, p. 23.
32. Derrida 1982b, p. 7.
33. Derrida 1991, pp. 26–27.
34. Derrida 1976, p. 24. I will come back to this notion in Chapter 11.

35. Derrida 1988, p. 18.

36. Stengers 1987.

Chapter 6

1. Bachelard 1968, p. 94.

2. Zamecnik 1979, p. 296.

3. Stephenson, interview, 1991.

4. I will come back to the *E. coli* system in Chapter 12.

5. MGHR, Committee on Research, Executive Committee Minutes, book 2, January 1951–December 1952.

6. Gale and Folkes 1953c.

7. Gale and Folkes 1953b, p. 728. For a more extended account of Gale's work, see Rheinberger 1996.

8. St. Aubin and Bucher 1951.

9. Frantz, letter to Rheinberger, July 7, 1994.

10. A buffer system highly fortified with cofactors and metabolic substrates.

11. Bucher, interview, 1993.

12. Bucher 1953; Frantz and Bucher 1954.

13. Originally, it had been used to stabilize suspensions of mitochondria. See Chapter 4.

14. Siekevitz 1952.

15. MGHR, Committee on Research, Executive Committee Minutes, book 3, January 1953–December 1954.

16. Zamecnik and Keller 1954, p. 338.

17. For a detailed account, see Rheinberger 1995.

18. They were not alone in observing such stimulation. Winnick (1950) had reported a stimulatory effect of ATP on amino acid incorporation in fetal liver homogenates and Greenberg on particles sedimented at low speed (Peterson and Greenberg 1952; Kit and Greenberg 1952).

19. Lipmann 1941, 1949.

20. It amounted to $10,406.88 for the period July 1, 1954 to June 30, 1955. MGHR, Committee on Research, book 2, February 1951–December 1953.

21. Letter of Zamecnik to Dean A. Clark, MGHR, Committee on Research, Executive Committee Minutes, book 3, January 1953–December 1954.

22. ATP, phosphocreatine, and the enzyme creatine phosphokinase were included.

23. Keller and Zamecnik 1954, p. 240.

24. Zamecnik and Keller 1954, p. 337.

25. Ibid., p. 351.

26. Stephenson, interview, 1991.

27. See, e.g., Peterson and Greenberg 1952.

28. See, e.g., Zamecnik and Keller 1954, p. 347.

29. Gale 1955, p. 183.

30. Loftfield 1957a, p. 351.

31. Ibid., p. 352.

32. Collins 1985, pp. 83–84, 147.

33. Rheinberger 1989.

34. Loftfield 1954.

35. Ibid., p. 465.

36. Simpson, Farber, and Tarver 1950.

37. Brachet 1947a and b; Caspersson 1947. See also Chapter 4 of this volume.

38. Gale and Folkes 1953c, 1954; Allfrey, Daly, and Mirsky 1953.

39. Zamecnik and Keller 1954, p. 352.

40. The ratio was 1:100 for mitochondria, 2.2:100 for the soluble fraction; "good" microsomes, for comparison, had a ratio of 14:100.

41. Loftfield, correspondence.

42. Keller and Zamecnik 1955, p. 234.

43. ZRN, letters of November 22 and December 6, 1954.

44. Littlefield, Keller, Gross, and Zamecnik 1955a, 1955b.

45. Strittmatter and Ball 1952. Avery, MacLeod, and McCarty (1944) had already used deoxycholate for isolating their "transforming agent."

46. The value corresponded to that given by Schachmann, Pardee, and Stanier (1952) for *Pseudomonas fluorescens* and by Petermann, Hamilton, and Mizen (1954) for rat liver and spleen particles.

47. Claude and Fullam 1945.

48. Porter 1953; for a description of these studies, see Rheinberger 1995.

49. Palade 1955. See also Rasmussen, forthcoming, Stanford University Press.

50. Palade and Siekevitz 1956, pp. 171–72.

51. Littlefield, Keller, Gross, and Zamecnik 1955a.

52. Palade 1955.

53. Petermann and Hamilton 1952; Petermann, Mizen, and Hamilton 1953; Petermann, Hamilton, and Mizen 1954.

54. Latour 1988, p. 227.

55. I will come back to these changes in Chapter 8.

56. Keller, Zamecnik, and Loftfield 1954.

57. Littlefield, Keller, Gross, and Zamecnik 1955a, p. 121.

58. Zamecnik and Keller 1954.

59. Jacob 1988, p. 235.

60. Ibid., p. 279.

Chapter 7

1. Lynch 1994, p. 148.

2. Van Fraassen and Sigman 1993, p. 74.

3. Mitchell 1987, p. 17.

4. Peirce 1955, pp. 102–3.

5. Jacob 1974, pp. 203–4.

6. Latour and Woolgar 1979; Hacking 1983. For recent collections of articles covering the field, see Lynch and Woolgar 1990b; Levine 1993; Hart Nibbrig 1994; Rheinberger, Hagner, and Wahrig-Schmidt 1996; see also the special issue of *Culture technique* on "Les vues de l'esprit" (1985) and the special issue on "Pictorial Representation in Biology" in *Biology and Philosophy 6* (1991).

7. Lynch 1994.

8. Goodman 1968, p. 9.

9. Latour 1988, pp. 222–36.

10. Lynch and Woolgar 1990a, p. 13.

11. Saussure 1959.

12. Bernard 1954, p. 14.

13. For the positioning of representation in Derrida's thought, see Gasché 1986. Gasché insists with Derrida that representation has to be understood as a form of iterability (pp. 212–17). See also Bennington and Derrida 1991.

14. Latour 1993b, p. 213.

15. Derrida 1988, pp. 18–19.

16. Derrida 1976, p. 9.

17. Latour 1987; 1993a.

18. Latour and Woolgar 1986; Hacking 1992, p. 44.

19. Goodman 1968; Mitchell 1987, p. 71.

20. Latour 1990c, p. 26 and the following.

21. Scientists talk about "significant," not about "true," traces.

22. Bachelard 1933, p. 140; see also Strack 1989 on "standard procedures" and "habituation."

23. Jacob 1988, p. 284.

24. The translation speaks of "engraving" and of " 'reading' matter" (Bachelard 1984, p. 168).

25. Jacob 1988, p. 281.

26. Goethe 1988, p. 17.

27. Derrida 1982b, p. 7.

28. Bernard 1966, p. 19; 1974, p. 15.

29. Goodman 1968, p. 8.

30. Cf., e.g., Tarski 1946, pp. 120–25.

31. Jacob 1988, p. 248.

32. Cf., e.g., Amann 1994.

33. Lamborg and Zamecnik 1960, p. 210.

34. Lacan 1989, p. 9.

35. Derrida 1976, p. 145.

36. Derrida 1988, p. 8.

37. Latour and Woolgar 1986, p. 51; see also Latour 1987, p. 64 and the following.

38. Derrida 1976, p. 9.

39. Hayles 1993, pp. 28, 33.

40. Husserl 1954.

41. Hacking 1983, p. 132. For a paleontological perspective, see Leroi-Gourhan 1964–65.

42. Baudrillard 1983, pp. 31–32, 146.

43. Hacking 1983, p. 136.

44. Lynch 1994, p. 146.

Chapter 8

1. Littlefield, Keller, Gross, and Zamecnik 1955a, p. 121.

2. Maas and Novelli 1953.

3. Lipmann 1954, p. 602. The model was based on observations with co-enzyme A-linked activation of acetate and pantoate activation for the formation of a peptide bond.

4. Lipmann 1970, p. 60. For Lipmann's wanderings "from $\sim$P to CoA to protein biosynthesis," see also Novelli 1966.

5. Zamecnik, interview, 1991.

6. Grier, Hood, and Hoagland 1949; Hoagland 1952.

7. An expression used by Hoagland in his autobiography. See Hoagland 1990, p. 58.

8. Hoagland 1990, p. 59 and the following.

9. Hoagland and Novelli 1954.

10. Fritz Lipmann, first draft of memoirs, p. 63. Library of Congress, Manuscript division, no. 80-498171.

11. Hoagland 1990, p. 71. See Chapter 6 in this volume for Zamecnik's project.

12. Doudoroff, Barker, and Hassid 1947; Maas and Novelli 1953; see also Lipmann 1954.

13. Hoagland 1955a. The paper was received on December 4, 1954. It appeared in the February issue of 1955.

14. Loftfield 1993, correspondence.

15. Hoagland 1955a; see also Hoagland 1955b.

16. Maas and Novelli 1953.

17. Hoagland 1955a.

18. Hoagland, Keller, and Zamecnik 1956. The paper was received for publication in June 1955.

19. Hoagland, Keller, and Zamecnik 1956.

20. Zamecnik 1979, p. 275.

21. Davie, Koningsberger, and Lipmann 1956.

22. Hoagland 1990, p. 83.

23. DeMoss and Novelli 1955; their article was received September 28, 1955, and published December 1955.

24. Green Library, Stanford University (GLS), Berg papers, Box 11, notebook 6.

25. Ibid., notebook 4.

26. Berg 1955; see also Berg 1956.

27. Hoagland, Keller, and Zamecnik 1956, p. 356.

28. Hoagland 1990, p. 79.

29. Hoagland, Zamecnik, Sharon, Lipmann, Stulberg, and Boyer 1957.

30. Hoagland, Keller, and Zamecnik 1956, pp. 355–56.

31. Monod, Pappenheimer, and Cohen-Bazire 1952; Halvorson and Spiegelman 1952; Rotman and Spiegelman 1954.

32. Simpson and Velick 1954; Askonas, Campbell, and Work 1954; Straub, Ullmann, and Acs 1955.

33. Francis and Winnick 1953; Friedberg and Walter 1955.

34. Loftfield 1954, 1955; Loftfield and Harris 1956.

35. Haurowitz 1956; Borsook 1956b.

36. Loftfield 1957b; Loftfield and Eigner 1958.

37. Palade 1955; Palade and Siekevitz 1956.

38. Claude 1941, 1943a.

39. For in vivo protein synthesis, see Borsook, Deasy, Haagen-Smit, Keighley, and Lowy 1950a; Hultin 1950; Keller 1951; Lee, Anderson, Miller, and Williams 1951; Tyner, Heidelberger, and LePage 1953; Smellie, McIndoe, and Davidson 1953; Allfrey, Daly, and Mirsky 1953. For in vitro protein synthesis, see Siekevitz 1952; Allfrey, Daly, and Mirsky 1953; Zamecnik and Keller 1954.

40. Porter 1953; Palade and Porter 1954; Porter and Blum 1953. See also Rasmussen, forthcoming, Stanford University Press.

41. Palade and Siekevitz 1956, pp. 189–90.

42. Petermann and Hamilton 1952, 1955; Petermann, Mizen, and Hamilton 1953, 1954.

43. Littlefield, Keller, Gross, and Zamecnik 1955a.

44. Palade and Siekevitz 1956.

45. Stephenson, Thimann, and Zamecnik 1956.

46. On Stephenson's list were bactericidal agents, iodoacetate, DNP, o-phenanthroline, azide, fluoride, cyanide, arsenate, malonate, versene, ethionine, dicumarol, antimycin A, and pancreatic ribonuclease!

47. Stephenson, Thimann, and Zamecnik 1956, p. 207.

48. Stephenson, interview, 1991.

49. Sissakian 1956.

50. Fraenkel-Conrat and Williams 1955.

51. Keller and Zamecnik 1955, 1956.

52. Sanadi, Gibson, and Ayengar 1954.

53. ZRN, letter of November 22, 1954.

54. Keller and Zamecnik 1956, p. 57.

55. In that respect, the system behaved differently than the disrupted *Staphylococcus* system of Gale and Folkes (1955b).

56. Littlefield and Keller 1956, 1957.

57. MGHR, Recommendations of the Scientific Advisory Committee, December 16 and 17, 1955, pp. 1–2.

58. Selby, Biesele, and Grey 1956.

59. Jeener 1948; see also Simkin and Work 1957.

60. Petermann, Hamilton, and Mizen 1954; Petermann and Hamilton 1955.

61. Zamecnik, Keller, Littlefield, Hoagland, and Loftfield 1956, p. 82.

62. Cf. Figure 6.1 in this volume.

63. Zamecnik, Keller, Littlefield, Hoagland, and Loftfield 1956, p. 87.

64. Hoagland 1990, p. 82.

65. Zamecnik, Keller, Hoagland, Littlefield, and Loftfield 1956, pp. 166–67.

66. Haurowitz 1949, 1950, especially chapter 17 on "Protein Synthesis."

67. Dounce 1952.

68. Lipmann 1954; cf. Figure 8.1 at the beginning of this chapter.

69. Chantrenne 1951; Koningsberger and Overbeek 1953; Todd 1955.

70. Gamow 1954.

71. See Kay's encompassing study of the "coding problem," forthcoming.

72. Zamecnik, Keller, Hoagland, Littlefield, and Loftfield 1956, p. 167.

73. Hoagland 1990, p. 82.

Chapter 9

1. Hacking 1983, p. 150.

2. Zamecnik 1960, p. 263.

3. Kuhn 1962.

4. See, e.g., Root-Bernstein 1989 and Bechtel and Richardson 1993, in a vast literature that will not be surveyed here.

5. See, e.g., Remer 1964.   6. Roberts 1989, p. x.

7. Lipmann 1971, p. v.   8. See Chapter V.

9. *Meyers Großes Taschenlexikon* 1990.

10. Cf. also Grmek and Fantini 1982.

11. Latour 1993a.

12. Darden and Maull 1977; Darden 1991.

13. Stengers 1987.   14. Kuhn 1992, pp. 19–20.

15. Fleck 1979.   16. Hacking 1992b, p. 6.

17. Kuhn 1992, p. 14.   18. Dijksterhuis 1962, p. 182.

19. Bernard 1957, pp. 14–15.   20. Bachelard 1968, pp. 10–12.

21. White 1980, p. 23.   22. Pickering 1992, pp. 2–3.

23. Foucault 1972b, p. 230. The insertion has been omitted from the English translation.

24. Foucault 1972a, especially part 4.

25. Heidegger 1977a, pp. 19–35.

26. Heidegger 1977b, p. 118; see Chapter 2 in this volume.

27. For recent analyses of molecular biology's discourse on information, see Doyle 1993; Keller 1994; Kay, forthcoming.

Chapter 10

1. Zamecnik 1979, pp. 299–300.

2. For an earlier reference, see Brachet 1952; for a later one, see the introductory remarks of J. N. Davidson to the British Biochemical Society symposium on February 18, 1956 (Davidson 1957).

3. Monod, Pappenheimer, and Cohen-Bazire 1952, p. 659; see also Brachet 1952; Monod and Cohn 1953, p. 58; Cohen and Barner 1954; Pardee 1954; Spiegelman, Halvorson, and Ben-Ishai 1955. For a historical account, see Gaudillière 1992.

4. For a review of studies on enzyme formation, see Spiegelman 1956a. Protein synthesis in protoplasts is summarized in Spiegelman 1956b. For protein synthesis in enucleated cells, see Malkin 1954. For in vivo and in vitro protein synthesis systems, see Gale and Folkes 1953a and c; Allfrey, Daly, and Mirsky 1953; Kruh and Borsook 1955; Gale 1955. For phage replication, see Hershey 1953. For a review, see, e.g., Borsook 1956b.

5. Hoagland 1990, p. 112.

6. Gale and Folkes 1955a, p. 683.

7. Brues, Tracy, and Cohn 1944; Zamecnik, letter to Rheinberger, November 5, 1990.

8. Littlefield, letter to Rheinberger, November 9, 1993. See also Littlefield, Keller, Gross, and Zamecnik 1955a, 1955b; Zamecnik, letter to Rheinberger, November 5, 1990; Hoagland 1990, p. 86.

9. Zamecnik, Keller, Littlefield, Hoagland, and Loftfield 1956, pp. 92–93.

10. Keller, Zamecnik, and Loftfield 1954; cf. p. 381.

11. Zamecnik, interview, 1991.

12. Zamecnik, Keller, Littlefield, Hoagland, and Loftfield 1956, pp. 93–98.

13. Grunberg-Manago and Ochoa 1955; see also Grunberg-Manago, Ortiz, and Ochoa 1955. For Zamecnik's awareness of Grunberg-Manago's finding, see Zamecnik 1979, p. 279.

14. Zamecnik, Keller, Hoagland, Littlefield, and Loftfield 1956, p. 172.

15. Potter, Hecht, and Herbert 1956; for a full account, see Herbert, Potter, and Hecht 1957.

16. Hoagland 1989, p. 104.

17. Zamecnik, laboratory notebooks (ZLN), October 31, 1955, "Expt to inquire whether C^{14}-orotate and C^{14}-ATP are incorporated into nucleic acid and into certain acid soluble nucleotides in our system."

18. Judson 1979, p. 314.

19. Zamecnik 1979, p. 279; Hoagland 1990, pp. 86–87.

20. ZLN, annotation to the experiment of November 3, 1955, "Expt to repeat the essential conditions of 10/31/55, but to 1) improve the washing procedure; and 2) introduce various controls and variants."

21. ZLN, experiment of November 10, 1955.

22. Stephenson, interview, 1991; acknowledgment in Hoagland, Zamecnik, and Stephenson 1957.

23. Zamecnik, interview, 1991.

24. ZRN, Potter to Zamecnik, April 30, 1956; Zamecnik to Potter, May 12, 1956.

25. ZLN.

26. ZLN, experiments from June 12, 15, and 19, 1956. Davie, Koningsberger, and Lipmann 1956.

27. Overbeek, letter to Rheinberger, February 2, 1995.

28. See also Zamecnik 1979, p. 278; to conclude from the notebooks, the assays were done in 1956 rather than 1955.

29. See Chapter 8.

30. Stephenson, interview, 1991. According to Loftfield, too, this "was the key." Loftfield, interview, 1993.

31. ZLN, July 12, 1956.

32. ZLN, experiment from September 19, 1956.

33. Zamecnik, Stephenson, Scott, and Hoagland 1957.

34. Hoagland 1990, p. 88.

35. Potter and Dounce 1956.

36. Latour 1987, p. 88.

37. Hoagland, Zamecnik, and Stephenson 1957.

38. ZLN, remark regarding the experiment from September 27, 1956.

39. ZLN, calculation from October 2, 1956.

40. ZLN, undated; October 1956 or January 1957.

41. Hoagland, Zamecnik, and Stephenson 1957. The paper appeared in *Biochimica et Biophysica Acta* 24:1 (April 1957).

42. Hoagland and Zamecnik 1957. Hoagland, Zamecnik, and Stephenson 1957.

43. Potter, Hecht, and Herbert 1956; Heidelberger, Harbers, Leibman, Takagi and Potter 1956; Herbert, Potter, and Hecht 1957. For an even earlier report on the incorporation of AMP into the RNA of a cell-free pigeon liver homogenate, see Goldwasser 1955.

44. Holley 1957. For a preliminary communication, see Holley 1956.

45. The manuscript was received by the *Journal of the American Chemical Society* on August 3, 1956. It should have been in the hands of Hoagland in September. Although the evidence for the participation of an RNA in amino acid activation was indirect, Hoagland recommended it for publication (Hoagland, letter to Rheinberger, May 24, 1990). The paper appeared on February 5, 1957. A brief survey reveals that Holley's paper appeared without delay, together with the majority of research articles submitted in September 1956.

46. Hoagland 1990, p. 92.

47. GLS, Berg papers, box 11, laboratory notebooks, 1953–59.

48. Hultin 1956; Hultin and Beskow 1956.

49. Hultin 1955, p. 216.

50. Ogata, Ogata, Mochizuki, and Nishiyama 1956.

51. Ogata, letter to Rheinberger, November 16, 1993. The results were first presented at the symposium "Biosynthesis of Protein and Enzymes" at the twenty-ninth annual meeting of the Japanese Biochemical Society held in Fukuoka on October 31, 1956. See Ogata and Nohara 1957; Ogata, Nohara, and Morita 1957.

52. The first paper appeared in August; the second in December 1956. They are extensively quoted in Loftfield 1957a, p. 369.

53. Ogata and Nohara 1957, see note added in proof.

54. Bachelard 1957, p. 13.

55. Mentioned in Hoagland, Zamecnik, and Stephenson 1957.

56. Stephenson, interview, 1991.

57. Hoagland 1990, pp. 117–19.

58. Hoagland 1989, p. 104.

59. Crick 1988, p. 96; for the published short remark, see Crick 1957.

60. Zamecnik, letter to Rheinberger, November 5, 1990; see also Judson 1979, pp. 293, 327; and Hoagland 1990, pp. 94–96.

61. Hoagland, Zamecnik, and Stephenson 1957, p. 215. This paper (Hoagland, Zamecnik, and Stephenson 1957) together with Hoagland's paper on amino acid activation (Hoagland 1955a) was quoted slightly more than 500 times during the period between 1955 and 1964. For comparison, Watson and Crick's *Nature* paper on the structure of DNA (Watson and Crick 1953) had the same number of quotations between 1953 and 1964. But whereas reference to the latter continued to be high (e.g., slightly below 400 between 1980 and 1989), reference to the former declined (twenty quotations between 1980 and 1989).

62. Hoagland 1989, p. 105.   63. Judson 1979, p. 327.

64. Hoagland 1989, p. 105.   65. Jacob 1988, p. 294.

66. Hoagland 1990, p. 96.

67. An assumption that shortly thereafter had to be dropped in favor of an ester bond. See Hoagland, Stephenson, Scott, Hecht, and Zamecnik 1958.

68. Loftfield 1957a, p. 379. The paper mentioned in the quotation as being a preprint is Crick, Griffith, and Orgel 1957.

69. Loftfield 1957a, p. 380.

70. ZLN, laboratory note of Zamecnik, April 1, 1957.

71. Loftfield 1957a, pp. 377–82.

72. It appeared in 1958. See Hoagland, Stephenson, Scott, Hecht, and Zamecnik 1958.

73. Hoagland 1990, pp. 97–98.

74. ZLN, letter of Francis Crick to Mahlon Hoagland attached to Zamecnik's laboratory notebooks, written in Cambridge, Cavendish Laboratory, January 20, 1957.

75. Zamecnik, Stephenson, and Hecht 1958.

76. Ibid., p. 77; emphasis added.

77. Derrida 1976, p. 145.

78. Hoagland 1958, p. 630.

79. Crick, letter to Hoagland of January 20, 1957, see note 74, this chapter.

80. ZLN, quotation from Jesse Scott, laboratory note, February 18, 1957. Note that M-RNA designates microsomal RNA, not messenger RNA.

81. Hoagland 1990, pp. 99–116.

82. Ibid., p. 114.

83. Ibid., p. 103.

84. For the fate of the "trinucleotide-adaptor," see Chapter 12.

85. For "*chemical* association," see Hoagland, Stephenson, Scott, Hecht, and Zamecnik 1958, introduction, p. 241; my emphasis. For "*complementarity*," see ibid., conclusion, p. 256; my emphasis.

86. See Gierer and Schramm 1956; Kirby 1956.

87. ZLN, summary of the experiments between February 7 and February 25, 1957.

88. ZLN, experiments dating from July 12 and July 20, 1956, respectively.

89. ZLN, laboratory note dating from October 2, 1956.

90. Zamecnik, interview, 1991. Alexander Todd in Cambridge was considered to be *the* expert on nucleic acid chemistry at that time. See, e.g., Todd 1956, where the nature of the internucleotidic linkage is discussed.

91. Hoagland, Stephenson, Scott, Hecht, and Zamecnik 1958, p. 255.

92. Ibid., p. 256.

93. Ogata and Nohara 1957; Koningsberger, Van der Grinten, and Overbeek 1957; Berg and Ofengand 1958; Schweet, Bovard, Allen, and Glassman 1958; Weiss, Acs, and Lipmann 1958; Holley and Prock 1958.

94. Jacob 1988, p. 294.

95. See Chapter 8.

96. Zamecnik, interview, 1991.

97. ZLN, notebooks, March 2, 1957.

98. Interview with Liselotte I. Hecht-Fessler, July 11, 1994.

99. Hecht, Stephenson, and Zamecnik 1958a.

100. Zamecnik 1960, p. 264.

101. Hecht, Stephenson, and Zamecnik 1958b.

102. Zamecnik, interview, 1991.

103. Hecht, Stephenson, and Zamecnik 1958b.

104. Hecht-Fessler, interview, 1994.

105. Canellakis 1957; Paterson and LePage 1957; Edmonds and Abrams 1957; Herbert 1958.

106. Hecht, Zamecnik, Stephenson, and Scott 1958.

107. Ibid., p. 962.

108. Zamecnik, Stephenson, and Hecht 1958, p. 74.

109. Zamecnik, Hoagland, Stephenson, and Scott 1958, p. 63.

110. Simkin and Work 1957; Littlefield and Keller 1957.

111. DeMoss, Genuth, and Novelli 1956; Berg 1957.

112. Hoagland 1958, p. 632.

113. Ibid., p. 633.

114. Hoagland 1958.

115. Loftfield, Hecht, and Eigner 1959.

116. Zachau, Acs, and Lipmann 1958. The paper had been communicated to the *Proceedings of the National Academy of Sciences* by Lipmann himself on July 30, 1958, and it appeared in the September issue of that journal.

117. Zamecnik, letter to Rheinberger, November 5, 1990.

118. Hecht, Stephenson, and Zamecnik 1959.

119. Berg and Ofengand 1958.

120. Hecht, Stephenson, and Zamecnik 1959, p. 517; emphasis added.

Chapter 11

1. Derrida 1976, p. 24.

2. Rudolf Daber, paleontologist at the Humboldt University in Berlin.

3. Bachelard 1984, p. 8. I have changed the English translation because it retains little of the sense of the original sentence.

4. Canguilhem 1975, pp. 178–79.

5. Kubler 1962, p. 35. Note the unwitting irony of this triple reassurance!

6. Goethe 1982, p. 424; Goethe 1957, p. 149.

7. See White 1980 for a critical assessment of the constitution of historical narrativity. For recent accounts, see also Carrard 1992 and Berkhofer 1995.

8. Derrida 1976, p. 61.

9. Kuhn 1992, p. 19.

10. The "disunity of science" has become a topic of increasing interest. See, e.g., Dupré 1993; Rosenberg 1994; Galison and Stump 1996.

11. Prigogine and Stengers 1979, p. 251.

12. Kubler 1962, pp. 83, 85.   13. Foucault 1972b, p. 231.

14. Nägele 1987, p. 1.   15. Jacob 1982, p. 67.

16. These expressions are Louis Althusser's. Althusser and Balibar 1968, pp. 115, 117, 131.

17. Derrida 1976, p. 23.   18. Derrida 1982b, p. 7.

19. Goethe 1982, p. 547.   20. Derrida 1982, p. 202.

21. Elkana 1970.   22. Canguilhem 1981, p. 15.

23. Foucault 1972b, p. 224.   24. Lwoff 1957, p. 252.

25. Louis Althusser has developed the notion of a "spontaneous philosophy of the scientist," in analogy to which my expression is built. See Althusser 1974.

26. Clark 1995, p. 67.

Chapter 12

1. Zamecnik 1960, p. 256.

2. Hofmeister 1902; Fischer 1906; Lipmann 1941; Bergmann 1942; Borsook and Dubnoff 1940; Schoenheimer 1942; Rittenberg 1941, 1950; Brachet 1942; Caspersson 1941; Sanger and Tuppy 1951; Palade 1955.

3. Zamecnik 1960, 278; emphasis added.

4. Cf. Yearley 1990.

5. Hoagland 1960, p. 373.

6. Zamecnik 1960, p. 263.

7. Smith, Cordes, and Schweet 1959.

8. Zamecnik 1960, p. 268.

9. Hoagland, Zamecnik, and Stephenson 1959, p. 110; Zamecnik 1960, p. 268.

10. Hoagland, Zamecnik, and Stephenson 1959, figure 2; Zamecnik 1960, figure 5.

11. For the "microsome," see Claude 1943a; for "ribonucleoprotein particle," see Petermann, Hamilton, and Mizen 1954; Littlefield, Keller, Gross, and Zamecnik 1955a, 1955b; for electron microscopy, see the review by Palade 1958; see also Zamecnik 1958 for a general overview.

12. Roberts 1958. The term *ribosome* had been proposed for the first time by Howard M. Dintzis in 1957 (Wim Möller, personal communication, and letter of Howard Dintzis to Wim Möller, dated August 22, 1989). See also Roberts 1964, p. 148.

13. Crick 1958, p. 153.

14. Crick 1957, 1958; Crick, Griffith, and Orgel 1957.

15. Zamecnik 1960, p. 274.

16. Dunn 1959; Spahr and Tissières 1959; Dunn, Smith, and Spahr 1960.

17. Hecht, Zamecnik, Stephenson, and Scott 1958.

18. Zamecnik 1960, p. 275.

19. Ibid., p. 275 n.

20. Hoagland, Zamecnik, and Stephenson 1959, p. 110.

21. Zamecnik 1960, p. 275.

22. Hoagland, Zamecnik, and Stephenson 1959, p. 111.

23. Chantrenne 1948; Haurowitz 1949; Dounce 1952; Koningsberger and Overbeek 1953; Gamow 1954, among others.

24. Hoagland 1959a, p. 41.   25. Ibid., p. 44.

26. Zamecnik 1960, p. 276.   27. Hoagland 1959b, p. 55.

28. Ibid., pp. 56, 61.   29. Hoagland 1960, pp. 401–2.

30. Ibid., pp. 406–7.   31. Zamecnik, interview, 1990.

32. Gale 1955; Astrachan and Volkin 1958; Zamecnik, interview, 1990.

33. Zamecnik, interview, 1990.

34. Hoagland 1960, p. 403.

35. Schweet, Lamfrom, and Allen 1958.

36. Chao and Schachman 1956.

37. Petermann, Hamilton, Balis, Samarth, and Pecora 1958.

38. Tissières and Watson 1958; Tissières, Watson, Schlessinger, and Hollingworth 1959.

39. Cf. Chapter 8.

40. Ts'o and Squires 1959.

41. Kurland 1960.

42. Schweet, Lamfrom, and Allen 1958.

43. Webster 1959.

44. Gale 1959a, 1959b, p. 164, and Gale, letter to Rheinberger, January 17, 1994.

45. Waldo Cohn, citing Masson Gulland in a discussion of a paper by Gale 1956, p. 183.

46. For "junk," see Hoagland 1960, p. 375.

47. To mention only a few papers from 1959, see Preiss, Berg, Ofengand, Bergmann, and Dieckmann 1959; Lipmann, Hülsmann, Hartmann, Boman, and Acs 1959; Lacks and Gros 1959; Tissières 1959; Dunn 1959; Spahr and Tissières 1959; Yu and Allen 1959; Smith, Cordes, and Schweet 1959; Holley and Merrill 1959.

48. For a review, see Spiegelman 1959. For a history of bacterial genetics, see Brock 1990.

49. Tissières and Watson 1958; Berg and Ofengand 1958; Preiss, Berg, Ofengand, Bergmann, and Dieckmann 1959. For the absence of a suitable fractionated protein synthesizing system, see Simkin 1959.

50. See Chapter 6.

51. Lamborg 1960. In fact, the first cell-free E. coli system was reported by Dietrich Schachtschabel and Wolfram Zillig from the Max Planck Institute for

Biochemistry in Munich during the Fourth International Congress of Biochemistry in Vienna in 1958. The paper was published in German and so did not come to the attention of most American, British, and French research groups. It has almost never been quoted. See Schachtschabel and Zillig 1959.

52. Among others see Gale and Folkes 1955a; Beljanski and Ochoa 1958a and b; Spiegelman 1959; Hunter, Brookes, Crathorn, and Butler 1959; Rogers and Novelli 1959; Connell, Lengyel, and Warner 1959; Nisman 1959.

53. Lamborg and Zamecnik 1960, p. 210.

54. Zamecnik 1979, p. 297.

55. Zamecnik, interview, 1990.

56. Tissières, Schlessinger, and Gros 1960. Mary Stephenson recalls: "We all liked Alfred [Tissières]. And Alfred referred to Marv in his work." But, "Marv's work was never quoted after Alfred published his experiment. You know, the most involved person doesn't get quoted" (Stephenson, interview, 1991).

57. See Chapter 6.

58. Monier, Stephenson, and Zamecnik 1960, p. 1.

59. Zamecnik 1979, p. 287. The material derived from this simple, direct procedure was very similar to the S-RNA prepared by the other, more elaborate methods. In this way, 70 to 80 mg of RNA could be obtained from 100 grams of fresh-pressed baker's yeast.

60. Monier, Stephenson, and Zamecnik 1960; Zamecnik 1960.

61. Zamecnik and Stephenson 1960; Zamecnik, Stephenson, and Scott 1960.

62. Lipmann, Hülsmann, Hartmann, Boman, and Acs 1959; Holley and Merrill 1959; Smith, Cordes, and Schweet 1959.

63. Holley, Apgar, Doctor, Farrow, Marini, and Merrill 1961; Holley, Apgar, Everett, Madison, Marquisee, Merrill, Penswick, and Zamir 1965.

64. Zamecnik 1979, p. 298. See Von Portatius, Doty, and Stephenson 1961; Stephenson and Zamecnik 1961.

65. Among others, see Von der Decken and Hultin 1958; Hultin and Von der Decken 1959; Bosch, Bloemendal, and Sluyser 1959, 1960.

66. Zamecnik 1960, p. 276.

67. Hoagland and Comly 1960. To my knowledge, it was (one of) the first double-label experiments in the field.

68. Hoagland and Comly 1960, p. 1560.

69. Ibid.

Chapter 13

1. For more on these episodes, cf. Judson 1979; Morange 1994, especially chapters 12, 13, and 14; Gaudillière 1996; Kay, forthcoming.

2. Cf. Grmek and Fantini 1982; Morange 1990, Burian 1990; Gaudillière 1992.

3. Cf. Judson 1979, 400–446; Burian 1993a.

4. Pardee, Jacob, and Monod 1959, p. 175.

5. Riley, Pardee, Jacob, and Monod 1960, p. 225.

6. Jacob 1988, p. 311. In Copenhagen, the core group of the molecular biological avant-garde had gathered. Jacob mentions Ole Maaløe, Jim Watson, Francis Crick, Seymour Benzer, Sydney Brenner, Jacques Monod, and Niels Bohr.

7. For a more detailed description, see Judson 1979, pp. 427–35; Gros 1986, chapter 5; Jacob 1988, pp. 311–14; Crick 1988, pp. 118–20; Morange 1994, chapter 13.

8. Astrachan and Volkin 1958.

9. See Gale and Folkes 1955a; Volkin and Astrachan 1956a, 1956b; Spiegelman 1956a, 1956b.

10. For Gale's quotation, see Gale and Folkes 1955a, p. 683. For Monod's reaction, see Gaebler 1956, p. 93, 100.

11. Spiegelman 1956b, p. 193.

12. Chantrenne 1956, p. 429.

13. For "stable factory," see Hoagland 1990, p. 107. At the Fourth International Congress of Biochemistry in Vienna in September 1958, Hoagland stated, in a comment on Ernest Gale's paper, "the planning of *sequence* of amino acids in protein is presumably a function of the microsomal particles. Here must reside the genetically determined, relatively stable RNA template which has the information necessary to arrange amino acids in their proper order" (Hoagland 1959c, p. 169). For "making the same proteins," see Zamecnik, interview, 1990.

14. Crick 1958, p. 157.

15. Hoagland 1990, chapter 6.

16. See, e.g., the remarks in Loftfield's 1957 review (Loftfield 1957a, pp. 375–77).

17. Riley, Pardee, Jacob, and Monod 1960, p. 225; see also Burian 1993b.

18. Jacob 1988, p. 313.

19. This is Jacob's imagery. See Jacob 1988, p. 313.

20. Brenner, Jacob, and Meselson 1961.

21. Jacob and Monod 1961, p. 319.

22. Nomura, Hall, and Spiegelman 1960.

23. Nomura 1990, p. 5.

24. Gros, Hiatt, Gilbert, Kurland, Risebrough, and Watson 1961; Gros 1986, chapter 5. For the French context of messenger RNA research, see Gaudillière 1996.

25. Kameyama and Novelli 1960; Nathans and Lipmann 1961; Matsubara and Watanabe 1961; Ofengand and Haselkorn 1961–62. In 1962, there were at least six reports from five laboratories using the *E. coli* system in a rapid publication journal such as *Biochemical and Biophysical Research Communications* (Univer-

sity of Pennsylvania; Rutgers University; New York University; NIH; Kyoto University) and seven reports from five laboratories in the *Federation Proceedings* of this year (Oak Ridge National Laboratory; Stanford University; St. Louis University; Rockefeller Institute, NIH). The *Journal of Molecular Biology* published four research articles based on the *E. coli* protein synthesis system during 1962.

26. Nirenberg 1969, p. 2.

27. For a detailed description, see Judson 1979, 470–82. My account is based on an interview with J. Heinrich Matthaei, October 29, 1992, and Matthaei's laboratory notebooks (MLN). The first experiment dates from November 1, 1960. MLN, M1, p. 1.

28. I had no chance to consult Nirenberg's notebooks as well. I thank Lily Kay for some preliminary information. For a detailed account, see Kay, forthcoming.

29. These experiments date from November 1960. MLN, M1.

30. This is the abbreviation used in the protocols. It is not stated explicitly whether "mRNA" stands for "microsomal RNA" or for "messenger RNA." For published use of "mRNA" as a short hand for microsomal RNA, see, e.g., Bosch, Bloemendal, and Sluyser 1959.

31. Cf. MLN, M2, experimental series 26, beginning with February 18, 1961.

32. David Novelli, who at the same time tried to establish a cell-free *E. coli* system, recalls that the "ability to obtain active preparations from day to day was entirely unpredictable." His preparations as well as those from others were "unstable and could not be stored." Novelli 1966, pp. 191–92.

33. In the laboratory jargon it was referred to as "short Siekevitz method." Matthaei, interview 1992. Cf. also MLN, M1, seventh experiment, November 14–15, 1960, where "Siekevitz' procedure" describes a hot TCA precipitation.

34. See the discussion in Zamecnik 1979, pp. 299–301, where Robert Olby raises this point. As far as can be seen from the notebooks, Matthaei used preincubated "S30"-extract in the middle of February 1961 for the first time. "S30" denotes the supernatant of a low-speed centrifugation of broken bacteria.

35. Matthaei and Nirenberg 1961a; cf. also Matthaei and Nirenberg 1961b.

36. MLN, M1 and M2, ninth experiment, November 3 to December 1, 1960.

37. MLN, M2, experiment 27B of March 2, 1961.

38. The nature of these RNAs cannot be identified from the protocols.

39. He used threonine, methionine, phenylalanine, arginine, lysine, leucine in addition to valine. MLN, M2, experiment 27K, March 24, 1961.

40. MLN, M1, p. 104; M2, experiment 29G.

41. The quotation is from MLN, M1, p. 107. See also M2, experiment 27N, May 25, 1961; for the final experiment, see M2, 27Q, May 27, 1961.

42. Matthaei and Nirenberg 1961c; Nirenberg and Matthaei 1961; see also Nirenberg and Matthaei 1963a.

43. See Nirenberg 1969; Judson 1979, p. 470; Matthaei, interview, 1992.

44. Nirenberg and Matthaei 1963b. Abstract from the Fifth International Congress of Biochemistry in Moscow in 1961.

45. See Judson 1979, p. 473; Matthaei, interview, 1992.

46. Matthaei and Nirenberg 1961a, p. 407.

47. To give an example, François Gros used the term *messenger RNA* in the discussions at a symposium on protein biosynthesis held at Wassenaar (Holland), August 29–September 2, 1960, and so did Hubert Chantrenne in the concluding remarks to that meeting. See Harris 1961, pp. 205, 389.

48. Crick, Barnett, Brenner, and Watts-Tobin 1961; Crick 1988, 122–36; Wittmann 1961; Wittmann 1963; Fraenkel-Conrat and Tsugita 1963.

49. Nirenberg and Matthaei 1961; Lengyel, Speyer, and Ochoa 1961.

50. Zamecnik 1979, p. 298.

51. Crick 1963.

52. Recent experiments had reconfirmed this difference. Rendi 1959; Von der Decken and Hultin 1960; McCorquodale, Veach, and Mueller 1961.

53. Hoagland 1961, p. 155.   54. Ibid., pp. 153, 155.

55. Ibid., p. 153.   56. Ibid.

57. Hoagland and Askonas 1963.

58. Hoagland, Scornik, and Pfefferkorn 1964.

59. See the discussion in Hoagland 1961.

60. See Nathans and Lipmann 1960.

61. "The function of GTP in the process and its possible relationship to the transfer factor are in urgent need of explanation" (Nathans and Lipmann 1961, p. 502).

62. Hoagland, Scornik, and Pfefferkorn 1964, p. 1191.

63. Allende, Monro, and Lipmann 1964.

64. Nishizuka and Lipmann 1966, p. 213. For a review, see Lipmann 1971, pp. 91–112.

65. For "ribosomal clusters," see Warner, Rich, and Hall 1962; for "active complexes," see Gilbert 1963; for "ergosomes," see Wettstein, Staehelin, and Noll 1963; for "aggregated ribosomes," see Gierer 1963; for "polysomes," see Warner, Knopf, and Rich 1963.

66. Zamecnik, interview, 1990.

67. Wilson and Hoagland 1965.

68. For a systematic review, see Spirin 1990.

69. Hoagland himself has spoken in this respect of "integrated protein synthesis" (Hoagland 1966).

70. For a detailed history, see Brock 1990.

71. For a review, see Watson 1963; Lipmann 1963; Crick 1963.

72. Zamecnik 1962b.

73. Allen and Zamecnik 1962; Schweet, Lamfrom, and Allen 1958; Yarmolinsky and de la Haba 1959.

74. Allen and Zamecnik 1963.

75. For the purification of individual S-RNAs, see Stephenson and Zamecnik 1962; for the characterization of their amino acid–carrying properties, see Yu and Zamecnik 1963a, 1964; Sarin and Zamecnik 1964, 1965a; for the recognition properties of synthetases, see Yu and Zamecnik 1963b; Lamborg, Zamecnik, Li, Kägi, and Vallée 1965.

76. Lamborg and Zamecnik 1965; Sarin and Zamecnik 1965b.

77. See Holley, Apgar, Everett, Madison, Marquisee, Merrill, Penswick, and Zamir 1965. For "a clever team," see Zamecnik 1979, p. 298. Betty Keller was the first to draw the cloverleaf secondary structure of transfer RNA, although she was not a coauthor of the 1965 paper in which it was shown. Robert Holley, in an interview in 1972, later remarked, "the clover-leaf arrangement [that] I didn't think of. Elizabeth (Betty) Keller, who was on the staff at Cornell [is] the one who really got the cell-free system for protein synthesis reproducible. She wrote the clover-leaf arrangement down, she discovered it, and John Penswick, [a] graduate student, also discovered it independently. Actually, Betty Keller drew the diagram that appeared in the *Science* paper, as one of three. If I had any inkling that it would have been the one that turned out to be correct, I of course would have made acknowledgement of their discovering it rather than having all the authors in the paper 'discovering' it" (Portugal and Cohen 1977, pp. 282, 285).

78. Nirenberg and Leder 1964.

79. See Kay 1994.

Epilogue

1. Derrida 1988, p. 10.

2. Rotman 1987, pp. 2, 102; see also Pickering 1995, chapter 4.

3. For the notion of "resonance," cf. Wahrig-Schmidt and Hildebrandt 1993.

4. Crick 1988, p. 112.   5. Lacan 1989, pp. 10, 12.

6. Rheinberger 1992b, p. 421.   7. Grene 1995, p. 17.

8. Kauffman 1995, pp. 252–53.   9. Ibid., pp. 298, 303.

References

Abir-Am, Pnina. 1980. "From biochemistry to molecular biology: DNA and the acculturated journey of the critic of science Erwin Chargaff." *History and Philosophy of the Life Sciences* 2: 3–60.

———. 1985. "Themes, genres and orders of legitimation in the consolidation of new scientific disciplines: Deconstructing the historiography of molecular biology." *History of Science* 23: 74–117.

———. 1991. "Noblesse oblige: Lives of molecular biologists." *Isis* 82: 326–43.

———. 1992. "The politics of macromolecules: Molecular biologists, biochemists, and rhetoric." *Osiris* 7: 164–91.

Agrawal, Sudhir, Tohru Ikeuchi, Daisy Sun, Prem S. Sarin, Andrzej Konopka, Jacob Maizel, and Paul C. Zamecnik. 1989. "Inhibition of human immunodeficiency virus in early infected and chronically infected cells by antisense oligodeoxynucleotides and their phosphorothioate analogues." *Proceedings of the National Academy of Sciences of the United States* 86: 7790–94.

Allen, David W., and Paul C. Zamecnik. 1962. "The effect of puromycin on rabbit reticulocyte ribosomes." *Biochimica et Biophysica Acta* 55: 865–74.

———. 1963. "T2 ribonuclease inhibition of polyuridylic acid-stimulated polyphenylalanine synthesis." *Biochemical and Biophysical Research Communications* 11: 294–300.

Allende, Jorge E., Robin Monro, and Fritz Lipmann. 1964. "Resolution of the E. coli amino acyl sRNA transfer factor into two complementary fractions." *Proceedings of the National Academy of Sciences of the United States of America* 51: 1211–16.

Allfrey, Vincent, Marie M. Daly, and Alfred E. Mirsky. 1953. "Synthesis of protein in the pancreas. II. The role of ribonucleoprotein in protein synthesis." *Journal of General Physiology* 37: 157–75.

Allfrey, Vincent G., Alfred E. Mirsky, and Syozo Osawa. 1957. "Protein synthesis in isolated cell nuclei." *Journal of General Physiology* 40: 451–90.

Althusser, Louis. 1974. *Philosophie et philosophie spontanée des savants* (1967). Paris: Maspero.

Althusser, Louis, and Etienne Balibar. 1968. *Lire le Capital.* Vol. 1. Paris: Maspero.

Amann, Klaus. 1994. "Menschen, Mäuse und Fliegen: Eine wissenssoziologische Analyse der Transformation von Organismen in epistemische Objekte." In Michael Hagner, Hans-Jörg Rheinberger, and Bettina Wahrig-Schmidt (eds.), *Objekte, Differenzen, Konjunkturen: Experimentalsysteme im historischen Kontext,* 259–89. Berlin: Akademie Verlag.

Amann, Klaus, and Karin Knorr Cetina. 1990. "The fixation of (visual) evidence." In Michael Lynch and Steve Woolgar (eds.), *Representation in Scientific Practice,* 85–121. Cambridge, Mass.: MIT Press.

Anfinsen, Chris B., Anne Beloff, A. Baird Hastings, and Art K. Solomon. 1947. "The in vitro turnover of dicarboxylic amino acids in liver slice proteins." *Journal of Biological Chemistry* 168: 771–72.

Askonas, Brigitte A., Peter N. Campbell, and Thomas S. Work. 1954. "The distribution of radioactivity in goat casein after injection of radioactive amino acids and its bearing on theories of protein synthesis." *Biochemical Journal* 56: iv.

Astrachan, Lazarus, and Elliot Volkin. 1958. "Properties of ribonucleic acid turnover in T2-infected Escherichia coli." *Biochimica et Biophysica Acta* 29: 536–44.

Aub, Joseph C., Austin M. Brues, René Dubos, Seymour S. Kety, Ira T. Nathanson, Alfred Pope, and Paul C. Zamecnik. 1944. "Bacteria and the toxic factor in shock." *War Medicine* 5: 71–73.

Avery, Oswald T., Colin M. MacLeod, and Maclyn McCarty. 1944. "Studies on the chemical nature of the substance inducing transformation of pneumococcal types: Induction of transformation by a desoxyribonucleic acid fraction isolated from pneumococcus type III." *Journal of Experimental Medicine* 79: 137–58.

Bachelard, Gaston. 1933. *Les intuitions atomistiques: Essai de classification.* Paris: Vrin.

———. 1951. *L'activité rationaliste de la physique contemporaine.* Paris: Presses Universitaires de France.

———. 1957. *La formation de l'esprit scientifique.* Paris: Vrin.

———. 1968. *The Philosophy of No.* Translated by G. C. Waterston. New York: Orion Press.

———. 1984. *The New Scientific Spirit.* Translated by Arthur Goldhammer. Boston: Beacon Press.

Barnes, Barry. 1974. *Scientific Knowledge and Sociological Theory.* London: Routledge and Kegan Paul.

———. 1977. *Interests and the Growth of Knowledge.* London: Routledge and Kegan Paul.

Bartels, Ditta. 1983. "The multi-enzyme programme of protein synthesis—its neglect in the history of biochemistry and its current role in biotechnology." *History and Philosophy of the Life Sciences* 5: 187–219.

Baudrillard, Jean. 1983. *Simulations*. Translated by Paul Foss, Paul Patton, and Philip Beitchman. New York: Semiotext(e).

Bazerman, Charles. 1988. *Shaping Written Knowledge: The Genre and Activity of the Experimental Article in Science*. Madison: University of Wisconsin Press.

Bechtel, William, and Robert C. Richardson. 1993. *Discovering Complexity: Decomposition and Localization as Strategies in Scientific Research*. Princeton, N.J.: Princeton University Press.

Beljanski, Mirko, and Severo Ochoa. 1958a. "Protein biosynthesis by a cell-free bacterial system." *Proceedings of the National Academy of Sciences of the United States of America* 44: 494–501.

——. 1958b. "Protein biosynthesis by a cell-free bacterial system." Pt. 2, "Further studies on the amino acid incorporation enzyme." *Proceedings of the National Academy of Sciences of the United States of America* 44: 1157–61.

Bennington, Geoffrey, and Jacques Derrida. 1991. *Jacques Derrida*. Paris: Seuil.

Berg, Paul. 1955. "Participation of adenyl-acetate in the acetate-activating system." *Journal of the American Chemical Society* 77: 3163–64.

——. 1956. "Acyl adenylates: The interaction of adenosine triphosphate and L-methionine." *Journal of Biological Chemistry* 222: 1025–34.

——. 1957. "Chemical synthesis and enzymatic utilization of adenyl amino acids." *Federation Proceedings* 16: 152.

Berg, Paul, and E. James Ofengand. 1958. "An enzymatic mechanism for linking amino acids to RNA." *Proceedings of the National Academy of Sciences of the United States of America* 44: 78–86.

Bergmann, Max. 1942. "A classification of proteolytic enzymes." *Advances in Enzymology* 2: 49–68.

Berkhofer, Robert F., Jr. 1995. *Beyond the Great Story: History as Text and Discourse*. Cambridge, Mass.: Harvard University Press.

Bernard, Claude. 1954. *Philosophie: Manuscrit inédit*. Edited by Jacques Chevalier. Paris: Editions Hatier-Boivin.

——. 1957. *An Introduction to the Study of Experimental Medicine*. Translated by Henry Copley Greene. New York: Dover.

——. 1965. *Cahier de notes 1850–1860*. Edited with commentary by Mirko D. Grmek. Paris: Gallimard.

——. 1966. *Leçons sur les phénomènes de la vie communs aux animaux et aux végétaux*. Paris: Vrin.

——. 1967. *The Cahier Rouge*. Translated by Hebbel E. Hoff, Lucienne Guillemin, and Roger Guillemin. In F. Grande and M. B. Visscher (eds.), *Claude Bernard and Experimental Medicine*. Cambridge, Mass.: Schenkman.

——. 1974. *Lectures on the Phenomena of Life Common to Animals and Plants*.

Translated by Hebbel E. Hoff, Roger Guillemin, and Lucienne Guillemin. Springfield, Ill.: Charles C. Thomas.

Bloor, David. 1976. *Knowledge and Social Imagery*. London: Routledge and Kegan Paul.

Blumenberg, Hans. 1986. *Die Lesbarkeit der Welt*. Frankfurt am Main: Suhrkamp.

Borsook, Henry. 1950. "Protein turnover and incorporation of labeled amino acids into tissue proteins in vivo and in vitro." *Physiological Reviews* 30: 206–19.

———. 1953. "Peptide bond formation." *Advances in Protein Chemistry* 8: 127–74.

———. 1956a. "The biosynthesis of peptides and proteins." In Claude Liébecq (ed.), *Proceedings of the Third International Congress of Biochemistry*, Brussels 1955, 92–104. New York: Academic Press.

———. 1956b. "The biosynthesis of peptides and proteins." *Journal of Cellular and Comparative Physiology* 47, supplement 1: 35–80.

Borsook, Henry, Clara L. Deasy, Arie J. Haagen-Smit, Geoffrey Keighley, and Peter H. Lowy. 1948. "The degradation of L-lysine in Guinea pig liver homogenate: Formation of α-aminoadipic acid," and "The degradation of α-aminoadipic acid in Guinea pig liver homogenate." *Journal of Biological Chemistry* 176: 1383–93, 1395–1400.

———. 1949a. "The incorporation of labeled lysine into the proteins of Guinea pig liver homogenate." *Journal of Biological Chemistry* 179: 689–704.

———. 1949b. "Uptake of labeled amino acids by tissue proteins in vitro." *Federation Proceedings* 8: 589–96.

———. 1950a. "The uptake in vitro of C^{14}-labeled glycine, L-leucine, and L-lysine by different components of Guinea pig liver homogenate." *Journal of Biological Chemistry* 184: 529–43.

———. 1950b. "Metabolism of C^{14}-labeled glycine, L-histidine, L-leucine, and L-lysine." *Journal of Biological Chemistry* 187: 839–48.

Borsook, Henry, and Jacob W. Dubnoff. 1940. "The biological synthesis of hippuric acid in vitro." *Journal of Biological Chemistry* 132: 307–24.

Bosch, Leendert, Hans Bloemendal, and Mels Sluyser. 1959. "Metabolic interrelationships between soluble and microsomal RNA in rat-liver cytoplasm." *Biochimica et Biophysica Acta* 34: 272–74.

———. 1960. "Studies on cytoplasmic ribonucleic acid from rat liver." Pt. 1, "Fractionation and function of soluble ribonucleic acid," and pt. 2, "Fractionation and function of microsomal ribonucleic acid." *Biochimica et Biophysica Acta* 41: 444–53, 454–61.

Brachet, Jean. 1942. "La localisation des acides pentosenucléiques dans les tissus animaux et les oeufs d'Amphibiens en voie de développement." *Archives de Biologie* 53: 207–57.

———. 1947a. "Nucleic acids in the cell and the embryo." *Nucleic Acid: Symposia of the Society for Experimental Biology* 1: 207–24.

————. 1947b. "The metabolism of nucleic acids during embryonic development." *Cold Spring Harbor Symposia on Quantitative Biology* 12: 18–27.

————. 1949. "The localization and the role of ribonucleic acid in the cell." *Annals of the New York Academy of Sciences* 50: 861–69.

————. 1952. "Acides ribonucléiques et biogénèse des protéines." *IIᵉ Congrès International de Biochimie*, Paris 1952. Comptes rendus, symposium no. 2: 85–95.

Brachet, Jean, and Raymond Jeener. 1943–45. "Recherches sur des particules cytoplasmiques de dimensions macromoléculaires riches en acide pentosenucléique." Pt. 1, "Propriétés générales, relations avec les hydrolases, les hormones, les protéines de structure." *Enzymologia* 11: 196–212.

Brachet, Jean, and John Rodney Shaver. 1949. "The injection of embryonic microsomes into early Amphibian embryos." *Experientia* 5: 204–5.

Brenner, Sidney, François Jacob, and Matthew Meselson. 1961. "An unstable intermediate carrying information from genes to ribosomes for protein synthesis." *Nature* 190: 576–81.

Brock, Thomas D. 1990. *The Emergence of Bacterial Genetics*. New York: Cold Spring Harbor Laboratory Press.

Brues, Austin M., Marjorie M. Tracy, and Waldo E. Cohn. 1944. "Nucleic acids of rat liver and hepatoma: Their metabolic turnover in relation to growth." *Journal of Biological Chemistry* 155: 619–33.

Bucher, Nancy L. R. 1953. "The formation of radioactive cholesterol and fatty acids from C^{14}-labeled acetate by rat liver homogenates." *Journal of the American Chemical Society* 75: 498.

————. 1987. "Dr. Aub, Huntington Hospital, and Cancer Research." *Harvard Medical Alumni Bulletin* (fall-winter): 46–51.

Bucher, Nancy L. R., and Andre Glinos. 1948. "Phosphatase distribution in rat liver during regeneration and after p-dimethylaminoazobenzene administration." *Unio Internationalis Contra Cancrum Acta* 6: 273–80.

Bucher, Nancy L. R., Robert B. Loftfield, and Ivan D. Frantz, Jr. 1949. "The effect of regeneration on the rate of protein synthesis and degradation in rat liver." *Cancer Research* 9: 619.

Buchwald, Jed Z. (ed.). 1995. *Scientific Practice: Theories and Stories of Doing Physics*. Chicago: University of Chicago Press.

Burian, Richard M. 1990. "La contribution française aux instruments de recherche dans le domaine de la génétique moléculaire." In Jean-Louis Fischer and William H. Schneider (eds.), *Histoire de la Génétique*, 247–69. Paris: A.R.P.E.M et Editions Sciences en Situation.

————. 1993a. "On the cusp between biochemistry and molecular biology: The Pyjama (or PaJaMo) experiment." Manuscript.

————. 1993b. "Task definition, and the transition from genetics to molecular genetics: Aspects of the work on protein synthesis in the laboratories of J. Monod and P. Zamecnik." *Journal of the History of Biology* 26: 387–407.

——. 1995. "The role of technique: Some transformations wrought by use of RNase and staining techniques, 1938–1952." Manuscript.

——. 1996. "Underappreciated pathways toward molecular genetics as illustrated by Jean Brachet's cytochemical embryology." In Sahotra Sarkar (ed.), *The Philosophy and History of Molecular Biology: New Perspectives*, 67–85. Dordrecht: Kluwer.

Butterfield, Herbert. 1957. *The Origins of Modern Science*. New York: Macmillan.

Cairns, John, Gunther S. Stent, and James D. Watson (eds.). 1992. *Phage and the Origins of Molecular Biology*. Expanded ed. New York: Cold Spring Harbor Laboratory Press.

Campbell, Peter N., and Thomas S. Work. 1953. "Biosynthesis of proteins." *Nature* 171: 997–1001.

Canellakis, Evangelo S. 1957. "On the mechanism of incorporation of adenylic acid from adenosine into ribonucleic acid by soluble mammalian enzyme systems." *Biochimica et Biophysica Acta* 25: 217–18.

Canguilhem, Georges. 1975. "L'histoire des sciences dans l'oeuvre épistémologique de Gaston Bachelard." In G. Canguilhem, *Etudes d'histoire et de philosophie des sciences*, 173–86. Paris: Vrin.

——. 1981. *Idéologie et rationalité dans l'histoire des sciences de la vie*. Paris: Vrin.

Carrard, Philippe. 1992. *Poetics of the New History: French Historical Discourse from Braudel to Chartier*. Baltimore: Johns Hopkins University Press.

Caspersson, Torbjörn. 1941. "Studien über den Eiweißumsatz der Zelle." *Naturwissenschaften* 29: 33–43.

——. 1947. "The relations between nucleic acid and protein synthesis." *Nucleic Acid: Symposia of the Society for Experimental Biology* 1: 127–51.

Castleman, Benjamin, David C. Crockett, and S. B. Sutton (eds.). 1983. *The Massachusetts General Hospital 1955–1980*. Boston: Little, Brown.

Chadarevian, Soraya de. 1996. "Sequences, conformation, information: Biochemists and molecular biologists in the 1950s." *J. of the History of Biology* 29: 361–86.

Chantrenne, Hubert. 1943–45. "Recherches sur des particules cytoplasmiques de dimensions macromoléculaires riches en acide pentosenucléique." Pt. 2, "Relations avec les ferments respiratoires." *Enzymologia* 11: 213–21.

——. 1947. "Hétérogénéité des granules cytoplasmiques du foie de souris." *Biochimica et Biophysica Acta* 1: 437–48.

——. 1948. "Un modèle de synthèse peptidique: Propriétés du benzoylphosphate de phényle." *Biochimica et Biophysica Acta* 2: 286–93.

——. 1951. "Recherches sur le mécanisme de la synthèse des protéines." *Pubblicazioni della Stazione Zoologica di Napoli* 23 (supplemento): 70–86.

——. 1956. "Metabolic changes in nucleic acids during the induction of enzymes by oxygen in resting yeast." *Archives of Biochemistry and Biophysics* 65: 414–26.

——. 1991. "Souvenirs de mes premières années au laboratoire du Rouge Cloître." *Fondation Jean Brachet, Bulletin de Liaison*, no. 7: 3–4.

Chao, Fu-Chuan, and Howard K. Schachmann. 1956. "The isolation and characterization of a macromolecular ribonucleoprotein from yeast." *Archives of Biochemistry and Biophysics* 61: 220–30.

Clark, William. 1995. "Narratology and the history of science." *Studies in History and Philosophy of Science* 26: 1–71.

Claude, Albert. 1938. "A fraction from normal chick embryo similar to the tumor-producing fraction of chicken tumor I." *Proceedings of the Society for Experimental Biology and Medicine* 39: 398–403.

——. 1941. "Particulate components of cytoplasm." *Cold Spring Harbor Symposia on Quantitative Biology* 9: 263–71.

——. 1943a. "The constitution of protoplasm." *Science* 97: 451–56.

——. 1943b. "Distribution of nucleic acids in the cell and the morphological constitution of cytoplasm." In Normand L. Hoerr (ed.), *Frontiers in Cytochemistry: Biological Symposia*, 10: 111–29. Lancaster, Penn.: Jaques Cattell Press.

——. 1950. "Studies on cells: Morphology, chemical constitution, and distribution of biochemical function." *The Harvey Lectures 1947–48* 43: 121–64.

Claude, Albert, and Ernest F. Fullam. 1945. "An electron microscope study of isolated mitochondria: Method and preliminary results." *Journal of Experimental Medicine* 81: 51–61.

Cohen, Seymour S., and Hazel D. Barner. 1954. "Studies on unbalanced growth in Escherichia coli." *Proceedings of the National Academy of Sciences of the United States of America* 40: 885–93.

Cohn, P. 1959. "Incorporation in vitro of amino acids into ribonucleoprotein fractions of microsomes." *Biochimica et Biophysica Acta* 33: 284–85.

Collins, Harry M. 1985. *Changing Order: Replication and Induction in Scientific Practice*. London: SAGE Publications.

Connell, George E., Peter Lengyel, and Robert C. Warner. 1959. "Incorporation of amino acids into protein of Azotobacter cell fractions." *Biochimica et Biophysica Acta* 31: 391–97.

Creager, Angela. 1996. "Wendell Stanley's dream of a free-standing biochemistry department at the University of California, Berkeley." *Journal of the History of Biology* 29: 331–60.

Crick, Francis C. 1955. "On degenerate templates and the adaptor hypothesis." Note for the RNA Tie Club, undated and unpublished. [Original with Sidney Brenner.]

——. 1957. Discussion note. In E. M. Crook (ed.), *The Structure of Nucleic Acids and Their Role in Protein Synthesis: Biochemical Society Symposium*, 14 (February 18, 1956), 25–26. London: Cambridge University Press.

——. 1958. "On protein synthesis." *Symposia of the Society for Experimental Biology London* 12: 138–63.

——. 1963. "The recent excitement in the coding problem." *Progress in Nucleic Acid Research* 1: 163–217.

——. 1970. "Molecular biology in the year 2000." *Nature* 228: 613–15.

———. 1988. *What Mad Pursuit*. New York: Basic Books.

Crick, Francis H. C., Leslie Barnett, Sydney Brenner, and R. J. Watts-Tobin. 1961. "General nature of the genetic code for proteins." *Nature* 192: 1227–32.

Crick, Francis H. C., John S. Griffith, and Leslie E. Orgel. 1957. "Codes without commas." *Proceedings of the National Academy of Sciences of the United States of America* 43: 416–21.

Dagognet, François. 1984. Preface to Claude Bernard, *Introduction à l'étude de la médecine expérimentale*, 9–21. Paris: Flammarion.

Damerow, Peter, and Wolfgang Lefèvre (eds.). 1981. *Rechenstein, Experiment, Sprache*. Stuttgart: Klett-Cotta.

Darden, Lindley. 1991. *Theory Change in Science: Strategies from Mendelian Genetics*. Oxford: Oxford University Press.

Darden, Lindley, and Nancy Maull. 1977. "Interfield theories." *Philosophy of Science* 44: 43–64.

Davidson, James N. 1957. "Cytological aspects of the nucleic acids." *The Structure of Nucleic Acids and Their Role in Protein Synthesis: Biochemical Society Symposium*, 14 (February 18, 1956), 27–31. London: Cambridge University Press.

Davie, Earl W., Victor V. Koningsberger, and Fritz Lipmann. 1956. "The isolation of a tryptophane-activating enzyme from pancreas." *Archives of Biochemistry and Biophysics* 65: 21–38.

Deleuze, Gilles. 1994. *Difference and Repetition*. Translated by Paul Patton. New York: Columbia University Press.

DeMoss, John A., Saul M. Genuth, and G. David Novelli. 1956. "The enzymatic activation of amino acids via their acyl-adenylate derivatives." *Proceedings of the National Academy of Sciences of the United States of America* 42: 325–32.

DeMoss, John A., and G. David Novelli. 1955. "An amino acid dependent exchange between inorganic pyrophosphate and ATP in microbial extracts." *Biochimica et Biophysica Acta* 18: 592–93.

Derrida, Jacques. 1976. *Of Grammatology*. Translated by Gayatri Chakravorty Spivak. Baltimore: Johns Hopkins University Press.

———. 1978. "Structure, sign, and play in the discourse of the human sciences." In J. Derrida, *Writing and Difference*, translated by Alan Bass, 278–93. Chicago: University of Chicago Press.

———. 1982a. *Dissemination*. Translated by B. Johnson. Chicago: University of Chicago Press.

———. 1982b. "Différance." In J. Derrida, *Margins of Philosophy*, translated by Alan Bass, 1–27. Chicago: University of Chicago Press.

———. 1988. "Signature event context." In J. Derrida, *Limited Inc*, translated by Samuel Weber, 1–23. Evanston: Northwestern University Press.

———. 1991. "Une 'folie' doit veiller sur la pensée." Interview with François Ewald, *Magazine littéraire* (March): 18–30.

Dijksterhuis, Eduard J. 1962. "The origins of classical mechanics: From Aristotle to Newton." In Marshall Clagett (ed.), *Critical Problems in the History of Science*, 163–90. Madison: University of Wisconsin Press.

Doudoroff, Michael, Horace A. Barker, and William Z. Hassid. 1947. "Studies with bacterial sucrose phosphorylase: The mechanism of action of sucrose phosphorylase as a glucose-transferring enzyme (transglucosidase)." *Journal of Biological Chemistry* 168: 725–32.

Dounce, Alexander L. 1952. "Duplicating mechanism for peptide chain and nucleic acid synthesis." *Enzymologia* 15: 251–58.

Doyle, Richard. 1993. "On Beyond Living: Rhetorics of Vitality and Post-vitality in Molecular Biology." Ph.D. diss., University of California, Berkeley.

Dunn, D. B. 1959. "Additional components in ribonucleic acid of rat-liver fractions." *Biochimica et Biophysica Acta* 34: 286–87.

Dunn, D. B., J. D. Smith, and Pierre F. Spahr. 1960. "Nucleotide composition of soluble ribonucleic acid from Escherichia coli." *Journal of Molecular Biology* 2: 113–17.

Dupré, John. 1993. *The Disorder of Things: Metaphysical Foundations of the Disunity of Science*. Cambridge, Mass.: Harvard University Press.

Edmonds, Mary, and Richard Abrams. 1957. "Incorporation of ATP into polynucleotide in extracts of Ehrlich ascites cells." *Biochimica et Biophysica Acta* 26: 226–27.

Elkana, Yehuda. 1970. "Helmholtz' 'Kraft': An illustration of concepts in flux." *Historical Studies in the Physical Sciences* 2: 263–98.

———. 1981. "A programmatic attempt at an anthropology of knowledge." In Everett Mendelsohn and Yehuda Elkana (eds.), *Sciences and Cultures*, 1–76. Dordrecht and Boston: Reidel.

Ernster, Lars, and Gottfried Schatz. 1981. "Mitochondria: A historical review." *Journal of Cell Biology* 91: 227s–55s.

Faxon, Nathaniel W. 1959. *The Massachusetts General Hospital 1935–1955*. Cambridge, Mass.: Harvard University Press.

Feyerabend, Paul. 1975. *Against Method: Outline of an Anarchistic Theory of Knowledge*. London: New Left Books.

Fischer, Emil. 1906. *Untersuchungen über Aminosäuren, Polypeptide und Proteïne (1899–1906)*. Berlin: Julius Springer.

Fischer, Ernst Peter, and Carol Lipson. 1988. *Thinking About Science: Max Delbrück and the Origins of Molecular Biology*. New York: W. W. Norton.

Fleck, Ludwik. 1979. *Genesis and Development of a Scientific Fact*. Translated by Fred Bradley and Thaddeus Y. Trenn. Chicago: Univ. of Chicago Press.

Foucault, Michel. 1972a. *The Archaeology of Knowledge*. Translated by A. M. Sheridan Smith. New York: Pantheon Books.

———. 1972b. *The Discourse on Language*. Translated by Rupert Swyer. Published as an appendix to Foucault, *The Archaeology of Knowledge*. New York: Pantheon Books.

Fraenkel-Conrat, Heinz, and Akira Tsugita. 1963. "Biological and protein-structural effects of chemical mutagenesis of TMV-RNA." *Proceedings of the Fifth International Congress of Biochemistry*, August 10–16, 1961, Moscow, vol. 3, 242–44. New York: Macmillan.

Fraenkel-Conrat, Heinz, and Robley C. Williams. 1955. "Reconstitution of active tobacco mosaic virus from its inactive protein and nucleic acid component." *Proceedings of the National Academy of Sciences of the United States of America* 41: 690–98.

Francis, M. David, and Theodore Winnick. 1953. "Studies on the pathway of protein synthesis in tissue culture." *Journal of Biological Chemistry* 202: 273–89.

Franklin, Allan. 1986. *The Neglect of Experiment*. Cambridge, U.K.: Cambridge University Press.

———. 1990. *Experiment, Right or Wrong*. Cambridge, U.K.: Cambridge University Press.

Frantz, Ivan D., Jr., and Nancy L. R. Bucher. 1954. "The incorporation of the carboxyl carbon from acetate into cholesterol by rat liver homogenates." *Journal of Biological Chemistry* 206: 471–81.

Frantz, Ivan D., Jr., and Howard Feigelman. 1949. "Biosynthesis of amino acids uniformly labeled with radioactive carbon, for use in the study of growth." *Cancer Research* 9: 619.

Frantz, Ivan D., Jr., and Robert B. Loftfield. 1950. "Equilibrium and exchange reactions involving peptides, amino acids, and proteolytic enzymes." *Federation Proceedings* 9: 172–73.

Frantz, Ivan D., Jr., Robert B. Loftfield, and Warren W. Miller. 1947. "Incorporation of C^{14} from carboxyl-labeled dl-alanine into the proteins of liver slices." *Science* 106: 544–45.

Frantz, Ivan D., Jr., Robert B. Loftfield, and Ann S. Werner. 1949. "Observations on the equilibrium between glycine and glycylglycine in the presence of liver peptidase." *Federation Proceedings* 8: 199.

Frantz, Ivan D., Jr., Paul C. Zamecnik, John W. Reese, and Mary L. Stephenson. 1948. "The effect of dinitrophenol on the incorporation of alanine labeled with radioactive carbon into the proteins of slices of normal and malignant rat liver." *Journal of Biological Chemistry* 174: 773–74.

Freud, Sigmund. 1957a. *On Narcissism: An Introduction*. Standard edition of the complete psychological works, vol. 14, 73–102. Translated under the general editorship of James Strachey in collaboration with Anna Freud, assisted by Alix Strachey and Alan Tyson. London: Hogarth Press.

———. 1957b. *Instincts and Their Vicissitudes*. Standard edition of the complete psychological works, vol. 14, 117–40. Translated under the general editorship of James Strachey in collaboration with Anna Freud, assisted by Alix Strachey and Alan Tyson. London: Hogarth Press.

Friedberg, Felix, Theodore Winnick, and David M. Greenberg. 1947. "Incorporation of labeled glycine into the protein of tissue homogenates." *Journal of Biological Chemistry* 171: 441.

Friedberg, Wallace, and Harry Walter. 1955. "Metabolic fate of doubly labeled heterologous proteins." *Federation Proceedings* 14: 214.

Frost, Robert. 1964. *Complete Poems*. New York: Holt, Rinehart, and Winston.

Fruton, Joseph S. 1952. "The enzymatic synthesis of peptide bonds." *IIe Congrès International de Biochimie*, Paris 1952. Comptes rendus, symposium no. 2, 5–18. Paris: Masson.

Gaebler, Oliver H. 1956. *Enzyme's: Units of Biological Structure and Function*. New York: Academic Press.

Gale, Ernest F. 1955. "From amino acids to proteins." In William D. McElroy and H. Bentley Glass (eds.), *A Symposium on Amino Acid Metabolism* (June 14–17, 1954), 171–92. Baltimore: Johns Hopkins Press.

——. 1956. "Nucleic acids and amino acid incorporation." In G. E. W. Wolstenholme and C. M. O'Connor (eds.), *CIBA Foundation Symposium on Ionizing Radiations and Cell Metabolism*, 174–84. Boston: Little, Brown.

——. 1959a. "Incorporation factors, amino acid incorporation and nucleic acid synthesis." In G. Tunevall (ed.), *Recent Progress in Microbiology*, 101–14. Springfield, Ill.: Charles C. Thomas Publisher.

——. 1959b. "Protein synthesis in sub-cellular systems." *Proceedings of the Fourth International Congress of Biochemistry*, Vienna, 1958, vol. 6, 156–65. London: Pergamon Press.

Gale, Ernest F., and Joan P. Folkes. 1953a. "The assimilation of amino-acids by bacteria." Pt. 14, "Nucleic acid and protein synthesis in staphylococcus aureus." *Biochemical Journal* 53: 483–92.

——. 1953b. "The assimilation of amino acids by bacteria." Pt. 18, "The incorporation of glutamic acid into the protein fraction of Staphylococcus aureus." *Biochemical Journal* 55: 721–29.

——. 1953c. "Amino acid incorporation by fragmented staphylococcal cells." *Biochemical Journal* 55: xi.

——. 1954. "Effect of nucleic acids on protein synthesis and amino-acid incorporation in disrupted staphylococcal cells." *Nature* 173: 1223–27.

——. 1955a. "The assimilation of amino acids by bacteria." Pt. 20, "The incorporation of labeled amino acids by disrupted staphylococcal cells,"; pt. 21, "The effect of nucleic acids on the development of certain enzymic activities in disrupted staphylococcal cells." *Biochemical J.* 59: 661–75, 675–84.

——. 1955b. "Promotion of incorporation of amino-acids by specific di- and trinucleotides." *Nature* 175: 592–93.

Galison, Peter. 1987. *How Experiments End*. Chicago: Univ. of Chicago Press.

——. 1988. "History, philosophy, and the central metaphor." *Science in Context* 2: 197–212.

——. 1995. "Context and constraints." In Jed Z. Buchwald (ed.), *Scientific Practice: Theories and Stories of Physics*, 13–41. Chicago: University of Chicago Press.

Galison, Peter, and David J. Stump. 1996. *The Disunity of Science: Boundaries, Contexts and Power*. Stanford: Stanford University Press.

Gamow, George. 1954. "Possible relation between deoxyribonucleic acid and protein structures." *Nature* 173: 318.

Garland, Joseph E. 1961. *Every Man Our Neighbor: A Brief History of the Massachusetts General Hospital 1811–1961*. Boston: Little, Brown.

Gasché, Rodolphe. 1986. *The Tain of the Mirror: Derrida and the Philosophy of Reflection*. Cambridge, Mass.: Harvard University Press.

Gaudillière, Jean-Paul. 1991. *Biologie moléculaire et biologistes dans les années soixante: La naissance d'une discipline: Le cas français*. Thèse de doctorat, Université Paris VII.

———. 1992. "J. Monod, S. Spiegelman et l'adaptation enzymatique: Programmes de recherche, cultures locales et traditions disciplinaires." *History and Philosophy of the Life Sciences* 14: 23–71.

———. 1993. "Molecular biology in the French tradition? Redefining local traditions and disciplinary patterns." *Journal of the History of Biology* 26: 473–98.

———. 1994. "Wie man Labormodelle für Krebsentstehung konstruiert: Viren und Transfektion am (US) National Cancer Institute." In Michael Hagner, Hans-Jörg Rheinberger, and Bettina Wahrig-Schmidt (eds.), *Objekte, Differenzen, Konjunkturen: Experimentalsysteme im historischen Kontext*, 233–57. Berlin: Akademie Verlag.

———. 1996. "Molecular biologists, biochemists and messenger RNA: The birth of a scientific network." *Journal of the History of Biology* 29: 417–45.

Gierer, Alfred. 1963. "Function of aggregated reticulocyte ribosomes in protein synthesis." *Journal of Molecular Biology* 6: 148–57.

Gierer, Alfred, and Georg Schramm. 1956. "Infectivity of ribonucleic acid from tobacco mosaic virus." *Nature* 177: 702–3.

Gilbert, Walter. 1963. "Polypeptide synthesis in E. coli. I. Ribosomes and the active complex." *Journal of Molecular Biology* 6: 374–88.

Goethe, Johann Wolfgang von. 1957. "Materialen zur Geschichte der Farbenlehre." In Dorothea Kuhn (ed.), *Die Schriften zur Naturwissenschaft*, pt. 1, texts, vol. 6. Weimar: Hermann Böhlau.

———. 1982. "Maximen und Reflexionen." In J. Goethe, *Werke* (Hamburg ed.), vol. 12, 365–547. München: Deutscher Taschenbuchverlag.

———. 1988. "The experiment as mediator between object and subject." In D. Miller (ed.), *Johann Wolfgang von Goethe, Scientific Studies*, 11–17. New York: Suhrkamp.

Goldwasser, Eugene. 1955. "Incorporation of adenosine-5′-phosphate into ribonucleic acid." *Journal of the American Chemical Society* 77: 6083–84.

Gooding, David. 1990. *Experiment and the Making of Meaning: Human Agency in Scientific Observation and Experiment*. Dordrecht: Kluwer.

Gooding, David, Trevor Pinch, and Simon Schaffer (eds.). 1989. *The Uses of Experiment*. Cambridge, U.K.: Cambridge University Press.

Goodman, Nelson. 1968. *Languages of Art*. Indianapolis: Bobbs-Merrill.

Greenberg, David M., Felix Friedberg, Martin P. Schulman, and Theodore Winnick. 1948. "Studies on the mechanism of protein synthesis with radioactive carbon-labeled compounds." *Cold Spring Harbor Symposia on Quantitative Biology* 13: 113–17.

Grene, Marjorie. 1984. *The Knower and the Known*. Washington, D.C.: Center for Advanced Research in Phenomenology; University Press of America.

———. 1995. *A Philosophical Testament*. Chicago: Open Court.

Grier, Robert S., M. B. Hood, and Mahlon B. Hoagland. 1949. "Observations on the effect of Beryllium on alkaline phosphatase." *Journal of Biological Chemistry* 180: 289–98.

Griffin, A. Clark, William N. Nye, Lafayette Noda, and J. Murray Luck. 1948. "Tissue proteins and carcinogenesis." Pt. 1, "The effect of carcinogenic azo dyes on liver proteins." *Journal of Biological Chemistry* 176: 1225–35.

Grmek, Mirko D., and Bernardo Fantini. 1982. "Le rôle du hasard dans la naissance du modèle de l'opéron." *Revue d'Histoire des Sciences* 35: 193–215.

Gros, François. 1986. *Les secrets du gène*. Paris: Editions Odile Jacob.

Gros, François, Howard Hiatt, Walter Gilbert, Chuck G. Kurland, R. W. Risebrough, and James D. Watson. 1961. "Unstable ribonucleic acid revealed by pulse labeling of E. coli." *Nature* 190: 581–85.

Grunberg-Manago, Marianne, and Severo Ochoa. 1955. "Enzymatic synthesis and breakdown of polynucleotides; polynucleotide phosphorylase." *Journal of the American Chemical Society* 77: 3165–66.

Grunberg-Manago, Marianne, Priscilla J. Ortiz, and Severo Ochoa. 1955. "Enzymatic synthesis of nucleic acidlike polynucleotides." *Science* 122: 907–10.

Guggenberger, Bernd. 1991. "Zwischen Ordnung und Chaos." *Frankfurter Allgemeine Zeitung*, February 2, Beilage "Bilder und Zeiten."

Hacking, Ian. 1983. *Representing and Intervening: Introductory Topics in the Philosophy of Natural Science*. Cambridge, U.K.: Cambridge University Press.

———. 1992a. "The self-vindication of the laboratory sciences." In Andrew Pickering (ed.), *Science as Practice and Culture*, 29–64. Chicago: University of Chicago Press.

———. 1992b. "'Style' for historians and philosophers." *Studies in History and Philosophy of Science* 23: 1–20.

Hagner, Michael, Hans-Jörg Rheinberger, and Bettina Wahrig-Schmidt. 1994. "Objekte, Differenzen, Konjunkturen." In M. Hagner, H.-J. Rheinberger, and B. Wahrig-Schmidt (eds.), *Objekte, Differenzen, Konjunkturen: Experimentalsysteme im historischen Kontext*, 7–21. Berlin: Akademie Verlag.

Halvorson, Harlyn O., and Sol Spiegelman. 1952. "The inhibition of enzyme formation by amino acid analogues." *Journal of Bacteriology* 64: 207–21.

Harris, Robert J. C. (ed.). 1961. *Protein Biosynthesis*. London and New York: Academic Press.

Hart Nibbrig, Christiaan L. (ed.). 1994. *Was heißt "Darstellen"?* Frankfurt am Main: Suhrkamp.

Haurowitz, Felix. 1949. "Biological problems and immunochemistry." *Quarterly Review of Biology* 24: 93–101.

——. 1950. *Chemistry and Biology of Proteins.* New York: Academic Press.

——. 1956. "The mechanism of protein biosynthesis." In Claude Liébecq (ed.), *Proceedings of the Third International Congress of Biochemistry*, Brussels, 1955, 104–5. New York: Academic Press.

Hayles, N. Katherine. 1993. "Constrained constructivism: Locating scientific inquiry in the theater of representation." In George Levine (ed.), *Realism and Representation: Essays on the Problem of Realism in Relation to Science, Literature, and Culture*, 27–43. Madison: University of Wisconsin Press.

Hecht, Liselotte I., Mary L. Stephenson, and Paul C. Zamecnik. 1958a. "Formation of nucleotide end groups and incorporation of amino acids into soluble RNA." *Federation Proceedings* 17: 239.

——. 1958b. "Dependence of amino acid binding to soluble ribonucleic acid on cytidine triphosphate." *Biochimica et Biophysica Acta* 29: 460–61.

Hecht, Liselotte I., Mary L. Stephenson, and Paul C. Zamecnik. 1959. "Binding of amino acids to the end group of a soluble ribonucleic acid." *Proceedings of the National Academy of Sciences of the United States of America* 45: 505–18.

Hecht, Liselotte I., Paul C. Zamecnik, Mary L. Stephenson, and Jesse F. Scott. 1958. "Nucleoside triphosphates as precursors of ribonucleic acid end groups in a mammalian system." *Journal of Biological Chemistry* 233: 954–63.

Heidegger, Martin. 1971. "The nature of language." In M. Heidegger, *On the Way to Language*, translated by Peter D. Hertz, 57–108. New York: Harper and Row.

——. 1977a. "The question concerning technology." In M. Heidegger, *The Question Concerning Technology and Other Essays*, translated by William Lovitt, 3–35. New York: Harper and Row.

——. 1977b. "The age of the world picture." In M. Heidegger, *The Question Concerning Technology and Other Essays*, translated by William Lovitt, 115–54. New York: Harper and Row.

Heidelberger, Charles, Eberhard Harbers, Kenneth C. Leibman, Y. Takagi, and Van R. Potter. 1956. "Specific incorporation of adenosine-5'-phosphate-^{32}P into ribonucleic acid in rat liver homogenates." *Biochimica et Biophysica Acta* 20: 445–46.

Hentschel, Klaus. 1993. "The conversion of St. John: A case study on the interplay of theory and experiment." *Science in Context* 6: 137–94.

——. 1995. "Zum Zusammenspiel von Instrument: Experiment und Theorie am Beispiel der Rotverschiebung im Sonnenspektrum und verwandter spektraler Verschiebungseffekte von ca. 1880 bis etwa 1960." *Habilitationsschrift.* Manuscript.

Herbert, Edward. 1958. "The incorporation of adenine nucleotides into ribonucleic acid of cell-free systems from liver." *Journal of Biological Chemistry* 231: 975–86.

Herbert, Edward, Van R. Potter, and Liselotte I. Hecht. 1957. "Nucleotide Metabolism." Pt. 7, "The incorporation of radioactivity from orotic acid-6-C^{14} into ribonucleic acid in cell-free systems from rat liver." *Journal of Biological Chemistry* 225: 659–74.

Hershey, Alfred D. 1953. "Nucleic acid economy in bacteria infected with bacteriophage T2." Pt. 2, "Phage precursor nucleic acid." *Journal of General Physiology* 37: 1–23.

Hoagland, Mahlon B. 1952. "Beryllium and growth." Pt. 3, "The effect of Beryllium on plant phosphatase." *Archives of Biochemistry and Biophysics* 35: 259–67.

——. 1955a. "An enzymic mechanism for amino acid activation in animal tissues." *Biochimica et Biophysica Acta* 16: 288–89.

——. 1955b. "Enzymatic mechanism for amino acid activation in animal tissues." *Federation Proceedings* 14: 73.

——. 1958. "On an enzymatic reaction between amino acids and nucleic acid and its possible role in protein synthesis." *Recueil des Travaux Chimiques des Pays-Bas et de la Belgique* 77: 623–33.

——. 1959a. "The present status of the adaptor hypothesis." *Brookhaven Symposia in Biology* 12: 40–46.

——. 1959b. "Nucleic acids and proteins." *Scientific American* 201 (December): 55–61.

——. 1959c. "Discussion of Dr. Gale's paper" ['Protein synthesis in sub-cellular systems']. *Proceedings of the Fourth International Congress of Biochemistry*, Vienna, September 1–6, 1958, Vol. 6, 166–70. London: Pergamon Press.

——. 1960. "The relationship of nucleic acid and protein synthesis as revealed by studies in cell-free systems." In Erwin Chargaff and James N. Davidson (eds.), *The Nucleic Acids*, vol. 3, 349–408. New York: Academic Press.

——. 1961. "Some factors influencing protein synthetic activity in a cell-free mammalian system." *Cold Spring Harbor Symposia on Quantitative Biology* 26: 153–57.

——. 1966. "Views on integrated protein synthesis in liver." In Nathan O. Kaplan and Eugene P. Kennedy (eds.), *Current Aspects of Biochemical Energetics*, 199–212. New York: Academic Press.

——. 1989. "Commentary on 'Intermediate reactions in protein biosynthesis.' " *Biochimica et Biophysica Acta* 1000: 103–5.

——. 1990. *Toward the Habit of Truth: A Life in Science.* New York: W. W. Norton.

——. 1996. "Biochemistry or molecular biology? The discovery of 'soluble RNA.' " *Trends in Biological Sciences Letters (TIBS)* 21: 77–80.

Hoagland, Mahlon B., and Brigitte A. Askonas. 1963. "A cytoplasmic RNA-containing fraction that stimulates amino acid incorporation." *Proceedings of the National Academy of Sciences of the United States of America* 49: 130–37.

Hoagland, Mahlon B., and Lucy T. Comly. 1960. "Interaction of soluble ribonucleic acid and microsomes." *Proceedings of the National Academy of Sciences of the United States of America* 46: 1554–63.

Hoagland, Mahlon B., Elizabeth B. Keller, and Paul C. Zamecnik. 1956. "Enzymatic carboxyl activation of amino acids." *Journal of Biological Chemistry* 218: 345–58.

Hoagland, Mahlon B., and G. David Novelli. 1954. "Biosynthesis of coenzyme A from phosphopantetheine and of pantetheine from pantothenate." *Journal of Biological Chemistry* 207: 767–73.

Hoagland, Mahlon B., Oscar A. Scornik, and Lorraine C. Pfefferkorn. 1964. "Aspects of control of protein synthesis in normal and regenerating rat liver." Pt. 2, "A microsomal inhibitor of amino acid incorporation whose action is antagonized by guanosine triphosphate." *Proceedings of the National Academy of Sciences of the United States of America* 51: 1184–91.

Hoagland, Mahlon B., Mary L. Stephenson, Jesse F. Scott, Liselotte I. Hecht, and Paul C. Zamecnik. 1958. "A soluble ribonucleic acid intermediate in protein synthesis." *Journal of Biological Chemistry* 231: 241–57.

Hoagland, Mahlon B., and Paul C. Zamecnik. 1957. "Intermediate reactions in protein biosynthesis." *Federation Proceedings* 16: 197.

Hoagland, Mahlon B., Paul C. Zamecnik, Nehama Sharon, Fritz Lipmann, Melvin P. Stulberg, and Paul D. Boyer. 1957. "Oxygen transfer to AMP in the enzymic synthesis of the hydroxamate of tryptophan." *Biochimica et Biophysica Acta* 26: 215–17.

Hoagland, Mahlon B., Paul C. Zamecnik, and Mary L. Stephenson. 1957. "Intermediate reactions in protein biosynthesis." *Biochimica et Biophysica Acta* 24: 215–16.

———. 1959. "A hypothesis concerning the roles of particulate and soluble ribonucleic acids in protein synthesis." In R. E. Zirkle (ed.), *A Symposium on Molecular Biology*, 105–14. Chicago: University of Chicago Press.

Hofmeister, Franz. 1902. "Über den Bau des Eiweißmolecüls." *Naturwissenschaftliche Rundschau* 17: 529–33, 545–49.

Hogeboom, George H., Walter C. Schneider, and George E. Palade. 1948. "Cytochemical studies of mammalian tissues." Pt. 1, "Isolation of intact mitochondria from rat liver: Some biochemical properties of mitochondria and submicroscopic particulate material." *Journal of Biological Chemistry* 172: 619–35.

Holley, Robert W. 1956. "An alanine-dependent, ribonuclease-inhibited conversion of AMP to ATP, and its possible relationship to protein synthesis." *Abstracts of Papers, One Hundred-Thirtieth Meeting, American Chemical Society,* September 16–21, Atlantic City, New Jersey, 43C.

——. 1957. "An alanine-dependent, ribonuclease-inhibited conversion of AMP to ATP, and its possible relationship to protein synthesis." *Journal of the American Chemical Society* 79: 658–62.

Holley, Robert W., Jean Apgar, Bhupendra P. Doctor, John Farrow, Mario A. Marini, and Susan H. Merrill. 1961. "A simplified procedure for the preparation of tyrosine- and valine-acceptor fractions of Yeast 'soluble ribonucleic acid.'" *Journal of Biological Chemistry* 236: 200–202.

Holley, Robert W., Jean Apgar, George A. Everett, James T. Madison, Mark Marquisee, Susan H. Merrill, John Robert Penswick, and Ada Zamir. 1965. "Structure of a ribonucleic acid." *Science* 147: 1462–65.

Holley, Robert W., and Susan H. Merrill. 1959. "Countercurrent distribution of an active ribonucleic acid." *Journal of the American Chemical Society* 81: 753.

Holley, Robert W., and P. Prock. 1958. "Intermediates in protein synthesis: Alanine activation and an active ribonucleic acid fraction." *Federation Proceedings* 17: 244.

Holmes, Frederick L. 1985. *Lavoisier and the Chemistry of Life: An Exploration of Scientific Creativity*. Madison: University of Wisconsin Press.

Hultin, Tore. 1950. "Incorporation in vivo of ^{15}N-labeled glycine into liver fractions of newly hatched chicks." *Experimental Cell Research* 1: 376–81.

——. 1955. "The incorporation in vivo of labeled amino acids into subfractions of liver cytoplasm fractions." *Experimental Cell Research*, supplement 3: 210–17.

——. 1956. "The incorporation in vitro of 1-C^{14}-glycine into liver proteins visualized as a two-step reaction." *Experimental Cell Research* 11: 222–24.

Hultin, Tore, and G. Beskow. 1956. "The incorporation of C^{14}-L-leucine into rat liver proteins in vitro visualized as a two-step reaction." *Experimental Cell Research* 11: 664–66.

Hultin, Tore, and Alexandra von der Decken. 1959. "The transfer of soluble polynucleotides to the ribonucleic acid of rat liver microsomes." *Experimental Cell Research* 16: 444–47.

Hunter, G. D., P. Brookes, A. R. Crathorn, and John A. V. Butler. 1959. "Intermediate reactions in protein synthesis by the isolated cytoplasmic-membrane fraction of Bacillus megaterium." *Biochemical Journal* 73: 369–76.

Hurlbert, Robert B., and Van R. Potter. 1954. "Nucleotide metabolism." Pt. 1, "The conversion of orotic acid-6-C^{14} to uridine nucleotides." *Journal of Biological Chemistry* 209: 1–21.

Husserl, Edmund. 1954. "Vom Ursprung der Geometrie." In W. Biemel (ed.), *Collected Works*, vol. 6, 365–86. Den Haag: Kluwer.

"Immunology as a Historical Object." 1994. Special issue of the *Journal of the History of Biology* 27, no. 3.

Jacob, François. 1974. "Le modèle linguistique en biologie." *Critique* 322: 197–205.

———. 1982. *The Possible and the Actual*. Seattle: University of Washington Press.

———. 1988. *The Statue Within: An Autobiography*. Translated by Franklin Philip. New York: Basic Books.

Jacob, François, and Jacques Monod. 1961. "Genetic regulatory mechanisms in the synthesis of proteins." *Journal of Molecular Biology* 3: 316–56.

Jardine, Nicholas. 1991. *The Scenes of Inquiry*. Oxford: Oxford University Press.

Jeener, Raymond. 1948. "L'hétérogénéité des granules cytoplasmiques: Données complémentaires fournies par leur fractionnement en solution saline concentrée." *Biochimica et Biophysica Acta* 2: 633–41.

Jeener, Raymond, and Jean Brachet. 1943–45. "Recherches sur l'acide ribonucléique des levures." *Enzymologia* 11: 222–34.

Judson, Horace F. 1979. *The Eighth Day of Creation*. New York: Simon and Schuster.

Kalckar, Herman M. 1941. "The nature of energetic coupling in biological synthesis." *Chemical Reviews* 28: 71–178.

Kameyama, Tadanori, and G. David Novelli. 1960. "The cell-free synthesis of β-galactosidase by Escherichia coli." *Biochemical and Biophysical Research Communications* 2: 393–96.

Kant, Immanuel. 1987. *Critique of Judgment*. Translated by Werner S. Pluhar. Indianapolis: Hackett.

Kauffman, Stuart. 1995. *At Home in the Universe: The Search for the Laws of Self-Organization and Complexity*. Oxford: Oxford University Press.

Kay, Lily E. 1993. *The Molecular Vision of Life*. Oxford: Oxford University Press.

———. 1994. "Wer schrieb das Buch des Lebens? Information und Transformation der Molekularbiologie." In M. Hagner, H.-J. Rheinberger, and B. Wahrig-Schmidt (eds.), *Objekte, Differenzen, Konjunkturen: Experimentalsysteme im historischen Kontext*, 151–79. Berlin: Akademie Verlag.

———. Forthcoming. *Who Wrote the Book of Life? A History of the Genetic Code*. Chicago: University of Chicago Press.

Keller, Elizabeth B. 1951. "Turnover of proteins of cell fractions of adult rat liver in vivo." *Federation Proceedings* 10: 206.

Keller, Elizabeth B., and Paul C. Zamecnik. 1954. "Anaerobic incorporation of C^{14}-amino acids into protein in cell-free liver preparations." *Federation Proceedings* 13: 239–40.

———. 1955. "Effect of guanosine diphosphate on incorporation of labeled amino acids into proteins." *Federation Proceedings* 14: 234.

———. 1956. "The effect of guanosine diphosphate and triphosphate on the incorporation of labeled amino acids into proteins." *Journal of Biological Chemistry* 221: 45–59.

Keller, Elizabeth B., Paul C. Zamecnik, and Robert B. Loftfield. 1954. "The role of microsomes in the incorporation of amino acids into proteins." *Journal of Histochemistry and Cytochemistry* 2: 378–86.

Keller, Evelyn Fox. 1983. *A Feeling for the Organism: The Life and Work of Barbara McClintock.* New York: W. H. Freeman.

———. 1994. "Language and science: Genetics, embryology, and the discourse of gene action." *Encyclopedia Britannica,* Great Ideas Today, pt. 1, "Current Developments in the Arts and Sciences," 1–29.

Kirby, K. S. 1956. "A new method for the isolation of ribonucleic acids from mammalian tissues." *Biochemical Journal* 64: 405.

Kit, Saul, and David M. Greenberg. 1952. "Incorporation of isotopic threonine and valine into the protein of rat liver particles." *Journal of Biological Chemistry* 194: 377–81.

Kittler, Friedrich. 1990. *Discourse Networks, 1800/1900.* Translated by M. Metteer, with C. Cullens. Stanford: Stanford University Press.

Knorr Cetina, Karin. 1981. *The Manufacture of Knowledge: An Essay on the Constructivist and Contextual Nature of Science.* Oxford: Pergamon Press.

Knorr Cetina, Karin, Klaus Amann, Stefan Hirschauer, and Karl-Heinrich Schmidt. 1988. "Das naturwissenschaftliche Labor als Ort der 'Verdichtung' von Gesellschaft." *Zeitschrift für Soziologie* 17: 85–101.

Kohler, Robert E. 1991a. *Partners in Science: Foundations and Natural Scientists 1900–1945.* Chicago: University of Chicago Press.

———. 1991b. "Systems of production: Drosophila, Neurospora, and biochemical genetics." *Historical Studies in the Physical and Biological Sciences* 22: 87–130.

———. 1994. *Lords of the Fly: Drosophila Genetics and the Experimental Life.* Chicago: University of Chicago Press.

Koningsberger, Victor V., and J. Theo G. Overbeek. 1953. "On the rôle of the nucleic acids in the biosynthesis of the peptide bond." *Koninklijke Nederlandse Akademie van Wetenschappen, Proceedings of the Section of Sciences* 56 (series B, physical sciences): 248–54.

Koningsberger, Victor V., Christian Olav van der Grinten, and J. Theo G. Overbeek. 1957. "Possible intermediates in the biosynthesis of proteins." Pt. 1, "Evidence for the presence of nucleotide-bound carboxyl-activated peptides in baker's yeast." *Biochimica et Biophysica Acta* 26: 483–90.

Kornberg, Arthur. 1989. *For the Love of Enzymes.* Cambridge, Mass.: Harvard University Press.

Kruh, Jacques, and Henry Borsook. 1955. "In vitro synthesis of ribonucleic acid in reticulocytes." *Nature* 175: 386–87.

Kubler, George. 1962. *The Shape of Time: Remarks on the History of Things.* New Haven and London: Yale University Press.

Kuhn, Thomas S. 1962. *The Structure of Scientific Revolutions.* Chicago: University of Chicago Press.

———. 1979. *The Essential Tension: Selected Studies in Scientific Tradition and Change.* Chicago: University of Chicago Press.

———. 1992. *The Trouble with the Historical Philosophy of Science.* An occasional publication of the Department of the History of Science, Harvard University. Cambridge, Mass.: Harvard University.

Kurland, Chuck G. 1960. "Molecular characterization of ribonucleic acid from Escherichia coli ribosomes." Pt. 1, "Isolation and molecular weights." *Journal of Molecular Biology* 2: 83–91.

Lacan, Jacques. 1986. *Séminaire VII: L'éthique de la psychanalyse.* Paris: Seuil.

———. 1989. "Science and truth." *Newsletter of the Freudian Field* 3: 4–29.

Lacks, Sanford, and François Gros. 1959. "A metabolic study of the RNA-amino acid complexes in Escherichia coli." *Journal of Molecular Biology* 1: 301–20.

Lamborg, Marvin R. 1960. "Amino acid incorporation into protein by extracts of E. coli." *Federation Proceedings* 19: 346.

Lamborg, Marvin R., and Paul C. Zamecnik. 1960. "Amino acid incorporation into protein by extracts of E. coli." *Biochimica et Biophysica Acta* 42: 206–11.

———. 1965. "Optical rotatory dispersion of E. coli sRNA in the far ultraviolet region." *Biochemical and Biophysical Research Communications* 20: 328–33.

Lamborg, Marvin R., Paul C. Zamecnik, Ting-Kai Li, Jeremias Kägi, and Bert L. Vallee. 1965. "Anomalous rotatory dispersion of soluble ribonucleic acid and its relation to amino acid synthetase recognition." *Biochemistry* 4: 63–70.

Latour, Bruno. 1987. *Science in Action.* Cambridge, Mass.: Harvard University Press.

———. 1988. *The Pasteurization of France.* Translated by Alan Sheridan and John Law. Cambridge, Mass.: Harvard University Press.

———. 1990a. "Postmodern? No, simply amodern! Steps towards an anthropology of science." *Studies in History and Philosophy of Science* 21: 145–71.

———. 1990b. "The force and the reason of experiment." In Homer E. Le Grand (ed.), *Experimental Inquiries,* 49–80. Dordrecht: Kluwer.

———. 1990c. "Drawing things together." In Michael Lynch and Steve Woolgar (eds.), *Representation in Scientific Practice,* 19–68. Cambridge, Mass.: MIT Press.

———. 1993a. *We Have Never Been Modern.* Translated by Catherine Porter. Cambridge, Mass.: Harvard University Press.

———. 1993b. "Le 'pédofil' de Boa Vista—montage photo-philosophique." In B. Latour, *La Clef de Berlin,* 171–225. Paris: Editions La Découverte.

Latour, Bruno, and Steve Woolgar. 1979. *Laboratory Life.* London: SAGE Publications.

———. 1986. *Laboratory Life.* 2d ed. Princeton, N.J.: Princeton University Press.

Ledingham, J. C. G., and W. E. Gye. 1935. "On the nature of the filterable tumour-exciting agent in avian sarcomata." *Lancet* 228 (1): 376–77.

Lee, Norman D., Jean T. Anderson, Ruth Miller, and Robert H. Williams. 1951. "Incorporation of labeled cystine into tissue protein and subcellular structures." *Journal of Biological Chemistry* 192: 733–42.

Lee, Norman D., Norma M. MacRae, and Robert H. Williams. 1951. "Effect of p-dimethylaminoazobenzene on the incorporation of labeled cystine into protein of the subcellular components of rat liver." *Federation Proceedings* 10: 363.

Le Grand, Homer E. (ed.). 1990. *Experimental Inquiries*. Dordrecht: Kluwer.

Lengyel, Peter, Joseph F. Speyer, and Severo Ochoa. 1961. "Synthetic polynucleotides and the amino acid code." *Proceedings of the National Academy of Sciences of the United States of America* 47: 1936–42.

Lenoir, Timothy. 1988. "Practice, reason, context: The dialogue between theory and experiment." *Science in Context* 2: 3–22.

——. 1992. "Practical reason and the construction of knowledge: The lifeworld of Haber-Bosch." In Ernan McMullin (ed.), *The Social Dimensions of Science*, 158–97. South Bend, Ind.: University of Notre Dame Press.

——. 1993. "The discipline of nature and the nature of disciplines." In Ellen Messer-Davidow, David Sylvan, and David Shumway (eds.), *Knowledges: Historical and Critical Studies in Disciplinarity*, 70–102. Charlottesville: University Press of Virginia.

Leroi-Gourhan, André. 1964–65. *Le geste et la parole*. Paris: Albin Michel.

"Les vues de l'esprit." 1985. Special issue of *Culture technique* 14.

Levine, George (ed.). 1993. *Realism and Representation: Essays on the Problem of Realism in Relation to Science, Literature, and Culture*. Madison: University of Wisconsin Press.

Linderstrøm-Lang, Kai. 1952. "Lane Medical Lectures: Proteins and enzymes (Lecture 5: Biological synthesis of proteins)." Stanford University Publications, Medical Series 6: 1–115.

Lipmann, Fritz. 1941. "Metabolic generation and utilization of phosphate bond energy." *Advances in Enzymology* 1: 99–162.

——. 1949. "Mechanism of peptide bond formation." *Federation Proceedings* 8: 597–602.

——. 1954. "On the mechanism of some ATP-linked reactions and certain aspects of protein synthesis." In William D. McElroy and Bentley Glass (eds.), *The Mechanism of Enzyme Action*, 599–604. Baltimore: Johns Hopkins University Press.

——. 1963. "Messenger ribonucleic acid." *Progress in Nucleic Acid Research* 1: 135–61.

——. 1971. *Wanderings of a Biochemist*. New York: Wiley-Interscience.

Lipmann, Fritz, W. C. Hülsmann, G. Hartmann, Hans G. Boman, and George Acs. 1959. "Amino acid activation and protein synthesis." *Journal of Cellular and Comparative Physiology* 54, supplement 1: 75–88.

Littlefield, John W., and Elizabeth B. Keller. 1956. "Cell-free incorporation of C^{14}-amino acids into cytoplasmic ribonucleoprotein particles." *Federation Proceedings* 15: 302–3.

——. 1957. "Incorporation of C^{14}-amino acids into ribonucleoprotein particles from the Ehrlich mouse ascites tumor." *Journal of Biological Chemistry* 224: 13–30.

Littlefield, John W., Elizabeth B. Keller, Jerome Gross, and Paul C. Zamecnik. 1955a. "Studies on cytoplasmic ribonucleoprotein particles from the liver of the rat." *Journal of Biological Chemistry* 217: 111–23.

——. 1955b. "Studies on protein synthesis in the liver." *Journal of Clinical Investigation* 34: 950.

Loftfield, Robert B. 1947. "Preparation of C^{14}-labeled hydrogen cyanide, alanine, and glycine." *Nucleonics* 1: 54–57.

——. 1954. "In vivo and in vitro incorporation of C-14 leucine into ferritin." *Federation Proceedings* 13: 465.

——. 1955. "Participation of free amino acids in protein synthesis." *Federation Proceedings* 14: 246.

——. 1957a. "The biosynthesis of protein." *Progress in Biophysics and Biophysical Chemistry* 8: 348–86.

——. 1957b. "Speed of protein synthesis." *Federation Proceedings* 16: 82.

Loftfield, Robert B., and Elizabeth A. Eigner. 1958. "The time required for the synthesis of a ferritin molecule in rat liver." *Journal of Biological Chemistry* 231: 925–43.

Loftfield, Robert B., John W. Grover, and Mary L. Stephenson. 1953. "Possible role of proteolytic enzymes in protein synthesis." *Nature* 171: 1024–25.

Loftfield, Robert B., and Anne Harris. 1956. "Participation of free amino acids in protein synthesis." *Journal of Biological Chemistry* 219: 151–59.

Loftfield, Robert B., Liselotte I. Hecht, and Elizabeth A. Eigner. 1959. "Alloisoleucine as a competitor for isoleucine and valine in protein synthesis." *Federation Proceedings* 18: 276.

Loomis, William F., and Fritz Lipmann. 1948. "Reversible inhibition of the coupling between phosphorylation and oxidation." *Journal of Biological Chemistry* 173: 807–8.

Löwy, Ilana. 1992. "The strength of loose concepts: Boundary concepts, federative experimental strategies and disciplinary growth: The case of immunology." *History of Science* 30: 371–95.

Lubar, Steven, and W. David Kingery. 1993. *History from Things: Essays on Material Culture.* Washington, D.C.: Smithsonian Institution Press.

Luhmann, Niklas. 1990. *Die Wissenschaft der Gesellschaft.* Frankfurt am Main: Suhrkamp.

Lunardini, Rosemary. 1993. "DNA drama." *Dartmouth Medicine* 18: 16–22.

Luria, Salvador. 1984. *A Slot Machine, A Broken Test Tube: An Autobiography.* New York: Harper and Row.

Lwoff, André. 1957. "The concept of virus." *Journal of General Microbiology* 17: 239–53.

Lwoff, André, and Agnes Ullmann (eds.). 1979. *Origins of Molecular Biology: A Tribute to Jacques Monod.* New York: Academic Press.

Lynch, Michael. 1985. *Art and Artifact in Laboratory Science: A Study of Shop Work and Shop Talk in a Research Laboratory.* London: Routledge and Kegan Paul.

———. 1994. "Representation is overrated: Some critical remarks about the use of the concept of representation in science studies." *Configurations* 2: 137–49.

Lynch, Michael, and Steve Woolgar. 1990a. "Sociological orientations to representational practice in science." In M. Lynch and S. Woolgar (eds.), *Representation in Scientific Practice*, 1–18. Cambridge, Mass.: MIT Press.

Lynch, Michael, and Steve Woolgar (eds.). 1990b. *Representation in Scientific Practice.* Cambridge, Mass.: MIT Press.

Maas, Werner K., and G. David Novelli. 1953. "Synthesis of pantothenic acid by depyrohosphorylation of adenosine triphosphate." *Archives of Biochemistry and Biophysics* 43: 236–38.

MacColl, San. 1989. "Intimate observation." *Metascience* 7: 90–98.

Malkin, Harold M. 1954. "Synthesis of ribonucleic acid purines and protein in enucleated and nucleated sea urchin eggs." *Journal of Cellular and Comparative Physiology* 44: 105–12.

Matsubara, Kenichi, and Itaru Watanabe. 1961. "Studies of amino acid incorporation with purified ribosomes and soluble enzymes from Escherichia coli." *Biochemical and Biophysical Research Communications* 5: 22–26.

Matthaei, J. Heinrich, and Marshall W. Nirenberg. 1961a. "The dependence of cell-free protein synthesis in E. coli upon RNA prepared from ribosomes." *Biochemical and Biophysical Research Communications* 4: 404–8.

———. 1961b. "Some characteristics of a cell-free DNAase sensitive system incorporating amino acids into protein." *Federation Proceedings* 20: 391.

———. 1961c. "Characteristics and stabilization of DNAase-sensitive protein synthesis in E. coli extracts." *Proceedings of the National Academy of Sciences of the United States of America* 47: 1580–88.

Mayr, Ernst. 1990. "When is historiography whiggish?" *Journal of the History of Ideas* 51: 301–9.

McCarty, Maclyn. 1985. *The Transforming Principle: Discovering That Genes Are Made of DNA.* New York: W. W. Norton.

McCorquodale, Donald J., E. G. Veach, and Gerald C. Mueller. 1961. "The incorporation in vitro of labeled amino acids into the proteins of normal and regenerating rat liver." *Biochimica et Biophysica Acta* 46: 335–41.

McIntosh, James. 1935. "The sedimentation of the virus of Rous sarcoma and the bacteriophage by a high-speed centrifuge." *Journal of Pathology and Bacteriology* 41: 215–17.

"The Mechanism of Protein Synthesis." 1969. *Cold Spring Harbor Symposia on Quantitative Biology* 34.

Melchior, Jacklyn B., and Harold Tarver. 1947a. "Studies in protein synthesis in vitro." Pt. 1, "On the synthesis of labeled cystine (S^{35}) and its attempted use as a tool in the study of protein synthesis." *Archives of Biochemistry* 12: 301–8.

———. 1947b. "Studies in protein synthesis in vitro." Pt. 2. "On the uptake of labeled sulfur by the proteins of liver slices incubated with labeled methionine (S^{35})." *Archives of Biochemistry* 12: 309–15.

Meyers Großes Taschenlexikon. 1990. 3d ed. Mannheim, Vienna, and Zurich: BI-Taschenbuchverlag.

Miller, Warren W. 1947. "High-efficiency counting of long-lived radioactive carbon as CO_2." *Science* 105: 123–25.

Mitchell, W. J. Thomas. 1987. *Iconology: Image, Text, Ideology.* Chicago: University of Chicago Press.

Monier, Robert, Mary L. Stephenson, and Paul C. Zamecnik. 1960. "The preparation and some properties of a low molecular weight ribonucleic acid from baker's yeast." *Biochimica et Biophysica Acta* 43: 1–8.

Monod, Jacques, and E. Borek (eds.). 1971. *Of Microbes and Life.* Ithaca, N.Y.: Cornell University Press.

Monod, Jacques, and Melvin Cohn. 1953. "Sur le mécanisme de la synthèse d'une protéine bactérienne: La β-galactosidase d'E. coli." *Fourth International Congress of Microbiology*, symposium on microbial metabolism, 42–62. Rome: Fondazione Emanuele Paterno.

Monod, Jacques, Alvin M. Pappenheimer Jr., and Germaine Cohen-Bazire. 1952. "La cinétique de la biosynthèse de la β-galactosidase chez E. coli considérée comme fonction de la croissance." *Biochimica et Biophysica Acta* 9: 648–60.

Morange, Michel. 1990. "Le concept du gène régulateur." In Jean-Louis Fischer and William H. Schneider (eds.), *Histoire de la Génétique*, 271–91. Paris: A.R.P.E.M and Editions Sciences en Situation.

———. 1994. *Histoire de la biologie moléculaire.* Paris: Editions La Découverte.

Myers, Greg. 1990. *Writing Biology: Texts in the Social Construction of Scientific Knowledge.* Madison: University of Wisconsin Press.

Nägele, Rainer. 1987. *Reading After Freud.* New York: Columbia University Press.

Nathans, Daniel, and Fritz Lipmann. 1960. "Amino acid transfer from sRNA to microsome." Pt. 2, "Isolation of a heat-labile factor from liver supernatant." *Biochimica et Biophysica Acta* 43: 126–28.

———. 1961. "Amino acid transfer from aminoacyl-ribonucleic acids to protein on ribosomes of Escherichia coli." *Proceedings of the National Academy of Sciences of the United States of America* 47: 497–504.

Nathanson, Ira T., A. L. Nutt, Alfred Pope, Paul C. Zamecnik, Joseph C. Aub, Austin M. Brues, and Seymour S. Kety. 1945. "The toxic factor in experimental traumatic shock." Pt. 1, "Physiological effects of muscle ligation in the dog." *Journal of Clinical Investigation* 24: 829–34 [pts. 2–6: 835–63].

Nirenberg, Marshall W. 1969. "The genetic code." Nobel lecture, 1:21. Stockholm: Nobel Foundation.

Nirenberg, Marshall W., and Philip Leder. 1964. "RNA codewords and protein synthesis." *Science* 145: 1399–1407.

Nirenberg, Marshall W., and J. Heinrich Matthaei. 1961. "The dependence of cell-free protein synthesis in E. coli upon naturally occurring or synthetic polyribonucleotides." *Proceedings of the National Academy of Sciences of the United States of America* 47: 1588–1602.

———. 1963a. "The dependence of cell-free protein synthesis in E. coli upon naturally occurring or synthetic template RNA." In Vladimir A. Engelhardt (ed.), *Proceedings of the Fifth International Congress of Biochemistry*, August 10–16, 1961, Moscow, vol. 1, 184–95. New York: Macmillan.

———. 1963b. "Comparison of ribosomal and soluble E. coli systems incorporating amino acids into protein." *Proceedings of the Fifth International Congress of Biochemistry*, August 10–16, 1961, Moscow, vol. 9, Abstract 2.115, p. 102. New York: Macmillan.

Nishizuka, Yasutomi, and Fritz Lipmann. 1966. "Comparison of guanosine triphosphate split and polypeptide synthesis with a purified E. coli system." *Proceedings of the National Academy of Sciences of the United States of America* 55: 212–19.

Nisman, B. 1959. "Incorporation and activation of amino acids by disrupted protoplasts of Escherichia coli." *Biochimica et Biophysica Acta* 32: 18–31.

Nomura, Masayasu. 1990. "History of ribosome research: A personal account." In Walter E. Hill, Peter B. Moore, Albert Dahlberg, David Schlessinger, Roger A. Garrett, and Jonathan R. Warner (eds.), *The Ribosome: Structure, Function, and Evolution*, 3–55. Washington D.C.: American Society for Microbiology.

Nomura, Masayasu, Benjamin D. Hall, and Sol Spiegelman. 1960. "Characterization of RNA synthesized in Escherichia coli after bacteriophage T2 infection." *Journal of Molecular Biology* 2: 306–26.

Nomura, Masayasu, Alfred Tissières, and Peter Lengyel (eds.). 1974. *Ribosomes*. New York: Cold Spring Harbor Laboratory Press.

Novelli, G. David. 1966. "From ~P to CoA to protein biosynthesis." In Nathan O. Kaplan and Eugene P. Kennedy (eds.), *Current Aspects of Biochemical Energetics*, 183–97. New York: Academic Press.

Nowotny, Helga. 1994. *Time: The Modern and Postmodern Experience*. Oxford: Polity Press.

Ofengand, James, and Robert Haselkorn. 1961–62. "Viral RNA-dependent incorporation of amino acids into protein by cell-free extracts of E. coli." *Biochemical and Biophysical Research Communications* 6: 469–74.

Ogata, Kikuo, Masana Ogata, Yoshio Mochizuki, and Tadamoto Nishiyama. 1956. "The in vitro incorporation of C^{14}-glycine into antibody and other protein fractions by popliteal lymph nodes of rabbits following the local injection of crystalline ovalbumin." *The Journal of Biochemistry* 43: 653–68.

Ogata, Kikuo, and Hiroyoshi Nohara. 1957. "The possible role of the ribonucleic acid (RNA) of the pH 5 enzyme in amino acid activation." *Biochimica et Biophysica Acta* 25: 659–60.

Ogata, Kikuo, Hiroyoshi Nohara, and Tomi Morita. 1957. "The effect of ribonuclease on the amino acid-dependent exchange between labeled inorganic pyrophosphate ($^{32}P^{32}P$) and adenosine triphosphate (ATP) by the pH 5 enzyme." *Biochimica et Biophysica Acta* 26: 656–57.

Olby, Robert C. 1990. "The molecular revolution in biology." In R. C. Olby, G. N. Cantor, J. R. R. Christie, and M. J. S. Hodge (eds.), *Companion to the History of Modern Science*, 503–20. London: Routledge.

Paillot, André, and André Gratia. 1938. "Application de l'ultracentrifugation à l'isolement du virus de la grasserie des vers à soie." *Comptes Rendus Hebdomadaires de la Société de Biologie* 90: 1178–80.

Palade, George E. 1951. "Intracellular distribution of acid phosphatase in rat liver cells." *Archives of Biochemistry* 30: 144–58.

———. 1955. "A small particulate component of the cytoplasm." *Journal of Biophysical and Biochemical Cytology* 1: 59–68.

———. 1958. "Microsomes and ribonucleoprotein particles." In Richard B. Roberts (ed.), *Microsomal Particles and Protein Synthesis*, 36–61. London: Pergamon Press.

Palade, George E., and Keith R. Porter. 1954. "Studies on the endoplasmic reticulum." Pt. 1, "Its identification in cells in situ." *Journal of Experimental Medicine* 100: 641–56.

Palade, George E., and Philip Siekevitz. 1956. "Liver microsomes: An integrated morphological and biochemical study." *Journal of Biophysical and Biochemical Cytology* 2: 171–200.

Pardee, Arthur B. 1954. "Nucleic acid precursors and protein synthesis." *Proceedings of the National Academy of Sciences of the United States of America* 40: 263–70.

Pardee, Arthur B., François Jacob, and Jacques Monod. 1959. "The genetic control and cytoplasmic expression of 'inducibility' in the synthesis of β-galactosidase by E. coli." *Journal of Molecular Biology* 1: 165–78.

Paterson, Alan R. P., and Gerald A. LePage. 1957. "Ribonucleic acid synthesis in tumor homogenates." *Cancer Research* 17: 409–17.

Peirce, Charles Sanders. 1955. "Logic as semiotic: The theory of signs." In Justus Buchler (ed.), *Philosophical Writings of Peirce*, 98–119. New York: Dover.

Petermann, Mary L., and Mary G. Hamilton. 1952. "An ultracentrifugal analysis of the macromolecular particle from normal and leukemic mouse spleen." *Cancer Research* 12: 373–78.

———. 1955. "A stabilizing factor for cytoplasmic nucleoproteins." *Journal of Biophysical and Biochemical Cytology* 1: 469–72.

Petermann, Mary L., Mary G. Hamilton, M. Earl Balis, Kumud Samarth, and Pauline Pecora. 1958. "Physicochemical and metabolic studies on rat liver nucleoprotein." In Richard B. Roberts (ed.), *Microsomal Particles and Protein Synthesis*, 70–75. London: Pergamon Press.

Petermann, Mary L., Mary G. Hamilton, and Nancy A. Mizen. 1954. "Electrophoretic analysis of the macromolecular nucleoprotein particles of mammalian cytoplasm." *Cancer Research* 14: 360–66.

Petermann, Mary L., Nancy A. Mizen, and Mary G. Hamilton. 1953. "The macromolecular particles of normal and regenerating rat liver." *Cancer Research* 13: 372–75.

Peterson, Elbert A., and David M. Greenberg. 1952. "Characteristics of the amino acid-incorporating system of liver homogenates." *Journal of Biological Chemistry* 194: 359–75.

Pickering, Andrew (ed.). 1992. *Science as Practice and Culture*. Chicago: University of Chicago Press.

———. 1995. *The Mangle of Practice: Time, Agency, and Science*. Chicago: University of Chicago Press.

"Pictorial Representation in Biology." 1991. Special issue of *Biology and Philosophy* 6.

Polanyi, Michael. 1958. *Personal Knowledge: Towards a Post-Critical Philosophy*. London: Routledge and Kegan Paul.

———. 1965. *Duke University Lectures 1964*. Microfilm, University of California, Berkeley. Copy from Library Photographic Service.

———. 1967. *The Tacit Dimension*. New York: Anchor.

———. 1969. *Knowing and Being*. Edited by Marjorie Grene. Chicago: University of Chicago Press.

Popper, Karl. 1968. *The Logic of Scientific Discovery*. New York: Harper and Row.

Porter, Keith R. 1953. "Observations on a submicroscopic basophilic component of cytoplasm." *Journal of Experimental Medicine* 97: 727–49.

Porter, Keith R., and Joseph Blum. 1953. "A study in microtomy for electron microscopy." *The Anatomical Record* 117: 685–710.

Portugal, Franklin H., and Jack S. Cohen. 1977. *A Century of DNA*. Cambridge, Mass.: MIT Press.

Potter, Joseph L., and Alexander L. Dounce. 1956. "Nucleotide-amino acid complexes in alkaline digests of ribonucleic acid." *Journal of the American Chemical Society* 78: 3078–82.

Potter, Van R., Liselotte I. Hecht, and Edward Herbert. 1956. "Incorporation of pyrimidine precursors into ribonucleic acid in a cell-free fraction of rat liver homogenate." *Biochimica et Biophysica Acta* 20: 439–40.

Preiss, Jack, Paul Berg, E. James Ofengand, Fred H. Bergmann, and Marianne Dieckmann. 1959. "The chemical nature of the RNA-amino acid compound formed by amino acid-activating enzymes." *Proceedings of the National Academy of Sciences of the United States of America* 45: 319–28.

Prigogine, Ilya, and Isabelle Stengers. 1979. *La nonvelle alliance*. Paris: Gallimard.

Rabinow, Paul. 1996. *Making PCR: A Story of Biotechnology*. Chicago: University of Chicago Press.

Rasmussen, Nicolas. Forthcoming. *Picture Control: The Electron Microscope and the Transformation of American Biology, 1940–1959*. Stanford, Calif.: Stanford University Press.

Reichenbach, Hans. 1938. *Experience and Prediction: An Analysis of the Foundations and the Structure of Knowledge*. Chicago: University of Chicago Press.

Remer, Theodore G. 1964. "Serendipity—the last word." *Science* 143: 196–97.

Rendi, R. 1959. "Incorporation of C^{14}-glycine in soluble RNA and microsomes of normal and regenerating rat liver." *Biochimica et Biophysica Acta* 31: 266–67.

Rendi, R., and Tore Hultin. 1960. "Preparation and amino acid incorporating ability of ribonucleoprotein-particles from different tissues of the rat." *Experimental Cell Research* 19: 253–66.

Rheinberger, Hans-Jörg. 1989. "H. M. Collins, Changing Order." *History and Philosophy of the Life Sciences* 11: 388–90.

———. 1992a. *Experiment, Differenz, Schrift: Zur Geschichte epistemischer Dinge*. Marburg, Germany: Basiliskenpresse.

———. 1992b. "Experiment, difference, and writing." Pt. 1, "Tracing protein synthesis," and pt. 2, "The laboratory production of transfer RNA." *Studies in History and Philosophy of Science* 23: 305–31, 389–422.

———. 1993. "Experiment and orientation: Early systems of in vitro protein synthesis." *Journal of the History of Biology* 26: 443–71.

———. 1994. "Experimental systems: Historiality, narration, and deconstruction." *Science in Context* 7: 65–81.

———. 1995. "From microsomes to ribosomes: 'Strategies' of 'representation' 1935–1955." *Journal of the History of Biology* 28: 49–89.

———. 1996. "Comparing Experimental Systems: Protein synthesis in microbes and in animal tissue at Cambridge (Ernst F. Gale) and at the Massachusetts General Hospital (Paul C. Zamecnik), 1945–1960." *Journal of the History of Biology* 29: 387–416.

Rheinberger, Hans-Jörg, and Michael Hagner. 1993. "Experimentalsysteme." In H.-J. Rheinberger and M. Hagner (eds.), *Die Experimentalisierung des Lebens*, 7–27. Berlin: Akademie.

Rheinberger, Hans-Jörg, Michael Hagner, and Bettina Wahrig-Schmidt (eds.). 1996. *Räume des Wissens: Repräsentation, Codierung, Spur*. Berlin: Akademie.

Rich, Alexander, and Norman Davidson (eds.). 1968. *Structural Chemistry and Molecular Biology*. San Francisco: Freeman.

Riley, Monica, Arthur B. Pardee, François Jacob, and Jacques Monod. 1960. "On the expression of a structural gene." *Journal of Molecular Biology* 2: 216–25.

Rittenberg, David. 1941. "The state of the proteins in animals as revealed by the use of isotopes." *Cold Spring Harbor Symposia on Quantitative Biology* 9: 283–89.

——. 1950. "Dynamic aspects of the metabolism of amino acids." *The Harvey Lectures 1948–49* 44: 200–219.

Roberts, Richard B. 1958. Introduction to Richard B. Roberts (ed.), *Microsomal Particles and Protein Synthesis*, vii–viii. New York: Pergamon Press.

——. 1964. "Ribosomes. A. General properties of ribosomes." In Richard B. Roberts (ed.), *Studies of Macromolecular Biosynthesis*, 147–68. Washington, D.C.: Carnegie Institution.

Roberts, Royston M. 1989. *Serendipity: Accidental Discoveries in Science*. New York: John Wiley.

Rogers, Palmer, and G. David Novelli. 1959. "Cell free synthesis of ornithine transcarbamylase." *Biochimica et Biophysica Acta* 33: 423.

Rohbeck, Johannes. 1993. *Technologische Urteilskraft: Zu einer Ethik technischen Handelns*. Frankfurt am Main: Suhrkamp.

Root-Bernstein, Robert Scott. 1989. *Discovering: Inventing and Solving Problems at the Frontiers of Scientific Knowledge*. Cambridge, Mass.: Harvard University Press.

Rosenberg, Alexander. 1994. *Instrumental Biology or the Disunity of Science*. Chicago: University of Chicago Press.

Rotman, Brian. 1987. *Signifying Nothing*. New York: St. Martin's Press.

Rotman, Boris, and Sol Spiegelman. 1954. "On the origin of the carbon in the induced synthesis β-galactosidase in Escherichia coli." *Journal of Bacteriology* 68: 419–29.

Rous, Peyton. 1911. "A sarcoma of fowl transmissible by an agent separable from tumor cells." *Journal of Experimental Medicine* 13: 397–411.

Rouse, Joseph. 1991. "Philosophy of science and the persistent narratives of modernity." *Studies in History and Philosophy of Science* 22: 141–62.

——. 1996. *Engaging Science: How to Understand Its Practices Philosophically*. Ithaca, N.Y.: Cornell University Press.

Sachs, Howard. 1957. "A stabilized enzyme system for amino acid incorporation." *Journal of Biological Chemistry* 228: 23–39.

St. Aubin, P. M. G., and Nancy L. R. Bucher. 1951. "A study of binucleate cell counts in resting and regenerating rat liver employing a mechanical method for the separation of liver cells." *The Anatomical Record* 112: 797–809.

Sanadi, D. Rao, David M. Gibson, and Padmasini Ayengar. 1954. "Guanosine triphosphate, the primary product of phosphorylation coupled to the breakdown of succinyl coenzyme A." *Biochimica et Biophysica Acta* 14: 434–36.

Sanger, Frederick, and Hans Tuppy. 1951. "The amino-acid sequence in the phenylalanyl chain of insulin." Pt. 1, "The identification of lower peptides from partial hydrolysates," and pt. 2, "The investigation of peptides from enzymic hydrolysates." *Biochemical Journal* 49: 463–81, 481–90.

Sanger, Frederick, S. Nicklen, and A. R. Coulson. 1977. "DNA sequencing with chain-terminating inhibitors." *Proceedings of the National Academy of Sciences of the United States of America* 74: 5463–67.

Sapolsky, Harvey M. 1990. *Science and the Navy*. Princeton, N.J.: Princeton University Press.

Sarin, Prem S., and Paul C. Zamecnik. 1964. "On the stability of aminoacyl-s-RNA to nucleophilic catalysis." *Biochimica et Biophysica Acta* 91: 653–55.

———. 1965a. "Modification of amino acid acceptance and transfer capacity of s-RNA in the presence of organic solvents." *Biochemical and Biophysical Research Communications* 19: 198–203.

———. 1965b. "Conformational differences between s-RNA and aminoacyl s-RNA." *Biochemical and Biophysical Research Communications* 20: 400–405.

Saussure, Ferdinand de. 1959. *Course in General Linguistics*. Translated by Wade Baskin. New York: Philosophical Library.

Schachmann, Howard K., Arthur B. Pardee, and Roger Y. Stanier. 1952. "Studies on the macromolecular organization of microbial cells." *Archives of Biochemistry and Biophysics* 38: 245–60.

Schachtschabel, Dietrich, and Wolfram Zillig. 1959. "Untersuchungen zur Biosynthese der Proteine." Pt. 1, "Über den Einbau ^{14}C-markierter Aminosäuren ins Protein zellfreier Nucleoproteid-Enzym-Systeme aus Escherichia coli B." *Hoppe-Seyler's Zeitschrift für physiologische Chemie* 314: 262–75.

Schneider, Walter C., and George H. Hogeboom. 1950. "Intracellular distribution of enzymes." Pt. 5, "Further studies on the distribution of cytochrome c in rat liver homogenates." *Journal of Biological Chemistry* 183: 123–28.

Schoenheimer, Rudolf. 1942. *The Dynamic State of Body Constituents*. Cambridge, Mass.: Harvard University Press.

Schweet, Richard S., Freeman C. Bovard, Esther Allen, and Edward Glassman. 1958. "The incorporation of amino acids into ribonucleic acid." *Proceedings of the National Academy of Sciences of the United States of America* 44: 173–77.

Schweet, Richard S., Hildegarde Lamfrom, and Esther Allen. 1958. "The synthesis of hemoglobin in a cell-free system." *Proceedings of the National Academy of Sciences of the United States of America* 44: 1029–35.

Selby, Cecily Cannan, John J. Biesele, and Clifford E. Grey. 1956. "Electron microscope studies of ascites tumor cells." *Annals of the New York Academy of Sciences* 63: 748–73.

Serres, Michel. 1980. *Le passage du nord-ouest (Hermes 5)*. Paris: Les Editions du Minuit.

———. 1987. *Statues*. Paris: Bourin.

———. 1989. "Préface qui invite le lecteur à ne pas négliger de la lire pour entrer dans l'intention des auteurs et comprendre l'agencement de ce livre." In M. Serres (ed.), *Eléments d'histoire des sciences*, 1–15. Paris: Bordas.

Shapin, Steven, and Simon Schaffer. 1985. *Leviathan and the Air Pump: Hobbes, Boyle, and the Experimental Life.* Princeton, N.J.: Princeton University Press.

Shaver, John Rodney, and Jean Brachet. 1949. "The exposition of chorioallantoic membranes of the chick embryo to granules from embryonic tissue." *Experientia* 5: 235.

Siekevitz, Philip. 1952. "Uptake of radioactive alanine in vitro into the proteins of rat liver fractions." *Journal of Biological Chemistry* 195: 549–65.

Siekevitz, Philip, and Paul C. Zamecnik. 1951. "In vitro incorporation of 1-C^{14}-DL-alanine into proteins of rat-liver granular fractions." *Federation Proceedings* 10: 246.

——. 1981. "The ribosome and protein synthesis." *Journal of Cell Biology* 91 (no. 3, pt. 2): 53S–65S.

Simkin, Julius L. 1959. "Protein biosynthesis." *Annual Review of Biochemistry* 28: 145–70.

Simkin, Julius L., and Thomas S. Work. 1957. "Protein synthesis in Guinea-pig liver: Incorporation of radioactive amino acids into proteins of the microsome fraction in vivo." *Biochemical Journal* 65: 307–15.

Simpson, Melvin V., E. Farber, and Harold Tarver. 1950. "Studies on ethionine." Pt. 1, "Inhibition of protein synthesis in intact animals." *Journal of Biological Chemistry* 182: 81–89.

Simpson, Melvin V., and Sidney F. Velick. 1954. "The synthesis of aldolase and glyceraldehyde-3-phosphate dehydrogenase in the rabbit." *Journal of Biological Chemistry* 208: 61–71.

Sissakian, Norair M. 1956. "Biochemical properties of plastides." In Claude Liébecq (ed.), *Proceedings of the Third International Congress of Biochemistry, Brussels, 1955*, 18–23. New York: Academic Press.

Smellie, Robert M. S., W. M. McIndoe, and James N. Davidson. 1953. "The incorporation of ^{15}N, ^{35}S and ^{14}C into nucleic acids and proteins of rat liver." *Biochimica et Biophysica Acta* 11: 559–65.

Smith, Kendric C., Eugene Cordes, and Richard S. Schweet. 1959. "Fractionation of transfer ribonucleic acid." *Biochimica et Biophysica Acta* 33: 286–87.

Spahr, Pierre F., and Alfred Tissières. 1959. "Nucleotide composition of ribonucleoprotein particles from Escherichia coli." *Journal of Molecular Biology* 1: 237–39.

Spiegelman, Sol. 1956a. "The present status of the induced synthesis of enzymes." In Claude Liébecq (ed.), *Proceedings of the Third International Congress of Biochemistry, Brussels, 1955*, 185–95. New York: Academic Press.

——. 1956b. "Protein synthesis in protoplasts." In G. E. W. Wolstenholme and C. M. O'Connor (eds.), *CIBA Foundation Symposium on Ionizing Radiations and Cell Metabolism*, 185–95. Boston: Little, Brown.

——. 1959. "Protein and nucleic acid synthesis in subcellular fractions of bacterial cells." In G. Tunevall (ed.), *Recent Progress in Microbiology*, 81–103. Spring-field, Ill.: Charles C. Thomas.

Spiegelman, Sol, Harlyn O. Halvorson, and Ruth Ben-Ishai. 1955. "Free amino acids and the enzyme-forming mechanism." In William D. McElroy and H. Bentley Glass (eds.), *A Symposium on Amino Acid Metabolism* (June 14–17, 1954), 124–70. Baltimore: Johns Hopkins University Press.

Spirin, Alexander. 1990. "Ribosome preparation and cell-free protein synthesis." In Walter E. Hill, Peter B. Moore, Albert Dahlberg, David Schlessinger, Roger A. Garrett, and Jonathan R. Warner (eds.), *The Ribosome: Structure, Function, and Evolution*, 56–70. Washington, D.C.: American Society for Microbiology.

Staiger, Emil (ed.). 1966. *Der Briefwechsel zwischen Schiller und Goethe*. Frankfurt am Main: Insel.

Star, Susan Leigh. 1986. "Triangulating clinical and basic research: British lo-calizationists, 1870–1906." *History of Science* 24: 29–48.

Star, Susan Leigh, and James R. Griesemer. 1988. "Institutional ecology, 'transla-tions' and boundary objects: Amateurs and professionals in Berkeley's Mu-seum of Vertebrate Zoology 1907–39." *Social Studies of Science* 19: 387–420.

Stein, William H., and Stanford Moore. 1948. "Chromatography of amino acids on starch columns: Separation of phenylalanine, leucine, isoleucine, meth-ionine, tyrosine, and valine." *Journal of Biological Chemistry* 176: 337–65.

Steinberg, Daniel, and Chris B. Anfinsen. 1952. "Evidence for intermediates in ovalbumin synthesis." *Journal of Biological Chemistry* 199: 25–42.

Stengers, Isabelle. 1987. "La propagation des concepts." In I. Stengers (ed.), *D'une science à l'autre: Des concepts nomades*, 9–26. Paris: Seuil.

Stent, Gunther S. 1968. "That was the molecular biology that was." *Science* 160: 390–95.

Stephenson, Mary L., Kenneth V. Thimann, and Paul C. Zamecnik. 1956. "In-corporation of C^{14}-amino acids into proteins of leaf disks and cell-free frac-tions of tobacco leaves." *Archives of Biochemistry and Biophysics* 65: 194–209.

Stephenson, Mary L., and Paul C. Zamecnik. 1961. "Purification of valine trans-fer ribonucleic acid by combined chromatographic and chemical proce-dures." *Proceedings of the National Academy of Sciences of the United States of America* 47: 1627–35.

——. 1962. "Isolation of valyl-RNA of a high degree of purity." *Biochemical and Biophysical Research Communications* 7: 91–94.

——. 1978. "Inhibition of Rous sarcoma viral RNA translation by a specific oligodeoxyribonucleotide." *Proceedings of the National Academy of Sciences of the United States of America* 75: 285–88.

Stephenson, Mary L., Paul C. Zamecnik, and Mahlon B. Hoagland. 1959. "Conditions for transfer reactions between soluble RNA and microsomes involved in protein synthesis." *Federation Proceedings* 18: 331.

Strack, Hans-Bernd. 1989. "Experimental standard procedures, habituation, and strategies for obtaining evidence." In Peter Weingartner and Gerhard Schurz (eds.), *Philosophy of the Natural Sciences*, proceedings of the Thirteenth International Wittgenstein Symposium, 377–81. Vienna: Hölder, Pichler, and Tempsky.

Straub, Ferenc B., Agnes Ullmann, and George Acs. 1955. "Enzyme synthesis in a solubilized system." *Biochimica et Biophysica Acta* 18: 439.

Strittmatter, Cornelius F., and Eric G. Ball. 1952. "A hemochromogen component of liver microsomes." *Proceedings of the National Academy of Sciences of the United States of America* 38: 19–25.

Suchman, Lucy A. 1990. "Representing practice in cognitive science." In Michael Lynch and Steve Woolgar (eds.), *Representation in Scientific Practice*, 301–21. Cambridge, Mass.: MIT Press.

Tarski, Alfred. 1946. *Introduction to Logic and to the Methodology of Deductive Sciences*. Oxford: Oxford University Press.

Tarver, Harold. 1954. "Peptide and protein synthesis: Protein turnover." In H. Neurath, and K. Bailey (eds.), *The Proteins*, vol. 2, pt. B, 1199–1296. New York: Academic Press.

Tissières, Alfred. 1959. "Some properties of soluble ribonucleic acid from Escherichia coli." *Journal of Molecular Biology* 1: 365–74.

———. 1974. "Ribosome research: Historical background." In Masayasu Nomura, Alfred Tissières, and Peter Lengyel (eds.), *Ribosomes*, 3–12. New York: Cold Spring Harbor Laboratory Press.

Tissières, Alfred, David Schlessinger, and François Gros. 1960. "Amino acid incorporation into proteins by E. coli ribosomes." *Proceedings of the National Academy of Sciences of the United States of America* 46: 1450–63.

Tissières, Alfred, and James D. Watson. 1958. "Ribonucleoprotein particles from Escherichia coli." *Nature* 182: 778–80.

Tissières, Alfred, James D. Watson, David Schlessinger, and B. R. Hollingworth. 1959. "Ribonucleoprotein particles from Escherichia coli." *Journal of Molecular Biology* 1: 221–33.

Todd, Alexander. 1955. "Nucleic acid structure and function." *Chemistry and Industry*, no. 37: 1139–44.

———. 1956. "Nucleic acids." In A. Todd (ed.), *Perspectives in Organic Chemistry*, 245–64. New York: Interscience.

Ts'o, Paul O. P., and R. Squires. 1959. "Quantitative isolation of intact RNA from microsomal particles of pea seedlings and rabbit reticulocytes." *Federation Proceedings* 18 (abstract 1351): 341.

Turnbull, David, and Terry Stokes. 1990. "Manipulable systems and laboratory strategies in a biomedical institute." In Homer E. Le Grand (ed.), *Experimental Inquiries*, 167–92. Dordrecht: Kluwer.

Tyner, Evelyn Pease, Charles Heidelberger, and Gerald A. LePage. 1953. "Intracellular distribution of radioactivity in nucleic acid nucleotides and proteins following simultaneous administration of P^{32} and glycine-2-C-[14]." *Cancer Research* 13: 186–203.

Van Fraassen, Bas C. and Jill Sigman. 1993. "Interpretation in science and in the arts." In George Levine (ed.): *Realism and Representation*, 73–99. Madison: University of Wisconsin Press.

Volkin, Elliot, and Lazarus Astrachan. 1956a. "Phosphorus incorporation in Escherichia coli ribonucleic acid after infection with bacteriophage T2." *Virology* 2: 149–61.

——. 1956b. "Intracellular distribution of labeled ribonucleic acid after phage infection of Escherichia coli." *Virology* 2: 433–37.

Von der Decken, Alexandra, and Tore Hultin. 1958. "A metabolic isotope transfer from soluble polynucleotides to microsomal nucleoprotein in a cell-free rat liver system." *Experimental Cell Research* 15: 254–56.

——. 1960. "The enzymatic composition of rat liver microsomes during liver regeneration." *Experimental Cell Research* 19: 591–604.

Von Portatius, Hans, Paul Doty, and Mary L. Stephenson. 1961. "Separation of L-valine acceptor 'soluble ribonucleic acid' by specific reaction with polyacrylic acid hydrazide." *Journal of the American Chemical Society* 83: 3351–52.

Wahrig-Schmidt, Bettina, and Friedhelm Hildebrandt. 1993. "Pathologische Erythrozytendeformation und renale Hämaturie." In Hans-Jörg Rheinberger and Michael Hagner (eds.), *Die Experimentalisierung des Lebens*, 74–96. Berlin: Akademie Verlag.

Warner, Jonathan R., Paul M. Knopf, and Alexander Rich. 1963. "A multiple ribosomal structure in protein synthesis." *Proceedings of the National Academy of Sciences of the United States of America* 49: 122–29.

Warner, Jonathan R., Alexander Rich, and Cecil E. Hall. 1962. "Electron microscope studies of ribosomal clusters synthesizing hemoglobin." *Science* 138: 1399–1403.

Watson, James D. 1963. "Involvement of RNA in the synthesis of proteins." *Science* 140: 17–26.

——. 1968. *The Double Helix: A Personal Account of the Discovery of the Structure of DNA*. London: Weidenfeld and Nicolson.

Watson, James D., and Francis H. C. Crick. 1953. "The structure of DNA." *Cold Spring Harbor Symposia on Quantitative Biology* 18: 123–31.

Weber, Samuel. 1989. "Upsetting the set up: Remarks on Heidegger's questing after technics." *Modern Language Notes* 104: 977–92.

Webster, George C. 1959. "Studies on the mechanism of protein synthesis by isolated nucleoprotein particles." *Federation Proceedings* 18 (abstract 1379): 348.

Weiss, Samuel B., George Acs, and Fritz Lipmann. 1958. "Amino acid incorporation in pigeon pancreas fractions." *Proceedings of the National Academy of Sciences of the United States of America* 44: 189–96.

Wettstein, Felix O., Theophil Staehelin, and Hans Noll. 1963. "Ribosomal aggregate engaged in protein synthesis: Characterization of the ergosome." *Nature* 197: 430–35.

White, Hayden. 1980. "The value of narrativity in the representation of reality." In W. J. Thomas Mitchell (ed.), *On Narrative*, 1–23. Chicago: University of Chicago Press.

Wilson, Samuel H., and Mahlon B. Hoagland. 1965. "Studies on the physiology of rat liver polyribosomes: Quantitation and intracellular distribution of ribosomes." *Proceedings of the National Academy of Sciences of the United States of America* 54: 600–607.

Wimsatt, William C. 1986. "Developmental constraints, generative entrenchment, and the innate-acquired distinction." In William Bechtel (ed.), *Integrating Scientific Disciplines*, 185–208. Dordrecht: Nijhoff.

Winnick, Theodore. 1950. "Studies on the mechanism of protein synthesis in embryonic and tumor tissues." Pt. 2, "Inactivation of fetal rat liver homogenates by dialysis, and reactivation by the adenylic acid system." *Archives of Biochemistry* 28: 338–47.

Winnick, Theodore, Felix Friedberg, and David M. Greenberg. 1947. "Incorporation of C[14]-labeled glycine into intestinal tissue and its inhibition by azide." *Archives of Biochemistry* 15: 160–61.

Winnick, Theodore, Felix Friedberg, and David M. Greenberg. 1948. "The utilization of labeled glycine in the process of amino acid incorporation by the protein of liver homogenate." *Journal of Biological Chemistry* 175: 117–26.

Winnick, Theodore, Ingrid Moring-Claesson, and David M. Greenberg. 1948. "Distribution of radioactive carbon among certain amino acids of liver homogenate protein, following uptake experiments with labeled glycine." *Journal of Biological Chemistry* 175: 127.

Winnick, Theodore, Elbert A. Peterson, and David M. Greenberg. 1949. "Incorporation of C[14] of glycine into protein and lipide fractions of homogenates." *Archives of Biochemistry* 21: 235–37.

Wise, Norton. 1992. "Mediations: Enlightenment balancing acts, or the technologies of rationalism." Manuscript.

Wittgenstein, Ludwig. 1953. *Philosophical Investigations.* Oxford: Basil Blackwell.

Wittmann, Heinz-Günter. 1961. "Ansätze zur Entschlüsselung des genetischen Codes." *Die Naturwissenschaften* 48: 729–34.

——. 1963. "Studies on the nucleic acid-protein correlation in tobacco mosaic virus." *Proceedings of the Fifth International Congress of Biochemistry*, Moscow, August 10–16, 1961, vol. 1, 240–54. New York: Macmillan.

Yarmolinsky, Michael B., and Gabriel L. de la Haba. 1959. "Inhibition by puromycin of amino acid incorporation into protein." *Proceedings of the National Academy of Sciences of the United States of America* 45: 1721–29.

Yearley, Steven. 1990. "The dictates of method and policy: Interpretational structures in the representation of scientific work." In Michael Lynch and

Steve Woolgar (eds.), *Representation in Scientific Practice*, 337–55. Cambridge, Mass.: MIT Press.

Yu, Chuan-Tao, and Frank W. Allen. 1959. "Studies on an isomer of uridine isolated from ribonucleic acids." *Biochimica et Biophysica Acta* 32: 393–406.

Yu, Chuan-Tao, and Paul C. Zamecnik. 1963a. "Effect of bromination on the amino acid accepting activities of transfer ribonucleic acid." *Biochimica et Biophysica Acta* 76: 209–22.

——. 1963b. "On the aminoacyl-tRNA synthetase recognition sites of yeast and E. coli transfer RNA." *Biochemical and Biophysical Research Communications* 12: 457–63.

——. 1964. "Effect of bromination on the biological activities of transfer RNA of E. coli." *Science* 144: 856–59.

Zachau, Hans Georg, George Acs, and Fritz Lipmann. 1958. "Isolation of adenosine amino acid esters from a ribonuclease digest of soluble, liver ribonucleic acid." *Proceedings of the National Academy of Sciences of the United States of America* 44: 885–89.

Zamecnik, Paul C. 1950. "The use of labeled amino acids in the study of the protein metabolism of normal and malignant tissues: A review." *Cancer Research* 10: 659–67.

——. 1953. "Incorporation of radioactivity from DL-leucine-1-C¹⁴ into proteins of rat liver homogenates." *Federation Proceedings* 12: 295.

——. 1958. "The microsome." *Scientific American*, 198 (March): 118–24.

——. 1960. "Historical and current aspects of the problem of protein synthesis." *The Harvey Lectures 1958–1959* 54: 256–81.

——. 1962a. "History and speculation on protein synthesis." *Proceedings of the Symposium on Mathematical Problems in the Biological Sciences* (New York, Rand Corporation) 14: 47–53.

——. 1962b. "Unsettled questions in the field of protein synthesis." *Biochemical Journal* 85: 257–64.

——. 1969. "An historical account of protein synthesis, with current overtones—a personalized view." *Cold Spring Harbor Symposia on Quantitative Biology* 34: 1–16.

——. 1974. "Joseph Charles Aub, 1890–1973." *Transactions of the Association of American Physicians* 87: 12–14.

——. 1976. "Protein synthesis—early waves and recent ripples." In A. Kornberg, B. L. Horecker, L. Cornudella, and J. Oro (eds.), *Reflections in Biochemistry*, 303–8. New York: Pergamon Press.

——. 1979. "Historical aspects of protein synthesis." *Annals of the New York Academy of Sciences* 325: 269–301.

——. 1983. "Cancer Research: Joseph Charles Aub." In Benjamin Castleman, David C. Crockett, and S. B. Sutton (eds.), *The Massachusetts General Hospital*, 343–48. Boston: Little, Brown.

———. 1984. "The machinery of protein synthesis: Biochemistry and the birth of molecular biology." *Trends in Biological Sciences Letters (TIBS)* 9: 464–66.

Zamecnik, Paul C., and Sudhir Agrawal. 1991. "The hybridization inhibition, or antisense, approach to the chemotherapy of AIDS." *AIDS Research Review* 1: 301–13.

Zamecnik, Paul C., Lydia E. Brewster, and Fritz Lipmann. 1947. "A manometric method for measuring the activity of the Cl. welchii lecithinase and a description of certain properties of this enzyme." *Journal of Experimental Medicine* 85: 381–94.

Zamecnik, Paul C., and Ivan D. Frantz Jr. 1949. "Peptide bond synthesis in normal and malignant tissue." *Cold Spring Harbor Symposia on Quantitative Biology* 14: 199–208.

Zamecnik, Paul C., Ivan D. Frantz Jr., Robert B. Loftfield, and Mary L. Stephenson. 1948. "Incorporation in vitro of radioactive carbon from carboxyl-labeled DL-alanine and glycine into proteins of normal and malignant rat livers." *Journal of Biological Chemistry* 175: 299–314.

Zamecnik, Paul C., Ivan D. Frantz Jr., and Mary L. Stephenson. 1949. "Use of starch column chromatography in study of amino acid composition and distribution of radioactivity in proteins of normal rat liver and hepatoma." *Cancer Research* 9: 612–13.

Zamecnik, Paul C., Mahlon B. Hoagland, Mary L. Stephenson, and Jesse F. Scott. 1958. "Studies on intermediates in protein synthesis." *Unio Internationalis Contra Cancrum Acta* 14: 63.

Zamecnik, Paul C., and Elizabeth B. Keller. 1954. "Relation between phosphate energy donors and incorporation of labeled amino acids into proteins." *Journal of Biological Chemistry* 209: 337–54.

Zamecnik, Paul C., Elizabeth B. Keller, Mahlon B. Hoagland, John W. Littlefield, and Robert B. Loftfield. 1956. "Studies on the mechanism of protein synthesis." In G. E. W. Wolstenholme and C. M. O'Connor (eds.), *CIBA Foundation Symposium on Ionizing Radiations and Cell Metabolism*, 161–73. Boston: Little, Brown.

Zamecnik, Paul C., Elizabeth B. Keller, John W. Littlefield, Mahlon B. Hoagland, and Robert B. Loftfield. 1956. "Mechanism of incorporation of labeled amino acids into protein." *Journal of Cellular and Comparative Physiology* 47, supplement 1: 81–101.

Zamecnik, Paul C., and Fritz Lipmann. 1947. "A study of the competition of lecithin and antitoxin for Cl. welchii lecithinase." *Journal of Experimental Medicine* 85: 395–403.

Zamecnik, Paul C., Robert B. Loftfield, Mary L. Stephenson, and Carroll M. Williams. 1949. "Biological synthesis of radioactive silk." *Science* 109: 624–26.

Zamecnik, Paul C., and Mary L. Stephenson. 1960. "Enrichment of specific activity of aminoacyl RNA." *Federation Proceedings* 19: 346.

————. 1978. "Inhibition of Rous sarcoma virus replication and cell transformation by a specific oligodeoxynucleotide." *Proceedings of the National Academy of Sciences of the United States of America* 75: 280–84.

Zamecnik, Paul C., Mary L. Stephenson, and Jesse Scott. 1960. "Partial purification of soluble RNA." *Proceedings of the National Academy of Sciences of the United States of America* 46: 811–22.

Zamecnik, Paul C., Mary L. Stephenson, Jesse F. Scott, and Mahlon B. Hoagland. 1957. "Incorporation of C^{14}-ATP into soluble RNA isolated from 105,000 × g supernatant from rat liver." *Federation Proceedings* 16: 275.

Zamecnik, Paul C., Mary L. Stephenson, and Liselotte I. Hecht. 1958. "Intermediate reactions in amino acid incorporation." *Proceedings of the National Academy of Sciences of the United States of America* 44: 73–78.

Index

In this index an "f" after a number indicates a separate reference on the next page, and an "ff" indicates separate references on the next two pages. A continuous discussion over two or more pages is indicated by a span of page numbers, e.g., "57–59." *Passim* is used for a cluster of references in close but not consecutive sequence.

Library of Congress Cataloging-in-Publication Data

Rheinberger, Hans-Jörg.
  Toward a history of epistemic things : synthesizing proteins in
the test tube / Hans-Jörg Rheinberger.
      p.   cm. — (Writing science)
  Includes bibliographical references and index.
  ISBN 0-8047-2785-6 (cloth). — ISBN 0-8047-2786-4 (pbk.)
    1. Proteins—Synthesis—Methodology—Research—History.
2. Molecular biology—History.   3. Science—Philosophy.   I. Title.
II. Series.
QP551.R47   1997
572'.645'072—dc21   96-47145
                                                              CIP

Original printing 1997
Last figure below indicates year of this printing:
06   05   04   03   02   01   00   99   98   97

Made in the USA
Columbia, SC
10 April 2022

58774833R00202